E X P E R T   A D V I S O R

AutoCAD®

D1545962

EXPERT ADVISOR

AutoCAD®

David S. Cohn

Addison-Wesley Publishing Company, Inc.
Reading, Massachusetts   Menlo Park, California   New York
Don Mills, Ontario   Wokingham, England   Amsterdam   Bonn
Sydney   Singapore   Tokyo   Madrid   San Juan

Library of Congress Cataloging-in-Publication Data
Cohn, David S.
    Expert advisor: AutoCAD.
    Includes indexes.
    1. AutoCAD (Computer program)   I. Title.
T385.C542 1988   604.2'4'02855369   88-3346
ISBN 0-201-05954-1 (pbk.)

Series Editor: Tanya Kucak
Technical reviewers: Christine Christianson, Kevin Mack, Tim Orr, Alan Getz, and Gary Hordemann
Cover design by Corey & Company: Designers
Text design by Joyce Weston
Set in 10.5-point Palatino by Publication Services Inc.

BCDEFGHIJ - HA - 898
Second Printing, December 1988

To my wife, Genevieve,
without whose constant love and support
I could not have become a computer nerd.

And to my children, Bruce and Andrea,
who have to put up with me
and who wouldn't forgive me
if their names weren't in here somewhere.

Acknowledgments

The author wishes to thank the editors, Carole Alden and Tanya Kucak, as well as Alexandra McDowell for their support and overall contribution to the creation of this book.

Thanks to Sandra Boulton at Autodesk for her continued help over the years and for the assistance in acquiring the necessary copies of Auto-CAD used in the preparation of this book. Thanks also to the others at Autodesk who have provided assistance and support in this and my other endeavors.

The author wants to thank Lionel Johnston of *CADalyst* magazine for providing the example that brought me to the point that I could write this book.

This book has been reviewed for its technical accuracy by SPOCAD, with reviews by Christine Christianson, Kevin Mack, Tim Orr, Alan Getz, and Gary Hordemann.

Thanks to Arnie Graber and Dennis Williamson at the Computer Center in Memphis, Tennessee, for the use of their laser printer. Thanks also to Jerry Norman at Nth Graphics Ltd. for the use of the Nth Engine graphics device and to Gene Sumrall of Cheetah International for his advice on purchasing a computer system.

A thank-you to Patt Hicks and Michaelle Tyler for putting up with all of the hassles in the office. Thanks also to Michael Haldeman for his index.

Finally, the author wishes to thank the members of the Memphis Auto-CAD Users Group for their contributions of questions resolved in this book and tips on the use of AutoCAD commands. Thanks also for the friendship and camaraderie provided by this dedicated group of individuals. Thanks specifically to group members Barry Bowen, David Minkin, and Jim Prewett for their thoughtful and careful review of the manuscript and routines therein. Thanks also to David and the Memphis engineering firm of Reaves and Sweeney for the use of their plotter, to Barry for the use of his AutoLISP routines, and to Jim for his continued friendship over the years.

Contents

Introduction

What Is AutoCAD?

AutoCAD is the largest selling computer-aided drafting (CAD) program in the world. It is, by many estimations, the largest program in terms of computer code ever written for a personal computer. That fact does not mean AutoCAD is a difficult program to learn or to master. It is, however, a complex program, offering numerous methods of accomplishing the same task.

Ultimately, the result of using AutoCAD should be increased speed and accuracy in the production of drawings whether your application is architecture, medical engineering, clothing design, mechanical engineering, technical illustration, land management, structural engineering, graphics, electrical engineering, or boat design. If you can't use the program efficiently, AutoCAD won't help you in your professional endeavors. To use AutoCAD effectively, you must have an understanding of how the commands operate. This understanding will enable you to develop methods for accomplishing your drawing tasks.

AutoCAD Compatibility and AutoCAD Devices   The AutoCAD program operates on many different computers and supports hundreds of different input devices, graphics cards, plotters, and printers. No differentiation is made in this book for different systems or devices, with a few exceptions.

This book is based on the MS-DOS version of AutoCAD running on IBM and IBM-compatible systems. Although you can use this book equally well with other versions of AutoCAD, there may be some slight variations, particularly with regard to the function keys and the DOS-level instructions.

Some features of AutoCAD, most notably the Advanced User Interface of AutoCAD release 9, require specific graphics cards or other peripheral devices. Again, you can make excellent use of this book even if a particular feature is not supported by your system. In the case of the Advanced User Interface restrictions, these are clearly pointed out throughout the text.

AutoCAD requires a certain minimum system configuration to run. You must have at least 512K bytes of RAM installed. To use the AutoLISP features of ADE-3, 640K bytes are required. If you have more memory, great.

AutoCAD can use up to 4 megabytes of either expanded or extended memory. In addition, AutoCAD release 9 requires an Intel 8087 or 80287 math coprocessor.

Your system must have at least one floppy disk drive and one hard disk. You must have a compatible graphics display card and monitor. AutoCAD can often be configured to use two monitors, one for text and one for graphics. If you are using only one monitor, it will be used for both text and graphics. You simply toggle between the two. In this case, when the book makes reference to the text screen and the graphics screen it is simply referring to toggling between these two different display modes.

AutoCAD requires some form of pointing device. You can use a digitizing tablet (with a stylus or multibutton cursor), a mouse, a light pen, a joy stick, or a track ball, as long as it can be configured to work with AutoCAD. If you have none of these devices, you can use the arrow keys on the keyboard as your pointing device. In any case, the function of the various keys is shown in the following table.

AUTOCAD FUNCTION	IBM PC KEY			
Ctrl	CTRL			
Flip Screen	F1			
Menu Cursor	INS			
Screen Cursor	HOME			
Abort Cursor	END			
Fast Cursor	PG UP			
Slow Cursor	PG DN			
Up Cursor	⬆			
Down Cursor	⬇			
Left Cursor	⬅			
Right Cursor	➡			
Toggle Coord	F6	or	CTRL	D
Toggle Grid	F7	or	CTRL	G
Toggle Ortho	F8	or	CTRL	O
Toggle Snap	F9	or	CTRL	B
Toggle Tablet	F10	or	CTRL	T

In addition to these required devices, you will probably want to have available a plotter and/or a printer plotter for converting your electronic drawings into printed form.

What Is *Expert Advisor: AutoCAD*?

Expert Advisor: AutoCAD is designed to provide ready answers to your questions about any AutoCAD command, feature, or variable. This book is not a tutorial. You could use this book to learn how to use AutoCAD, but it is more of a reference book or an encyclopedia of AutoCAD. Although I don't expect that you will read this book from cover to cover, you will want to spend time studying a particular command or feature. I am sure that you will discover something about practically every command that you didn't realize before. I have been using AutoCAD daily for over three years. During that time I have used every version of the program. I thought I knew how to use just about every command and yet still discovered new features while creating this book.

How the Book Is Organized I have tried to make every facet of *Expert Advisor: AutoCAD* useful. Every AutoCAD command, program feature, and system variable is documented. Entries are provided in alphabetical order so that you can quickly find the information you need. (A short section of numerical entries follows the alphabetical entries.) Cross-references are provided in the text so that you will be able to find related information quickly and easily. Because of this careful organization, it is easy to flip to appropriate topics if you need additional information.
Here is what you will find for each entry covered in this volume:

Overview   This section provides a description of what the command does, in simple terms. If you have forgotten a command's function or simply want to learn how commands unfamiliar to you work, this section will be helpful to you.

Procedure   In this section, you will find step-by-step directions on how to use the feature or command most effectively. If you remember a command but have forgotten how to use it, this section will give you a review.

Examples   Here you will find practical examples for the use of a given command or feature. You will also find other potential uses for the command, many of which you may not have thought of. Most examples contain illustrations, and many include short macros or AutoLISP routines.

Warnings   AutoCAD is not free of pitfalls and problems. In this section you will find out where you are most likely to encounter problems. If you can foresee these potential problems, you will probably be able to avoid them.

Tips   This section will make the most interesting reading for users who want to rapidly increase their expertise with AutoCAD. The tips may also contain additional techniques and shortcuts, as well as cross-references to other entries in the book.

In the back of this volume you will find four appendices: a summary list of AutoCAD system variables, a scale factors chart, a quick reference to the entries, and selected AutoCAD drawings. There are also two indexes. First, an Applications Index lists common uses of the commands and features and where to find them. Second, a General Index lists all of the commands, functions, and topics covered in this book. Even though you will be able to find most of these subjects listed alphabetically in the book, some topics are embedded within the text. The General Index will help you find these topics.

This book addresses all of the features of AutoCAD versions 2.5 and 2.6 as well as release 9. This includes all intermediate versions, such as 2.52 and 2.62. In addition, you will find information on the Advanced Drafting Extension (ADE) numbers 1 to 3. At the beginning of each entry, the applicable ADE number, if any, is listed. Only one ADE number will appear, since ADE-3 includes ADE-2 and ADE-1, and ADE-2 includes ADE-1. The version number(s) will also be listed. If the command or feature was included in versions prior to 2.5, you will see

[all versions]

If the entry was available beginning with version 2.5, you will see

[v2.5, v2.6, r9]

and so on. The ADE and version information is summarized in Appendix C. Within each entry, features or options that are available only in later versions or with higher ADE numbers are noted.

Typographical and Other Conventions   Some standard conventions are used throughout this book, especially when listing command keystrokes. If the text says to press Return, this is the same as pressing the Enter key. Most AutoCAD commands allow the Space bar to duplicate the function of the Return key. This is true for all commands except those requesting text information. You can use whichever key is convenient. In this book, Return is always used, and you will see

$\boxed{\text{RETURN}}$

in command sequences. If not listed explicitly, a Return is expected at the end of any command keystroke line.

Press Ctrl-B or

$\boxed{\texttt{Ctrl-B}}$

means you should hold down the key labeled Ctrl at the same time as holding down the B key. Any other Ctrl key sequences are listed in the same fashion; for example, Ctrl-X or

$\boxed{\texttt{Ctrl-X}}$

If the text lists the keystrokes as ^C, this means to actually type the caret and then the letter C.

In the key sequences for AutoCAD commands, all of the actual AutoCAD prompts appear. (Occasionally, I have omitted responses such as the 1 selected, 1 found message displayed during object selection.) The Auto-CAD prompts appear unaltered. User response is always in **bold** or ***bold-face italic***. If your response is meant to be entered exactly as provided, the response is simply in **boldface**. For example,

Command: **LINE**

means that you should enter the letters L I N E (the LINE command) in response to the AutoCAD command prompt. You must press Return after entering the command, and the Return is understood (not shown). This same convention is used for responses made at the DOS prompt level.

If you are being instructed to make a selection that is your choice, the response will be shown in ***bold italics*** and will also be enclosed in parentheses. For example,

From point: ***(select a point)***

means that you should select any point you wish. Generally, you will need to complete your entry by pressing Return (except when selecting a point using the pointing device).

Only when you must press Return without making any other entry is Return actually included in the text, such as pressing Return to complete the LINE command:

```
To point: ( RETURN )
```

Some examples are accompanied by explanations of each step. These explanations appear in parentheses immediately following the user input:

```
Second corner: (select point C) (objects will be highlighted)
```

When AutoCAD prompts you to respond, a default value is often displayed. This default value appears within angle brackets < > and is selected by pressing Return. When the actual value displayed is shown in an example, the default will appear in monospace within the angle brackets:

```
Rotate objects as they are copied? <Y>:
```

When the actual default value will necessarily vary depending on the particular situation, the value or an explanation of the value will appear in *italics* within the angle brackets:

```
Object snap target height (1-50 pixels) <current value>:
```

All AutoCAD commands are shown entered in capital letters, although AutoCAD is not case-sensitive and accepts uppercase and lowercase spellings, and combinations of both, as legitimate input. For clarity, uppercase is used for commands in this book. AutoLISP routines are illustrated in lowercase, except that AutoCAD commands and variables within AutoLISP routines are capitalized. Again, this is done for clarity. AutoLISP is not case-sensitive.

Production Notes   For those people, like me, who are interested in these things, I'd like to mention the hardware and software used in the preparation of this book. All of the text was written using PC Write, an excellent shareware word processing program written by Bob Wallace (Quicksoft, Seattle, WA). All of the drawings and illustrations were created using Auto-CAD versions 2.5 and 2.6 and release 9 (Autodesk, Sausalito, CA). The computer screens were captured with HotShot, a RAM-resident text and graphics screen capture program (SymSoft, Mountain View, CA). These screens were captured as PC Paintbrush .PCX format files and were then gray-shaded and imported into Xerox Ventura Publisher desktop publishing software, where they were rescaled for a truer screen ratio. The screen borders were also added at this point.

The AutoCAD drawings were created using a SummaSketch digitizer with a four-button cursor. The drawings were plotted on a Hewlett-Packard HP7585-B plotter. The Ventura Publisher illustrations were printed on a Hewlett-Packard LaserJet II. The remaining portions of the computer system consisted of a Standard 286 (PC Source, Austin, TX), running at 8 MHz, 0 wait states, with 1 megabyte of RAM and an Intel 80287 math coprocessor. A Vega Deluxe EGA card (Video-7, Fremont, CA) was used to drive the text screen in AutoCAD, for both the graphics and text screens with AutoCAD release 9 and for all of the other programs used. An Nth Engine (Nth Graphics, Austin, TX) was used to drive the graphics screen for AutoCAD versions 2.5 and 2.6. Both of these cards were used with a Mitsubishi AUM-1371A Diamond Scan monitor.

Alphabetical Entries

ACAD

[all versions]

Overview ACAD is the DOS-level command that starts the AutoCAD program. From the DOS prompt, typing *ACAD* causes the AutoCAD program to load and AutoCAD's main menu to be displayed.

Procedure To start AutoCAD, type *ACAD* at the DOS prompt. When AutoCAD is loaded, the main menu is displayed. If you want to automatically load a drawing, enter the ACAD command followed by a space and the name of the drawing (the name may include a device and a subdirectory name if you desire). In addition, you can cause AutoCAD to automatically load a drawing file and run a prewritten script file (that is, a text file consisting of a series of AutoCAD commands and having the file extension .SCR). To do this, enter the ACAD command, a space, the name of the drawing, a space, and the name of the script file. Again, the drawing name and the script file name may include a device and a subdirectory name if you want.

Examples To load the AutoCAD program from the DOS prompt, type

C>**ACAD**

AutoCAD's main menu will be displayed.

   If you want to load an existing drawing as soon as you start AutoCAD, you can include the name of the drawing on the DOS command line, leaving a space between the ACAD command and the drawing name, as in

C>**ACAD** *Drawing-name*

where *Drawing-name* is replaced by the name of your drawing. You can include a drive designation and a subdirectory path. Do not include the .DWG file extension.

   If you wish to start AutoCAD, load an existing drawing, and display a named view that has been previously saved, use the same procedure but include the view name after the drawing name, separating them with a comma:

C>**ACAD** *Drawing-name,view-name*

If the view name you included does not exist, AutoCAD displays an error message and loads your drawing as it appeared when you last saved it.

You can also load AutoCAD, call up an existing drawing, and run a predetermined script file by including the script file name after the drawing name, separating the two with a space:

C>**ACAD** Drawing-name Script

As with the drawing name, you can include a drive designation and subdirectory but must not include the file extension.

Warnings You must either start AutoCAD from within the subdirectory that you have copied the AutoCAD program into (usually called \ACAD) or include the AutoCAD subdirectory on your computer's path using the DOS PATH command. To avoid adding your actual drawing files to the subdirectory containing the AutoCAD program files, set a path to the AutoCAD subdirectory and start AutoCAD from the subdirectory containing your drawing.

When loading a drawing, you can use the method of providing a view name for AutoCAD to utilize only if your version of AutoCAD includes ADE-1 (Advanced Drafting Extension number 1; see ADE).

Tips The ACAD command is often included in a DOS batch file (a text file with the file extension .BAT). This batch file, which you might want to call DRAW.BAT, can be set up to not only load AutoCAD but also to set any necessary parameter for the system's path and any AutoCAD environment variables that you might want to set. (See Environment Variables.)

```
PATH C:\ACAD;C:\DOS
set lispheap=40000
set lispstack=5000
ACAD
```

ACADPREFIX

[v2.6, r9]

Overview The ACADPREFIX system variable stores the directory name specified by the ACAD environment variable (if one has been set). ACADPREFIX has a value only if the ACAD environment variable has been set prior to loading AutoCAD. ACADPREFIX is a read-only variable. Its value cannot be altered. (See Environment Variables.)

Procedure   Instead of using the DOS PATH command, you can use the DOS SET command to determine the subdirectory within which AutoCAD always searches when looking for a drawing file. When the ACAD environment variable is set, AutoCAD searches within that subdirectory before it searches in any other subdirectories that may be specified in the computer's path.

Examples   To cause AutoCAD to always search the subdirectory \DRAWINGS \DETAILS before searching any other subdirectory, the ACAD environment variable needs to be set before loading AutoCAD. From the DOS prompt, enter

```
C>SET ACAD=c:\drawings\details
```

Once AutoCAD is loaded, you can observe that the ACADPREFIX system variable is now set to \DRAWINGS\DETAILS by using the SETVAR command or the (getvar) AutoLISP function. In this example, the SETVAR command is used from AutoCAD's command prompt:

```
Command: SETVAR
Variable name or ?: ACADPREFIX
ACADPREFIX = "\DRAWINGS\DETAILS" (read only)
```

The (read only) notation indicates that you cannot change the value from within the drawing.

Warnings   ACADPREFIX is not available in AutoCAD version 2.5.

Tips   To make efficient use of the ACAD environment variable feature, place the SET ACAD= command in a DOS batch file, which you will execute prior to loading AutoCAD. You could use this batch file to set other environment variables as well.

ACADVER   [r9]

Overview   The ACADVER system variable stores the AutoCAD version number. ACADVER is a read-only variable.

Procedure The version of AutoCAD that you are currently running can be returned using the SETVAR command or the (getvar) AutoLISP function.

Examples In this example, the SETVAR command returns the current version number.

```
Command: SETVAR
Variable name or ?: ACADVER
ACADVER = "9.0" (read only)
```

Warnings ACADVER is not available in versions of AutoCAD prior to release 9.

Tips The ACADVER value is completely dependent on the version of AutoCAD that you are currently running.

ADE
[all versions]

Overview ADE (Advanced Drafting Extension), an AutoCAD program feature, is the way Autodesk refers to the various enhancements to the basic AutoCAD package. These enhancements are available as three distinct options numbered ADE-1, ADE-2, and ADE-3. Each higher ADE package requires the lower package or packages to be present. This book assumes that the user's version of AutoCAD includes all three ADE packages. Where appropriate, notation has been made for commands that are available only with a particular ADE package.

- ADE-1 adds the dimensioning features and the BREAK, CHAMFER, FILLET, HATCH, SKETCH, and UNITS commands to the basic Auto-CAD program.
- ADE-2 adds the attributes feature, the ability to drag objects on screen, the isometric drawing aids, and the MIRROR, MSLIDE, OSNAP, and VIEW commands.
- ADE-3 adds the ability to use IGES interchange, draw polylines, and use the SHELL command to go to the operating system while in the middle of an AutoCAD drawing, as well as the DIVIDE, DOUGHNUT, DTEXT, DXBIN, ELEV, ELLIPSE, EXPLODE, EXTEND, HIDE, MEASURE, OFFSET, POLYGON, ROTATE, SCALE, STRETCH, and TRIM commands.

Procedure   To determine which ADE packages are included in your AutoCAD package, start AutoCAD and look at the main menu. The ADE level appears at the top of the screen, as shown in Figure 1.

Tips   The ADE packages provide AutoCAD with all of the powerful features that have made it the best-selling CAD program in the world. If you don't already have all three packages included in your copy of AutoCAD, you should purchase the necessary upgrade.

ADI   [v2.5, v2.6, r9]

Overview   The ADI (Autodesk Device Interface), an AutoCAD program feature, is actually a device driver standard provided by Autodesk to third-party developers so that they can write their own device drivers for peripherals such as graphics cards, digitizers, and plotters. Most AutoCAD drivers are written by Autodesk and are designed to run on the most popular devices. To take advantage of advanced features present in many new peripherals,

```
              A U T O C A D
        Copyright (C) 1982,83,84,85,86,87 Autodesk, Inc.
        Version 2.6 (4/3/87) IBM PC
        Advanced Drafting Extensions 3
        Serial Number:   xx-xxxxxx

        Main Menu

            0.   Exit AutoCAD
            1.   Begin a NEW drawing
            2.   Edit an EXISTING drawing
            3.   Plot a drawing
            4.   Printer Plot a drawing

            5.   Configure AutoCAD
            6.   File Utilities
            7.   Compile shape/font description file
            8.   Convert old drawing file

        Enter selection:
```

Figure 1

the peripheral developer can make use of the ADI feature to write his or her own driver. If you are using one of the newer high-resolution screens, you may be using an ADI driver.

Procedure

When you configure AutoCAD, you may choose to use an ADI driver if one has been copied onto your computer. The steps needed to configure the individual driver depend on the driver itself. (Refer to the instructions supplied by the peripheral manufacturer.) The steps needed to configure AutoCAD to use the ADI driver are exactly the same as configuring Auto-CAD to use any other device. Simply choose the ADI driver from the proper configuration menu. (See Configuring AutoCAD.)

Examples

AutoCAD's standard drivers usually do not take advantage of advanced features available with most peripherals. Many graphics cards, particularly the newest Enhanced Graphics Adapter (EGA) cards, are sold with an ADI driver. These drivers provide resolutions up to 800 × 600 pixels when used with the new multiscanning monitors. Other ADI drivers enable almost instantaneous panning and zooming within AutoCAD drawings or can display more than one view on the screen at the same time, often at differing magnifications. Changes in the computer field take place rapidly. Check with your computer dealer or read the advertisements in computer and AutoCAD magazines. Most manufacturers advertise their product's enhanced capabilities when used with AutoCAD. These enhanced capabilities are almost always accomplished using an ADI driver. In the future, Autodesk will provide fewer standard drivers for new peripherals and depend instead on the individual manufacturers to write ADI drivers.

Warnings

You must load the RAM-resident ADI driver provided by the peripheral manufacturer prior to starting AutoCAD. Otherwise, AutoCAD displays an error message when you enter the drawing editor.

The Advanced User Interface features of AutoCAD release 9 are available only when you use the IBM Color/Graphics Adapter (CGA; in monochrome mode only), the IBM Enhanced Graphics Adapter (EGA), the IBM Video Graphics Array (VGA), the Hercules Graphics Card (monochrome), and ADI version 3.0 or above. To use the Advanced User Interface with an ADI driver, the individual ADI driver must be capable of implementing these features. (See Advanced User Interface.)

Tips   To ensure that your RAM-resident ADI driver is loaded prior to starting AutoCAD, include it in the DOS batch file that you use to start AutoCAD or in your computer's AUTOEXEC.BAT file.

Advanced User Interface   [ADE-3] [r9]

Overview   The Advanced User Interface provides additional methods of interacting with AutoCAD. A menu bar, pull-down menus, icon menus, and pop-up dialog boxes are provided in addition to AutoCAD's standard screen, tablet, and button menus. You can customize the menu bar, pull-down menus, and icon menus through the provisions of AutoCAD's menu files. Dialog boxes are called by specific AutoCAD commands.

Procedure   **Menu Bar**   (See Menu Bar.) The menu bar is accessed by moving the screen cursor to the top of the graphics screen. When the cursor reaches the status line area, the status line is replaced by the menu bar. Moving the cursor left or right causes the different menu-bar titles to be highlighted. In Figure 2, the menu bar is displayed and the Display title is highlighted.

```
┌─────────────────────────────────────────────────────────────┬──────────┐
│ Tools  Draw  Edit  Display  Modes  Options  File             │AutoCAD   │
│                                                              │* * * *   │
│                                                              │SETUP     │
│                                                              │          │
│                                                              │BLOCKS    │
│                                                              │DIM:      │
│                                                              │DISPLAY   │
│                                                              │DRAW      │
│                                                              │EDIT      │
│                                                              │INQUIRY   │
│                                                              │LAYER:    │
│                                                              │SETTINGS  │
│                                                              │PLOT      │
│                                                              │UTILITY   │
│                                                              │          │
│                                                              │3D        │
│                                                              │          │
│                                                              │ASHADE:   │
│                                                              │          │
│                                                              │SAVE:     │
│                                                              │          │
│ ─────────────────────────────────────────────────────────── │          │
│ Loaded menu D:\ACAD9\ACAD.mnx                                │          │
│ Command:                                                     │          │
└─────────────────────────────────────────────────────────────┴──────────┘
```

Figure 2

9

Pull-Down Menu   (See Pull-Down Menu.) When the menu bar is displayed, you can access a pull-down menu by moving the screen cursor until the desired menu-bar item is highlighted and then pressing the pick button. Moving the cursor down then highlights the various items in the pull-down menu. Pressing the pick button when one of the pull-down menu items is highlighted executes that command. The pull-down menu remains displayed until you pick an item; type on the keyboard; press the pick button while over another area of the graphics screen, menu bar, regular screen menu, or tablet menu; or make a selection using the button menu. Figure 3 shows a pull-down menu.

Icon Menu   (See Icon Menu.) Icon menus are made up of AutoCAD slide files. You activate icon menus by selecting a menu macro that calls them. When an icon menu is activated, it covers the graphics screen. The cross-hair screen cursor is replaced by an arrow cursor. Up to sixteen icons are presented on a menu, each with an associated pick-button box. Moving the screen cursor until the arrow is in one of the button boxes causes the button box to be highlighted and that icon selection to appear with a

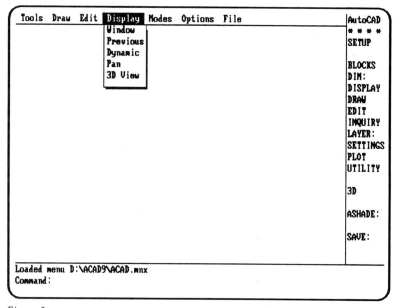

Figure 3

box around it for confirmation.To select that icon, press the pick button. The only way to exit from an icon menu is to pick something from it. Figure 4 is an example of an icon menu.

Dialog Box   (See Dialog Box.) Dialog boxes pop up over the graphics screen in response to several AutoCAD commands. (See DDRMODES, DDEMODES, DDLMODES, and DDATTE.) A dialog box also appears during the INSERT command if the block being inserted has attributes included in its definition and if the system variable ATTDIA has a value of 1. When a dialog box appears, the cross-hair screen cursor is replaced by an arrow cursor. The cursor is moved by moving the pointing device.

Dialog boxes have several selections available, each within a box. All dialog boxes have an OK selection. Most also have a Cancel selection. Dialog boxes may present different types of selections.

- Check buttons are small rectangles that either are blank or contain a check mark. This type of selection is used for selecting the current isoplane or turning AutoCAD's Snap, Grid, Axis, Ortho, or Blip modes on and off.

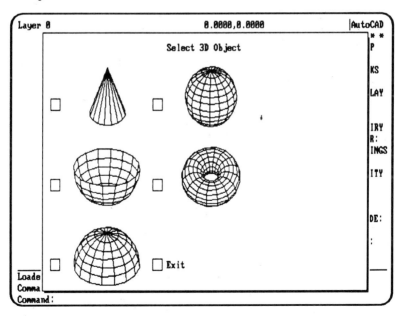

Figure 4

- Action buttons simply perform an operation, such as selecting OK or Cancel.
- Input buttons allow you to type in a value. When you move the arrow cursor over an input button, the input button is highlighted. You can then type in a value and press the Return key or select the OK button. If the value you enter is invalid, no action is taken. You must either backspace and enter the value again or cancel the input.
- Requestor button selections within dialog boxes cause a second dialog box to overlay the first. You must enter a response or cancel the top dialog box before you are returned to the underlying box. At other times, a dialog box selection may cause a warning to be displayed, such as when you freeze the current layer. You must acknowledge the warning message by clicking the OK button displayed within the alert dialog box before AutoCAD will continue. Figure 5 is an example of a dialog box.

A dialog box remains on the screen until an acceptable selection is made or the dialog box is canceled. Pressing the Return key has the same effect as

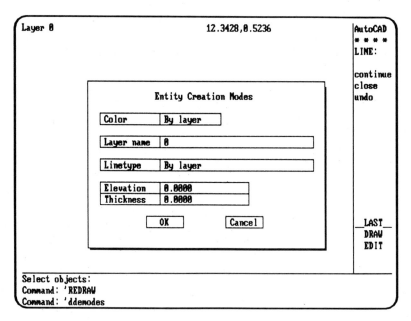

Figure 5

pressing the pick button when the OK selection is highlighted. Ctrl-C or Escape has the same effect as Cancel.

Examples The settings for SNAPMODE, GRIDMODE, AXISMODE, ORTHOMODE, and BLIPMODE can all be set from within the Drawing Aids dialog box, called by the DDRMODES command, by simply clicking (that is, pressing the pick button) on the appropriate check buttons. In addition, the snap spacing, snap angle, grid spacing, axis spacing, and isometric drawing features can all be accessed while within this one dialog box. Figure 6 shows the Drawing Aids dialog box.

Warnings The availability of the Advanced User Interface features depends on the version of AutoCAD and the type of graphics card in use. These features are available only in AutoCAD release 9 with ADE-3. In addition, the Advanced User Interface features are available only when using the IBM Color/Graphics Adapter (CGA; in monochrome mode only), the IBM Enhanced Graphics Adapter (EGA), the IBM Video Graphics Array (VGA), the Hercules Graphics Card (monochrome), or the Autodesk Device Interface

```
Layer 0                         12.3428,0.5236                    AutoCAD
                                                                  * * * *
                    Snap                                          LINE:

         X Spacing   1.0000              Snap                     continue
         Y Spacing   1.0000              Grid                     close
                                         Axis                     undo
         Snap angle  0                   Ortho
         X Base      0.0000              Blips  ✓
         Y Base      0.0000
                                                Isoplane
                    Grid
                                           ✓  Left
         X Spacing   0.0000                   Top
         Y Spacing   0.0000                   Right

                    Axis
                                         Isometric
         X Spacing   0.0000                                        LAST
         Y Spacing   0.0000                                        DRAW
                                                                   EDIT
                      OK             Cancel

Command: '
Command: '
Command: 'ddrmodes
```

Figure 6

13

(ADI) version 3.0 or above. If the currently configured graphics card supports the Advanced User Interface, the system variable POPUPS is set to 1. Otherwise, the POPUPS value is 0.

Tips
If the graphics card does not support the Advanced User Interface features, commands that call these features, if included in menu files, are simply ignored.

You can move the arrow cursor within icon menus and dialog boxes by using either the keyboard arrow keys or the pointing device. You can access menu bars and pull-down menus only by using the pointing device.

AutoCAD's Advanced User Interface features are included as part of the AutoCAD standard menu file, ACAD.MNU.

Menu bars, pull-down menus, and icon boxes are all user-customizable through the facilities of AutoCAD's menu file. (See Menu Bar, Pull-Down Menu, and Icon Menu.) Dialog boxes are not user-customizable. (See Dialog Box.)

AFLAGS
[ADE-2] [v2.5, v2.6, r9]

Overview
The system variable AFLAGS (Attribute Flags) determines the setting of the Invisible, Constant, and Verify options in effect when using the ATTDEF command to create an attribute definition. (See ATTDEF.)

Procedure
The AFLAGS value is normally set by the ATTDEF command when you respond to the following prompt:

```
Attribute modes -- Invisible:N  Constant:N  Verify:N
Enter (ICV) to change, RETURN when done:
```

This controls the display mode of each attribute. The mode is the sum of the AFLAGS value, where 1 equals invisible, 2 equals constant, and 4 equals verify.

If you are using AutoCAD release 9, an additional attribute mode, Preset, is available. In this case, the ATTDEF command prompts

```
Attribute modes -- Invisible:N  Constant:N  Verify:N  Preset:N
Enter (ICVP) to change, RETURN when done:
```

A value of 8 indicates that the current mode is preset. In this release of AutoCAD, a value of 9 would mean preset and invisible.

Examples An AFLAGS value of 3 means that the attribute is invisible and constant. The AFLAGS variables can be changed using the SETVAR command or the (getvar) and (setvar) AutoLISP functions. To simplify matters, a complete listing of the available values is provided in the following table.

VALUE	INVISIBLE	CONSTANT	VERIFY	PRESET
0	N	N	N	N
1	Y	N	N	N
2	N	Y	N	N
3	Y	Y	N	N
4	N	N	Y	N
5	Y	N	Y	N
6	N	Y	Y	N
7	Y	Y	Y	N
8	N	N	N	Y
9	Y	N	N	Y
10	N	Y	N	Y
11	Y	Y	N	Y
12	N	N	Y	Y
13	Y	N	Y	Y
14	N	Y	Y	Y
15	Y	Y	Y	Y

Tips The AFLAGS value is always initially set to 0. If you change the value, that change remains in effect only until the end of the current editing session. When you return to the main menu, the current value is discarded.

It is generally unnecessary to change the AFLAGS variable using any method other than the ATTDEF command.

If you are using AutoCAD release 9, AFLAGS values over 15 simply repeat the same sequence again (in other words, a value of 16 is the same as a value of 0). For AutoCAD versions 2.5 and 2.6, the values repeat after 7 (in other words, for these versions, a value of 8 is the same as a value of 0).

ANGBASE   [ADE-1] [v2.5, v2.6, r9]

Overview   The ANGBASE (Angle Base) system variable stores the angle for the 0 direction.

Procedure   Normally you select the ANGBASE angle when you are using the UNITS command, but you can also change it by using the SETVAR command or AutoLISP's (getvar) and (setvar) functions. The value of ANGBASE is a real number. (See SETVAR and UNITS.)

Examples   When you use UNITS, you see the prompt

`Enter direction for angle 0 <0>:`

Changing this value from within the UNITS command has the same effect as using AutoLISP or the SETVAR command to change the ANGBASE value. In this way, you can change the value while another command is active.

Tips   The initial ANGBASE value is determined from the prototype drawing (default value is 0 degrees). The current value is saved with the drawing. ANGBASE should normally remain set to 0. This will result in the North or 90-degree direction being toward the top of the drawing.

Changing the ANGBASE value transparently using the 'SETVAR command results in an immediate change in AutoCAD's interpretation of the 0 direction.

In release 9 only, you can also change the ANGBASE value by changing the X base and/or Y base values within the dialog box presented by the DDRMODES command. (See DDRMODES.)

ANGDIR

[ADE-1] [v2.5, v2.6, r9]

Overview The ANGDIR (Angle Direction) system variable controls whether Auto-CAD considers increasing positive angle values to be drawn in a clockwise or counterclockwise direction.

Procedure The value for the ANGDIR variable is either 1 or 0. A value of 1 indicates that positive angle values represent a clockwise direction. A value of 0 indicates a counterclockwise direction.

Examples When using the UNITS command, if you answer yes when the command prompts

```
Do you want angles measured clockwise?
```

the ANGDIR value will be changed to 1. You can also change the ANGDIR value by using AutoLISP's (getvar) and (setvar) functions. (See SETVAR and UNITS.)

Tips Being able to change the direction in which angles are evaluated is a relatively new enhancement. Thus, many existing macros and AutoLISP routines assume that positive angles represent counterclockwise angles. Unless you have a definite need to vary from this, you should accept AutoCAD's default setting.

The initial ANGDIR value is determined from the prototype drawing (default value is 0; counterclockwise). The current value is saved with the drawing.

Changing the ANGDIR value transparently using the 'SETVAR command results in an immediate change in AutoCAD's interpretation of angle direction.

Angles

[all versions]

Overview AutoCAD allows angles to be represented in several ways. You can choose the method of angle representation appropriate to your particular application. AutoCAD's conventional methods are shown in the following table.

METHOD	EXAMPLE	VERSION
Decimal degrees	42.5	
Degrees/minutes/seconds	42d30'0.00"	[ADE-1]
Grads	47.2222g	[ADE-1]
Radians	0.7418r	[ADE-1]
Surveyor's units	N 47d30'0" E	[ADE-1]

Procedure

Normally you choose the method of angle representation with the UNITS command. The initial method of angle representation is determined by the prototype drawing. In addition to using the UNITS command, you can change the method in which angles are represented by changing the value of the system variable AUNITS, using either the SETVAR command or AutoLISP's (getvar) and (setvar) functions.

The method you choose determines how angles are displayed on the STATUS line and within command prompts and how AutoCAD will allow angles to be entered. When you are using decimal degrees, grads, or radians, only that method of angle measure is acceptable. When you are using degrees/minutes/seconds, you can enter angles in decimal degrees also. When you are using surveyor's units, all three methods (decimal, degrees/minutes/seconds, and surveyor's units) are acceptable.

Any number of decimal places are acceptable for input, although Auto-CAD will display only the number of decimal places selected when you use the UNITS command or change the system variable AUPREC. Typically, positive angles are drawn in a counterclockwise direction, but you can change this by using the UNITS command or by changing the system variable ANGDIR. An angle value of 0 normally is drawn at 3 o'clock, but you can alter this by using the UNITS command or by changing the system variable ANGBASE. AutoCAD's default angle orientation is shown in the following drawing.

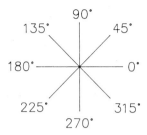

A

Examples   So that AutoCAD recognizes when you are providing an angle in response to a command prompt, use the < symbol. For example, to draw a line from the 0,0 point to a point 6 units away at a 45-degree angle, you would enter the following:

```
From point: 0,0
To point: @6<45
```

Notice in this example that a *d* is not necessary after the 45. If a *d* is not provided, AutoCAD understands that the angle entered is in degrees (if you selected that method of angle representation). The same would be true if you had selected radians or grads. It is not necessary to include the angle representation symbol except in the case of degrees/minutes/seconds or surveyor's units.

Tips   Unless you are entering angles in grads or radians, you probably will find it useful to set the method of angle display to degrees/minutes/seconds (or surveyor's units when you are drawing site plans), since these methods also allow angles to be entered in decimal degrees.

APERTURE

[ADE-2] [all versions]

Overview   APERTURE is an AutoCAD command and a system variable. Whenever an Object Snap mode is used, a special target box is added to the cross-hairs. The size of the target box is measured in screen pixels or dots and can be changed via the APERTURE command. When you use an Object Snap mode, any objects that are within or that cross the target box during entity

selection are considered in AutoCAD's evaluation of the object-entity selection. The size of the target box (in pixels) is stored as the APERTURE
system variable value and saved in AutoCAD's configuration file. The
value of APERTURE can be changed with the SETVAR command.

Procedure

```
Command: APERTURE
Object snap target height(1-50 pixels)<current value>:
```

Enter the number of pixels that you wish the aperture box to extend from
the cross-hairs as shown in the following drawing. When APERTURE is
used as an AutoCAD command, values from 1 to 50 pixels are valid. When
it is a system variable, any value greater than or equal to 1 is valid. The
larger the number, the bigger the aperture box.

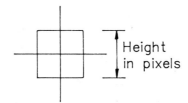

Examples

You can vary the size of the aperture box while another command is active
by using the transparent SETVAR command. This is accomplished by preceding the command with an apostrophe. For example, to change the
aperture box size while in the middle of the LINE command, you could do
the following:

```
To point: 'SETVAR
>>Variable name or ?: APERTURE
>>New value for APERTURE <5>: 10
Resuming LINE command.
To point:
```

In this example, the aperture box was changed from a value of 5 pixels
to double that size, 10 pixels. Changing the size of the aperture box
using either of the previous methods results in an immediate change in
the box's size.

Tips Set the aperture box size so that it is neither too large nor too small. Remember, the Object Snap feature must scan all the entities that fall within the aperture box before making a selection (subject to Quick Snap mode). Thus, the larger the box the slower the Object Snap feature. On the other hand, you cannot use the Object Snap feature with entities that do not fall within the box. The Object Snap feature will not find the intersection of dashed lines unless a portion of those lines falls within the aperture box. (For more information, see Object Snap.)

The initial APERTURE system variable value is determined from the prototype drawing (default value is 10 pixels). The current value is saved with the drawing.

ARC

[all versions]

Options

A	Included angle
C	Center point
D	Starting direction
E	End point
L	Length of chord
R	Radius
Blank	(As a reply to start point) Sets start point and direction as end of last line or arc

Overview The ARC command is one of the basic AutoCAD drawing commands. It is used for drawing partial circles, or arcs, of any size. Arcs are one of the eleven basic AutoCAD drawing entities. There are eight different methods available for drawing arcs:

1. Three points on an arc
2. Start point, center (of circle), end point
3. Start point, center, included angle
4. Start point, center, length of chord
5. Start point, end point, radius
6. Start point, end point, included angle

7. Start point, end point, starting direction
8. As a continuation of the previous arc or line

In the case of an arc described by specifying its start point and the center point of the circle of which the arc is part, either the start point or the center point may be specified first. This results in eleven different methods of defining an arc; each method is described under "Examples."

Procedure

```
Command: ARC
Center/<Start point>:
```

At this step, you can begin by entering points or an appropriate individual initial at each AutoCAD prompt. You can enter points by using a pointing device, entering exact coordinates, or entering relative coordinates.

Examples

Three-Point Arcs   Three-point arcs can be drawn clockwise or counterclockwise.

```
Command: ARC
Center/<Start point>: (pick first point)
Center/End/<Second point>: (pick second point)
End point: (pick third point)
```

The following drawing shows a typical three-point arc.

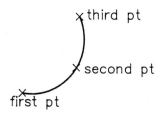

Start Point, Center, End Point   An arc that you specify using the start point, the center, and the end point will not necessarily pass through the end point. The arc is defined by the radius, which is the distance from the start point to the center. The end point simply provides the angle at which the arc ends. Arcs specified in this fashion are always drawn counterclockwise. The next three examples illustrate the methods that allow you to select the start point, the center point, and the end point in different order.

```
Command: ARC
Center/<Start point>: (pick start point)
Center/End/<Second point>: C
Center: (pick center point)
Angle/Length of chord/<End point>: (pick end point)
```

or

```
Command: ARC
Center/<Start point>: C
Center: (pick center point)
Start: (pick start point)
Angle/Length of chord/<End point>: (pick end point)
```

or

```
Command: ARC
Center/<Start point>: (pick start point)
Center/End/<Second point>: E
End point: (pick end point)
Angle/Direction/Radius/<Center point>: (pick center point)
```

Regardless of the order in which the points are selected, the same arc will be drawn. The following drawing shows a typical arc resulting from any of these three methods.

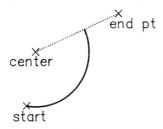

Start Point, Center, Included Angle   The radius is set to the distance from the start point to the center, and the arc subtends the included angle. Specifying a positive angle causes the arc to be drawn counterclockwise. Specifying a negative angle results in the arc being drawn in a clockwise direction.

You can select the start point first:

```
Command: ARC
Center/<Start point>: (pick start point)
Center/End/<Second point>: C
```

```
Center: (pick center point)
Angle/Length of chord/<End point>: A
Included angle: (specify the angle)
```

You can also select the center point followed by the start point:

```
Command: ARC
Center/<Start point>: C
Center: (pick center point)
Start: (pick start point)
Angle/Length of chord/<End point>: A
Included angle: (specify the angle)
```

In either case, the same arc will be drawn, as shown in the following drawing.

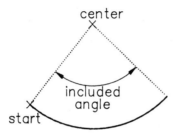

Start Point, Center, Length of Chord   The distance from the start point to the center determines the radius of the arc. The chord length — the length of the line connecting the start point and the end point of the arc — determines the ending angle of the arc. A positive chord length causes the arc to be drawn counterclockwise. A negative chord length results in the arc being drawn clockwise. You can first select the start point followed by the center point:

```
Command: ARC
Center/<Start point>: (pick start point)
Center/End/<Second point>: C
Center: (pick center point)
Angle/Length of chord/<End point>: L
Length of chord: (specify the chord length)
```

You can also select the center point and then the start point:

```
Command: ARC
Center/<Start point>: C
Center: (pick center point)
Start: (pick start point)
Angle/Length of chord/<End point>: L
Length of chord: (specify the chord length)
```

The following drawing shows the resulting arc when the chord length is a positive value.

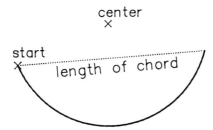

Start Point, End Point, Radius   The resulting arc always passes through the start and the end points. When you enter a positive radius, the arc is drawn counterclockwise. A negative radius causes a clockwise arc direction.

```
Command: ARC
Center/<Start point>: (pick start point)
Center/End/<Second point>: E
End point: (pick end point)
Angle/Direction/Radius/<Center point>: R
Radius: (enter radius)
```

An arc drawn with a positive radius is shown in the following drawing.

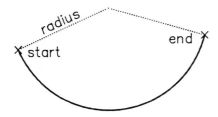

Start Point, End Point, Included Angle   The resulting arc always passes through the start and the end points. The resulting arc is drawn counterclockwise when you specify a positive angle and clockwise when you specify a negative angle.

```
Command: ARC
Center/<Start point>: (pick start point)
Center/End/<Second point>: E
End point: (pick end point)
Angle/Direction/Radius/<Center point>: A
Included angle: (specify angle)
```

The following drawing illustrates a typical arc where the included angle is a positive value.

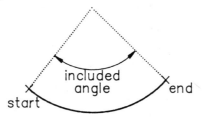

Start Point, End Point, Starting Direction   The resulting arc always passes through the start and the end points. The direction can be entered as an angle or simply as a single point. The resulting arc can be drawn clockwise or counterclockwise.

```
Command: ARC
Center/<Start point>: (pick start point)
Center/End/<Second point>: E
End point: (pick end point)
Angle/Direction/Radius/<Center point>: D
Direction from start point: (specify direction or point)
```

The following drawing shows an arc drawn by selecting the start point, end point, and starting direction.

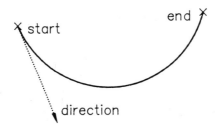

Tips   If DRAGMODE is on, you can enter *Drag* before you specify the third parameter, and the arc will be visually dragged. If DRAGMODE is set to Auto, the visual dragging of the arc will always occur (see DRAGMODE).

If you are using AutoCAD release 9 (with the standard menu ACAD.MNU), you can select the ARC command from the pull-down menu under the Draw menu bar. This selection displays an options menu in the regular screen menu area. The menu item repetition in release 9 causes these options to repeat automatically until you cancel them by pressing Ctrl-C or by selecting another command from the menu (only with AutoCAD release 9). (See Advanced User Interface and Repeating Commands.)

AREA
[all versions]

Options

E   Entity (prompts for selection of a circle or polyline)
A   Add (adds each calculated area to a running total)
S   Subtract (subtracts each subsequent area from a running total)

Overview   AREA is both an AutoCAD command and a system variable. The AREA command allows you to quickly determine the area and the perimeter of a polygon, circle, or polyline. The AREA system variable stores the last area value returned by the AREA command.

Procedure   The AREA command allows you to select the method of area calculation to be used.

```
Command: AREA
<First point>/Entity/Add/Subtract:
```

You can then simply begin to select points. AutoCAD will automatically calculate the perimeter and the area of the space that would be enclosed by straight lines connecting each point. Pressing Return closes the space by making the last side a line from the last point selected back to the first point.

```
Command: AREA
<First point>/Entity/Add/Subtract: (select a point)
Next point: (select a point)
Next point: (select a point), continuing to select points
Next point: [ RETURN ]
Area = <area>, Perimeter = <perimeter>
```

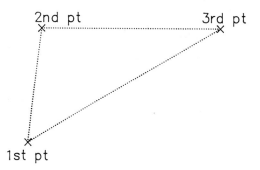

You can use the AREA command to calculate the area and the circumference of a circle or the area and the perimeter of a polyline. Enter *E* or *Entity* in response to the AREA command prompt. You will be prompted to select a circle or a polyline, and AutoCAD's pick box will replace the cross-hairs. When you select the desired circle or polyline entity, the area and the perimeter, circumference, or length will be displayed as appropriate. AutoCAD uses the centerline of the polyline to determine the area and the perimeter. Areas and perimeters of open polylines are calculated as if the polyline were closed by a straight segment from the start point to the end point. If the Add or Subtract option has been selected, the values are added or subtracted from the running totals and the prompt reissued.

```
Command: AREA
<First point>/Entity/Add/Subtract: E
Select circle or polyline: (do so)
Area = <area>, Perimeter = <perimeter>
```

Select the entity by moving the pick box to any point on the entity and pressing the pick button, as illustrated in the following drawing.

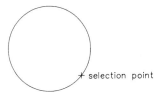

selection point

You can place the AREA command into Add mode by entering *A* in response to the AREA command prompt. Areas calculated will be displayed and added to a running total, after which the previous prompt will be repeated. A message will be displayed on the prompt line to let you know you are in Add mode. If the last area calculated was that of a circle or a polyline, the Select circle or polyline prompt will be displayed. To select individual points, press Return. When you are in Add mode, your only options are to select individual points, to select entities, or to toggle to Subtract mode.

When you are in Subtract mode (having responded to the AREA command subprompt with *S*), each calculated area will be deducted from the running total. The prompt will indicate that you are in Subtract mode.

Pressing Return in response to the options prompt completes the AREA command and returns you to AutoCAD's command prompt.

The area and the perimeter calculated will be displayed in the units format selected by the UNITS command. (See UNITS.) For example, if architectural units have been selected, the AREA command will return values in the format

```
Area = 288 square inches (2 square feet), Perimeter = 6'-0"
```

Each time you activate the AREA command, the total area is initially reset to zero.

Examples   By using a combination of Add and Subtract modes, you can calculate the area of complex shapes. Consider, for example, the gasket in the following drawing. To calculate the area of the gasket minus the area of the holes, you could draw the gasket as a closed polyline and the holes as circles. Then, using the AREA command, you could quickly calculate the area of the gasket.

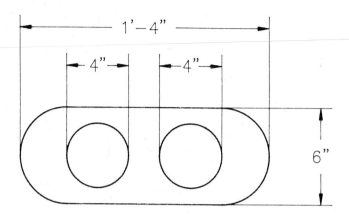

```
Command: AREA
<First point>/Entity/Add/Subtract: A

<First point>/Entity/Subtract: E
(ADD mode) Select circle or polyline: (point to polyline)
Area = 88.27 square in. (0.6130 square ft.), Perimeter = 3'-2 7/8"
Total area = 88.27 square in. (0.6130 square ft.)

(ADD mode) Select circle or polyline: ( RETURN )
<First point>/Entity/Subtract: S

<First point>/Entity/Subtract: E
(SUBTRACT mode) Select circle or polyline: (point to circle)
Area = 12.57 square in. (0.0873 square ft.), Circumference = 1'-0 9/16"
Total area = 75.71 square in. (0.5257 square ft.)

(SUBTRACT mode) Select circle or polyline: (point to other circle)
Area = 12.57 square in. (0.0873 square ft.), Circumference = 1'-0 9/16"
Total area = 63.14 square in. (0.4385 square ft.)

(SUBTRACT mode) Select circle or polyline: ( RETURN )
<First point>/Entity/Subtract: ( RETURN )
```

Warnings   The AREA command will calculate the area and the perimeter only of spaces enclosed by straight-line segments, not by arcs. To obtain the area of a space enclosed by arcs or a combination of arcs and lines, use the PLINE command to enclose the space with a polyline or use the PEDIT command to convert existing lines and arcs into polylines. Then use the AREA command to return the area enclosed. (See PLINE and PEDIT.)

If you are using a version of AutoCAD prior to version 2.6, the AREA command does not have Entity, Add, or Subtract modes.

In release 9, the AREA command calculates the area inside a spline curve. It does not "see" the frame. (See PEDIT.)

Tips

The area and the perimeter calculated by the program are stored in the AREA and PERIMETER system variables. These values are updated with each use of the AREA command. The values can be read (but not changed) using the SETVAR command or with AutoLISP.

The AREA command returns the area and circumference of circle entities. The area and perimeter are returned for closed polylines, and the area and length are returned for open polylines. For open polylines, the area is calculated as if the open polyline were closed by a straight polyline segment connecting the end point of the polyline back to the start point. But the length of the closing segment is not included in the length reported by the AREA command.

ARRAY—Polar. *See also* **ARRAY—Rectangular** [all versions]

Overview

The ARRAY—Polar command allows you to quickly make multiple copies of any number of objects. The objects are copied in a circular (polar) pattern and can be rotated as they are copied. The original objects are counted as one array item, but each object copied can be edited individually.

Procedure

After entering the command, you must select the objects you want to combine in the array. You can add to or remove from the selection set using AutoCAD's standard methods (see Entity Selection). Indicate the end of object selection by pressing Return at the Select objects: prompt. AutoCAD then asks if you want to generate a rectangular array or a polar array. You respond with a *P*, to indicate polar. You must next select the center point around which the selected objects will be arrayed. You will be prompted for three parameters: the number of times the objects are to be copied, the angle that the array will fill, and the angle between each copy of your objects. You must specify any two of the three parameters.

Once you have entered the parameters, AutoCAD asks if you want to have the objects rotated as they are copied. The default is yes and causes the objects to be rotated in relation to the rotation of the array itself each time they are copied. AutoCAD determines the distance from the array's center to the center point of each object, depending on the type of entity. Attributes, points, blocks, shapes, solids, and text use their insertion or start points. Circles and arcs use their center points. Lines, polylines, and traces are measured to their closest end points.

Examples

Number of Items and Angle to Fill   Enter the number of times the objects should be copied (including the original) and how far around the center point the array should rotate. A positive angle causes the objects to be arrayed in a counterclockwise direction. A negative angle copies them clockwise.

```
Command: ARRAY
Select objects: (select what you want to array)
Rectangular/Polar array (R/P): P (or polar)
Center point of array: (select center point)
Number of items: (enter the number)
Angle to fill (+=CCW,-=CW)<360>: (enter angle)
Rotate objects as they are copied? <Y>: (Y or N)
```

In the following drawing the original object is arrayed 4 times, the angle to fill is 180, and the object is rotated as it is copied.

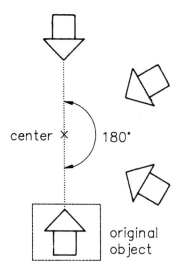

center ✕   180°

original
object

Number of Items and Angle between Items   Enter the number of times the objects should be copied (including the original). Enter zero in response to the angle to fill. Then specify the angle between each copy of the objects. A positive angle causes the objects to be arrayed in a counter-clockwise direction. A negative number arrays them clockwise.

```
Command: ARRAY
Select objects: (select what you want to array)
Rectangular/Polar array (R/P): P (or polar)
Center point of array: (select center point)
Number of items: (enter the number) (in this case, 4)
Angle to fill (+=CCW,-=CW)<360>: 0
Angle between items (+=CCW,-=CW): (enter angle) (in this case, 60)
Rotate objects as they are copied? <Y>: (Y or N) (in this case, Y)
```

The following drawing shows the resulting array.

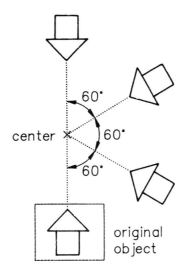

Angle to Fill and Angle between Items   Skip the number of times the objects should be copied by pressing Return at this prompt. Specify the angle to fill and the angle between each copy of the objects. If you specify one of these angles as negative, the objects will be copied in a clockwise direction. If both angles are positive, the objects will be arrayed in a counterclockwise direction.

```
Command: ARRAY
Select objects: (select what you want to array)
```

```
Rectangular/Polar array (R/P): P (or polar)
Center point of array: (select center point)
Number of items: ( RETURN )
Angle to fill (+=CCW,-=CW)<360>: (enter angle) (in this case, 180)
Angle between items (+=CCW,-=CW): (enter angle) (in this case, 60)
Rotate objects as they are copied? <Y>: (enter Y or N) (in this case, Y)
```

AutoCAD determines the number of times to copy the objects, as shown in the following drawing.

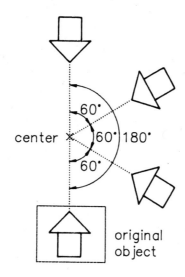

Warnings

AutoCAD determines the distance of an object from the center point of the array based on different criteria. Thus, in the construction of polar arrays, individual objects may be shifted into different relationships without being rotated as they are copied. To avoid this problem, first combine the objects into a block and then array the block. The individual copies of the block can be exploded later to allow you to edit them individually. (See BLOCK and EXPLODE.)

Tips

If you make an error, you can restore the drawing to its condition prior to the ARRAY command by entering *U* at the next command prompt (see UNDO).

ARRAY—Rectangular   *See also* ARRAY—Polar   [all versions]

Overview
The ARRAY—Rectangular command allows you to quickly make multiple copies of any number of objects. The objects will be copied in a rectangular pattern consisting of a specified number of rows and columns. The original objects are counted as one array item, but each object copied can be edited individually.

Procedure
Activate the ARRAY command and type *R* to indicate rectangular. After entering the command, you must select the objects you want to combine in the array. You can add to or remove from the selection set using Auto-CAD's standard methods (see Entity Selection). Indicate the end of object selection by pressing Return at the Select objects: prompt. AutoCAD then asks if you want to generate a rectangular array or a polar array. Respond with an *R*, to indicate rectangular.

AutoCAD then prompts for the number of rows (the number of times the objects are to be copied up or down) and the number of columns (the number of times the objects are to be copied left or right). The original objects are included in this count. Although the default value is 1, a value of 1 is meaningless as a response to both prompts. Either the number of rows or the number of columns must be greater than 1. Finally, you are asked for the unit cell, or distance between rows, and the distance between columns. If distances are entered, a positive number causes the objects to be copied to the right and/or upward. A negative value results in the objects being copied downward and/or to the left. It is also possible to indicate the distance between rows and columns simply by pointing on the drawing or by specifying points. This method specifies the unit cell, the corners of a rectangle that defines the location of the first copy of the objects in relation to the original. In this case, the second prompt for the distance between columns would be skipped.

Examples
```
Command: ARRAY
Select objects: (select the desired objects)
Rectangular/Polar array (R/P): R
Number of rows (---)<1>: (enter number) (in this case, 3)
Number of columns (¦¦¦)<1>: (enter number) (in this case, 4)
Unit cell or distance between rows (---): (enter distance or point)
Distance between columns (¦¦¦): (enter distance)
```

A typical rectangular array is shown in the following drawing.

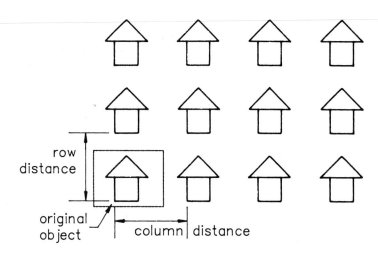

A rectangular array is drawn parallel to the cross-hair on the screen. Normally this means that the objects are arrayed along the X and Y axes. If the snap rotation angle is set to a value other than 0, however, the array can be rotated. This angle can be set with the SNAP—Rotate command or by changing the SNAPANG system variable (see SNAP).

In Figure 7, the rectangular array is generated at an angle parallel to the cross-hair.

When one or more objects must appear in the drawing in a regular repetitive fashion, the ARRAY command should be used. The ARRAY command was used to replicate the thirty-four circuit assemblies in the printed circuit board drawing in Appendix D (see Figure D-5).

Tips

If you have made an error, you can restore the drawing to its condition prior to the ARRAY command by entering *U* at the next command prompt (see UNDO).

Rectangular arrays are useful for many things. One use that is often overlooked is in the creation of charts and tables. By drawing one typical row of a chart, complete with horizontal lines, and then arraying the text and lines, you can generate a complete chart very quickly. Then all you have to do is use the CHANGE command to alter the nontypical text and draw in the vertical lines.

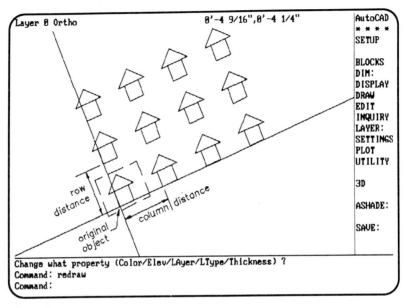

Figure 7

Associative Dimensions [ADE-1] [v2.6, r9]

Overview   The associative dimensioning feature allows you to place angular, diameter, linear, and radius dimensions into a drawing. AutoCAD's dimensioning mode "ties" those dimensions to the object that they annotate. If the object is subsequently edited, the associative dimension can be automatically updated accordingly.

Procedure   Dimensions are annotated to a drawing as either associative or nonassociative dimensions. The dimensioning system variable DIMASO controls the method used (see DIMASO). If associative dimensioning is enabled, AutoCAD automatically creates a new layer named DEFPOINTS. This layer contains the associative dimension's definition points. Although the DEFPOINTS layer is normally off, the definition points are still visible if the layer containing the dimension is visible. The definition points will not be plotted, however, unless the DEFPOINTS layer is on.

The positioning of the definition points depends on the dimension subcommand used (the individual dimensioning subcommands describe the placement of their definition points). The midpoint of the dimension text is always a definition point.

There is no difference in the operation of the dimensioning commands when you are using associative dimensions. There is a difference, however, in the editing of the dimension or the object to which it is associated. Associative dimensions, unlike nonassociative dimensions, behave as a single entity. Definition points are used to tie the object and its dimension together; thus, when editing an object and its associated dimension, you should select the object using a method that ensures that you select both the object and the definition points.

Only seven AutoCAD commands can be used to edit an object and its associative dimension. Other editing commands can be used but will not cause the associative dimension to be updated. The seven commands, along with certain restrictions in their use with associative dimensions, follow:

COMMAND	RESTRICTIONS
ARRAY	Works only with rotated polar arrays
EXTEND	Works only with linear dimensions
MIRROR	
ROTATE	
SCALE	
STRETCH	Works only with linear and angular dimensions
TRIM	Works only with linear dimensions

In addition to these commands, the dimensioning subcommands HOME-TEXT, NEWTEXT, and UPDATE permit editing of associative dimensions exclusively.

Examples   The following examples are commands used to edit associative dimensions.

ARRAY   The ARRAY—Polar command can be used to array objects along with their associative dimensions. The object must be rotated as it is arrayed.

```
Command: ARRAY
Select objects: (select point A)
Select objects: (select point B)
```

```
Select objects: ( RETURN )
Rectangular or Polar Array (R/P): P
Center point of array: ENDPOINT
of: (select point C)
Number of items: 3
Angle to fill (+=ccw -=cw) <360>: 180
Rotate objects as they are copied? <Y>: ( RETURN )
```

The following drawing illustrates what happens when an object and its associative dimension are not rotated as they are arrayed.

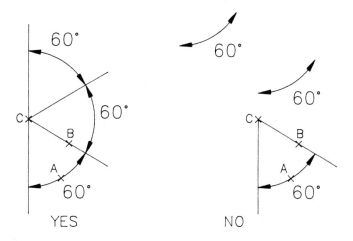

EXTEND [ADE-3]   The EXTEND command can be used to extend both an object and its associative dimension. Note, however, that if the object being extended does not lie parallel to the dimension's definition points, the result will not be correct.

```
Command: EXTEND
Select boundary edge(s)...
Select objects: (select point A)
Select objects: ( RETURN )
Select object to extend: (select point B)
Select object to extend: (select point C)
Select object to extend: ( RETURN )
```

The appearance of a typical drawing before and after the object is extended is shown in the following drawing.

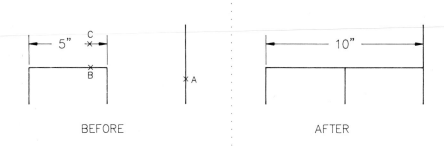

BEFORE ...... AFTER

MIRROR [ADE-2] You can use the MIRROR command to mirror an object and its associative dimension. The resulting dimension text will read in the same direction as the original dimension text, regardless of the setting of the MIRRTEXT variable. This is illustrated in the following drawing. The object on the left has been mirrored about the mirror line.

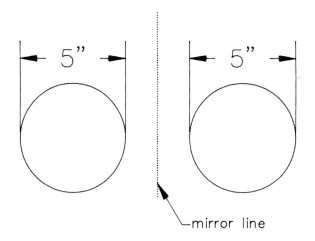

mirror line

ROTATE [ADE-3] An associative dimension that is rotated will change to reflect its new orientation.

```
Command: ROTATE
Select objects: W
First corner: (select point A)
Second corner: (select point B)
Select objects: RETURN
Base point: ENDPOINT
```

```
of: (select point C)
<Rotation angle>/Reference: 180
```

Notice in the following drawing that the dimension text retains the correct orientation.

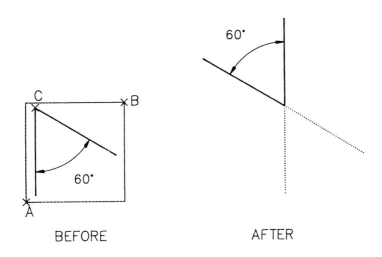

BEFORE   AFTER

SCALE [ADE-3] If both the object and the associative dimension are selected, the SCALE command will change the size of the object and correctly alter its dimensions.

```
Command: SCALE
Select objects: (select point A)
Select objects: (select point B)
Select objects: RETURN
Base point: CENTER
of: (select point A)
<Scale factor>/Reference: 2
```

The size of the arrows and the height of the dimension text are not altered, as shown in the following before and after drawing.

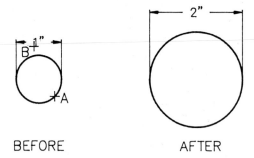

BEFORE   AFTER

STRETCH [ADE-3]   The STRETCH command is a useful command when used with associative dimensions. It can be used with any of the linear dimension types to change the object and yet maintain the original alignment of the dimension. The triangle in the following example is stretched to illustrate this feature. Notice that the aligned dimension is measured and realigned, while the vertical dimension remains vertical.

```
Command: STRETCH
Select objects to stretch by window...
Select objects: C
First corner: (select point A)
Second corner: (select point B)
Select objects: ( RETURN )
Base point: 0,0
New point: @1,1
```

This example is illustrated in the following drawing.

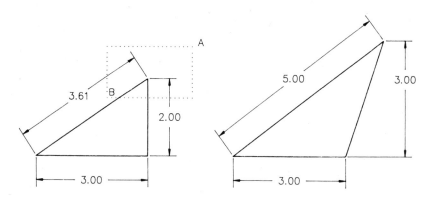

Whereas the ROTATE command cannot be used to rotate one line of an angular dimension and alter the dimension, the STRETCH command can.

```
Command: STRETCH
Select objects to stretch by window...
Select objects: C
First corner: (select point A)
Second corner: (select point B)
Select objects: [ RETURN ]
Base point: ENDPOINT
of: (select a point)
New point: (select point C)
```

The following drawing illustrates this example.

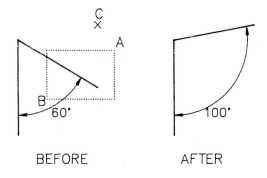

BEFORE   AFTER

TRIM [ADE-3] You can use the TRIM command to trim both an object and its associative dimension. Note, however, that if the object being trimmed does not lie parallel to the dimension's definition points, the result will not be correct (or the dimension may not be trimmed at all). In this example, the same line and dimension originally extended in the previous EXTEND example is trimmed back to its original position.

```
Command: TRIM
Select cutting edge(s)...
Select objects: (select point A)
Select objects: [ RETURN ]
Select object to trim: (select point B)
Select object to trim: (select point C)
Select object to trim: [ RETURN ]
```

This example is illustrated in the following drawing.

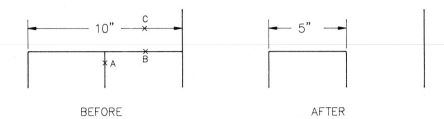

BEFORE   AFTER

Warnings   When you are editing an associative dimension, the result will take on the current values of all of the dimensioning system variables. That means the dimensioning text style, the units display mode, and even the visibility of extension lines will take on the current values. Therefore, the associative dimension may change if any of these values have changed.

Associative dimensioning is not available in AutoCAD version 2.5.

Tips   The EXPLODE command, when used on an associative dimension entity, converts the dimension into a nonassociative dimension. The individual components of the associative dimension (dimension line, arrows, and text) can then be edited individually. The individual entities will be changed to the 0 (zero) layer and will have color and linetype defined as BYBLOCK.

Associative dimension definition points can be used as object snap points with the Node object snap.

If the dimensioning system variable DIMSHO is on, associative dimensions will be constantly updated as they are dragged. This can cause very slow response. In most cases, you should set this variable to off.

Use the dimensioning subcommands HOMETEXT, NEWTEXT, and UPDATE to make changes to associative dimensions while you are in AutoCAD's Dimensioning mode. (See the individual variables by name for more information.)

Updating an associative dimension causes the dimension to be redrawn. It thus becomes the last entity, enabling you to use the BASELINE, CONTINUE, or LEADER command to append to it.

You can move the location of an associative dimension's text by using the STRETCH command (use the crossing-selection method and draw the crossing window around the dimension text). You can use the same method to change the location of the dimension line. If this results in a dimension line that no longer has to be split, the line will automatically "heal" itself.

The following drawing shows a dimension line that has been healed after the associative dimension text was stretched to a new location.

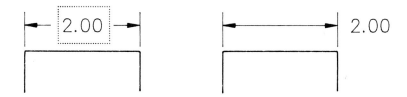

@ (At Symbol)

[all versions]

Overview The @ (at symbol) is used in AutoCAD to enter a relative coordinate. (See Coordinates.)

Procedure The easiest way to think of the @ is to imagine that it represents the last coordinate used in any drawing command. The @ tells AutoCAD to evaluate the coordinate relative to the last referenced point in the drawing.

Examples Say you have just finished drawing a line to the coordinate 6,12 and you now want to start the next line at a point exactly 14 units to the right and 3 units above that location. You could respond to the From point: prompt by entering @14,3. What you have told AutoCAD, in effect, is to start drawing the new line "at" a point 14 units to the right and 3 units above the last drawn point, the point stored by the @ key.

```
To point: 6,12
To point: ( RETURN )
Command: ( RETURN )
LINE From point: @14,3
```

Figure 8 illustrates this feature.

Tips You should become familiar with using relative rather than absolute coordinates. It is a much more flexible way to work with AutoCAD.

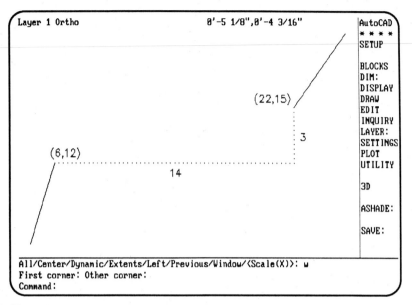

Figure 8

ATTDEF

Overview   The ATTDEF (Attribute Define) command allows you to create an attribute definition. Attributes are one of AutoCAD's basic drawing entities. An attribute is text information that can be saved as part of a block definition. When a block containing attributes is inserted into a drawing, you will be prompted to supply values (information) for each attribute. This information can be extracted later to a separate file for use by other programs.

Procedure   To define an attribute, enter the ATTDEF command.

```
Command: ATTDEF
Attribute modes -- Invisible:N  Constant:N  Verify:N
Enter (ICV) to change, RETURN when done:
```

AutoCAD will prompt for the attribute modes, indicated by the letters I (invisible), C (constant), and V (verify). Attributes can be visible or invisible, constant or variable. You can also define the attribute so that you must

46

verify its value each time the block containing it is later inserted. A visible attribute appears as text within your drawing. An invisible attribute is not seen, but the value that you enter later will be part of your drawing. You can turn attribute values on or off, regardless of their definition setting, by using the ATTDISP (Attribute Display) command. (See ATTDISP.) A constant attribute has the same value every time the block containing it is inserted into the drawing. If the attribute is variable, you are prompted to supply the value with each insertion of the block.

The attribute modes are toggles; that is, they switch from one mode to the next each time you enter the associated letter. For example, entering *I* causes the mode prompt to change to the following:

```
Attribute modes -- Invisible:Y  Constant:N  Verify:N
Enter (ICV) to change, RETURN when done:
```

When you are satisfied with the modes you have selected, press Return to indicate that you are ready to go on. You must then enter the attribute tag, which can contain any characters except blanks. The attribute tag is the name of the attribute and identifies the attribute in any disk file you extract it to. You can enter the tag name in uppercase or lowercase; AutoCAD will convert the tag name into uppercase. AutoCAD then asks for the attribute prompt. This must be worded exactly as you want to be prompted whenever the block containing the attribute is later inserted into a drawing. If you have selected Constant mode, the attribute prompt will be skipped, since it has no meaning. The default value for the attribute prompt is the attribute tag.

Finally, AutoCAD asks for a default attribute value. This is the value that will be displayed inside the default brackets < > whenever the block containing the attribute is later inserted into a drawing. If the attribute was defined as constant, you will be asked for the attribute value rather than the default attribute value. You should supply the constant value of the attribute that will be used each time the block containing the attribute is later inserted into a drawing.

If, for some reason, leading blank spaces are required in the attribute value, you can precede the value with a backslash character (\) followed by the required number of blank spaces.

Once the attribute is defined, AutoCAD issues prompts identical to those displayed when you are entering text. You can define the style, alignment, text height, and angle of the attribute text. The only difference is that you will not be prompted for the actual text string. The attribute value takes the place of the text string. Later, when you include the attribute definition in a block, the attribute value will disappear from the screen. When you insert the block containing the attribute into a drawing, the attribute value will appear in the same location with the specified text style, alignment, text height, and rotation.

```
Command: ATTDEF
Attribute modes -- Invisible:N  Constant:N  Verify:N
Enter (ICV) to change, RETURN when done: (select desired modes)
Attribute tag: (enter tag name) (remember, no blanks)
Attribute prompt: (enter text for prompt) (the tag is the default)
Default attribute value: (enter default value)
Start point or Align/Center/Fit/Middle/Right/Style: (select)
Height <default>: (select text height)
Rotation angle <default>: (select text rotation angle)
```

You can repeat the ATTDEF command by pressing Return again. When you reach the prompt for the start point, you can align the next attribute value under the previous one by pressing Return at that point.

Release 9 Only   If you are using AutoCAD release 9, the ATTDEF command will have one additional attribute mode, Preset.

```
Command: ATTDEF
Attribute modes -- Invisible:N  Constant:N  Verify:N  Preset:N
Enter (ICVP) to change, RETURN when done: (enter I, C, V, or P)
```

When an attribute is preset, AutoCAD will not prompt for its value at the time of block insertion. The preset value will be assigned automatically. You can use attribute editing commands later to change preset values (see ATTEDIT and DDATTE). All other operations of the ATTDEF command remain the same.

Examples   The following command sequence can be used to define attributes as part of a typical architectural detail marker.

```
Command: ATTDEF
Attribute modes — Invisible:N  Constant:N  Verify:N
Enter (ICV) to change, RETURN when done: ( RETURN )
Attribute tag: DETAIL
Attribute prompt: Enter detail number
Default attribute value: ?
Start point or Align/Center/Fit/Middle/Right/Style: M
```

```
Middle point: (select point A)
Height <0'-0 3/16">: ( RETURN )
Rotation angle <0>: ( RETURN )
Command: ( RETURN )
ATTDEF
Attribute modes — Invisible:N  Constant:N  Verify:N
Enter (ICV) to change, RETURN when done: ( RETURN )
Attribute tag: SHEET
Attribute prompt: Enter sheet number
Default attribute value: ?
Start point or Align/Center/Fit/Middle/Right/Style: ( RETURN )
```
The typical detail marker is shown in the following drawing.

Warnings

Once an attribute is defined as a constant and inserted into a drawing, its value cannot be changed.

If a block containing attributes is redefined, the attributes associated with any previous insertion of the block will be lost (see BLOCK).

The Preset attribute mode is not available in AutoCAD version 2.5 or 2.6.

Tips

The CHANGE command can be used to change the values and modes associated with an attribute definition before it is saved as part of a block (see CHANGE).

The current active display mode used by the ATTDEF command is saved as the system variable AFLAGS and can be changed using the SETVAR command or AutoLISP. The mode is the sum of the AFLAGS value, where 1 equals Invisible, 2 equals Constant, 4 equals Verify, and 8 equals Preset (for release 9 only). For example, an AFLAGS value of 3 would mean that the attribute was invisible and constant. (For more information, see AFLAGS.)

When a block containing several attributes is inserted into a drawing, you will be prompted to supply values for each attribute in the same order in which the attributes were originally defined. Plan your work so that you define attributes in the exact order that you wish to have them included in the block. The only way to change the order is to start over. (Note: In earlier versions of AutoCAD, using the CHANGE command to alter an attribute definition changed the order in which attributes were prompted as part of a block. Also, the attribute prompts occurred in the reverse order from the way they were defined. This is no longer the case.)

Plan the names that you will use for attribute tags to make it easier to edit attributes later by using wildcards (* and ?). For example, when you edit an attribute, specifying *ROOMN** gets both ROOMNAME and ROOMNUMBER but skips over ROOMFINISH. (See ATTEDIT.)

When you use AutoCAD's three-dimensional visualization mode, attribute definitions are assigned a zero thickness regardless of the current thickness set by the ELEV command. You can change their thicknesses later by using the CHANGE command. (See ELEV.)

ATTDIA

[ADE-2] [r9]

Overview

The ATTDIA (Attribute Dialog) system variable controls whether AutoCAD will automatically activate a dialog box for entering attribute values when you are using the INSERT command to insert a block that contains attributes. ATTDIA is an integer value.

Procedure

If the ATTDIA value is 0, attribute prompting occurs normally. In such situations, attribute prompting depends completely on the Attribute modes established by the ATTDEF command. If the value of ATTDIA is 1, however, AutoCAD automatically displays a dialog box when prompting for attribute information.

You can change the ATTDIA value with the SETVAR command or the (getvar) and (setvar) AutoLISP functions.

Examples

When the ATTDIA value equals 1, AutoCAD displays a dialog box during the INSERT command if the block contains attributes. The dialog box contains up to ten lines of attribute names and values, along with a scroll button if the block contains more than ten attributes. The attribute prompt or tag is displayed as the attribute name, and the default value is indicated.

```
Command: SETVAR
Variable name or ?: ATTDIA
New value for ATTDIA <0>: 1
```

A typical attribute dialog box is shown in Figure 9.

Warnings   Dialog boxes are displayed only if the graphics device supports the Advanced User Interface features (see Advanced User Interface).

During dialog-box editing of attributes, the Verify Attribute mode is ignored.

Tips   You disable the prompting for attribute information during the INSERT command by changing the ATTREQ system variable value.

During dialog-box editing of attributes, you must complete the block insertion by selecting the OK button inside the dialog box. Picking the Cancel button cancels the entire INSERT command and returns to the command prompt.

```
Layer 0                           7.1432,3.9447          AutoCAD
                                                         * * * *
                                                         INSERT:
                                                         ?
              Enter Attributes        ┌── Up ──┐         *
                                      │ Page up │        corner
                                                         xyz
   Enter room number....... │ 101                  │     drag
   Enter room name......... │ LIVING ROOM          │
   Enter floor material.... │ Carpet               │     Scale
   Enter base material..... │ 6" wood - profile A  │     Xscale
   Enter north wall materia │ Gypsum Board         │     Yscale
   Enter north wall finish. │ Latex - flat         │     Zscale
   Enter east wall material │ Gypsum Board         │     Rotate
   Enter east wall finish.. │ Latex - flat         │
   Enter south wall materia │ Gypsum Board         │
   Enter south wall finish. │ 1/4" plate mirror    │
                                                         __LAST__
                                      │ Page down │      DRAW
                                      │  Down │          EDIT

        ┌── OK ──┐          ┌── Cancel ──┐

  Y scale factor (default=X):
  Rotation angle <0>:
```

Figure 9

The initial ATTDIA value is determined from the prototype drawing (the default value is 0; use regular command-line prompting for attribute information). The current value is saved with the drawing file.

ATTDISP

[ADE-2] [all versions]

Options

On   All attributes visible
Off   All attributes invisible
N   Normal: visibility set individually

Overview   The ATTDISP (Attribute Display) command allows you to change the visibility of attributes in a drawing regardless of the original setting of the Invisible mode in the ATTDEF command (see ATTDEF).

Procedure   Command: **ATTDISP**
Normal/On/Off <current value>: **(enter desired visibility)**

Normal causes the attributes to display in whatever mode was assigned in the attribute definition. On makes all attributes visible regardless of their original definition. Off makes all attributes invisible regardless of their definition.

Examples   There may be times, particularly when editing attributes, that you may want to have all attribute values displayed, even those that are normally invisible. Use the ATTDISP command to make all attributes visible. Edit the necessary attributes. When you are finished, restore the Attribute Display mode to normal by executing the ATTDISP command again.

Command: **ATTDISP**
Normal/On/Off <current value>: **ON**
Command: **ATTDISP**
Normal/On/Off <current value>: **N** (restore normal display)

Warnings   Changing the attribute display mode causes the drawing to regenerate unless REGENAUTO is off (see REGENAUTO).

Tips   The current Attribute Display mode is held in the system variable ATT-MODE and is saved with the drawing.

If your drawing has many attributes in it, it will take longer to redraw or regenerate. Turning ATTDISP off while you are working on the drawing speeds up your work. Be sure you have reset the desired Attribute Display mode prior to plotting your drawing. Only attribute values that are actually visible on the screen will plot.

ATTEDIT   [ADE-2] [all versions]

Overview   The ATTEDIT (Attribute Edit) command allows you to edit attributes already present in a drawing independently of the blocks of which they are a part. The attributes can be edited individually or globally. When editing attributes individually, you can change any property (location, text height, orientation, or string value). When you are doing global edits, only the attribute string values can be altered. Global editing can include all attributes (including invisible attributes and attributes not on the screen) or can be restricted to visible attributes only.

Procedure   Command: **ATTEDIT**
Edit attributes one by one? <Y>

You are first prompted to select individual or global editing. Responding Y selects individual attribute editing. Only attributes currently visible on the screen can be selected, but the selection set can be restricted further. Responding N selects global editing. Only attribute values can be edited, but you can further limit the selected attributes.

After selecting individual or global editing, you can further restrict the attributes you will edit by specifying the block names, attribute tags, and/or specific attribute values that you want to edit. The ATTEDIT command prompts

Block name specification <*>:
Attribute tag specification <*>:
Attribute value specification <*>:

If you simply press Return, the wildcard * (asterisk) will cause all blocks, tags, and values to match. You can restrict the selection to specific block names, attribute tags, and values by entering only the name you wish to match. Or you can use wildcard combinations.

If you selected global editing, you are next asked if you wish to edit only visible attributes:

```
Global edit of Attribute values.
Edit only Attributes visible on screen? <Y>
```

If you answer Y, all of the selected attributes will be marked with an X on the screen. AutoCAD will prompt for the attribute string you wish to change and the new string:

```
String to change:
New string:
```

If you answer N, AutoCAD will flip to the text screen (on single-screen systems) and display the message

```
Drawing must be regenerated afterwards.
```

You will then be prompted for the string to change and the new string.

When editing individual attributes, you are prompted to select the attributes that you wish to edit. You can select them using any selection method. (See Entity Selection.) The first attribute found will be marked with an X. The ATTEDIT command then prompts you to select the attribute property you want to change. You can change any or all of the properties listed before selecting Next to move on to the next selected attribute.

```
Value/Position/Height/Angle/Style/Layer/Color/Next <N>:
```

You can change the attribute string value by selecting V (value). You are then prompted

```
Change or Replace? <R>
```

If you want to change only a few characters, select C (change). AutoCAD prompts for the string to change and the new string. To replace the entire value, select R (replace). AutoCAD then prompts

```
New Attribute value:
```

Examples You can use the block name, attribute tag, and attribute value specifications to limit the attributes you select for editing. In the following example, all attributes, both visible and invisible, are selected for editing. The number of one of the drawing sheets has been changed from A-5 to A-6. All of the details called out and referenced as being on sheet A-5 have to be changed to their new reference, A-6. The selection is limited to only the block named DETAIL. Selection is furthur limited to the attribute tag SHEET and the attribute value A-5. (The block that is being edited is the example from the ATTDEF command.)

```
Command: ATTEDIT
Edit Attributes one by one? <Y> N
Drawing must be regenerated afterwards.
Block name specification <*>: DETAIL
Attribute tag specification <*>: SHEET
Attribute value specification <*>: A-5
28 attributes selected.
String to change: 5
New string: 6
A-6
Regenerating drawing.
```

Thus, the sheet referenced in twenty-eight separate call-outs was changed in one step with the ATTEDIT command.

Warnings The position of the attribute can be changed but will depend on the original text placement definition (left-justified, centered, and so on).

Tips You can select attribute values that are null by entering a backslash (\) in response to the attribute value specification prompt.

   If you are using AutoCAD release 9 and your graphics card supports the new features, you can utilize the Advanced User Interface features to edit attributes from within a dialog box by using the DDATE command (see DDATE).

ATTEXT

[ADE-2] [all versions]

Options

C	CDF format
D	DXF format
S	SDF format
E	Extract selected objects only

Overview

The ATTEXT (Attribute Extract) command permits you to extract attribute entities from the drawing to a disk file for use by another program. The drawing itself is not affected. You can specify the format of the extracted file, as well as the device name, file name, and subdirectory the file will be written to. This is considered one of the most powerful AutoCAD commands because it allows other programs to use information contained in an AutoCAD drawing.

Procedure

Command: **ATTEXT**
CDF, SDF or DXF Attribute extract (or Entities)? <C>:

The attributes extracted may be limited to those entities you select by first entering an *E* and then selecting those entities by pointing, windowing, and so on. Press Return a final time to indicate that you have finished selecting entities.

Once you have selected the specific entities (or simply pressed Return to select all entities), you will need to select the format of the resulting attribute extract file. Enter *C* for Comma Delimited format (CDF), *S* for Space Delimited format (SDF), or *D* for Drawing Interchange file (DXF). CDF is the default format. CDF and SDF extracts require the use of a template file, which you must create before you use the ATTEXT command. By carefully creating the template file, you can extract specific information and format it so it can be easily used by another program or even as a text file all by itself. DXF files contain all the information associated with the selected entities, including insertion points and rotation angles. The information contained in a DXF file can be readily handled only by programs written for that specific purpose, but they are a powerful form of extract file nonetheless.

Examples   **Comma Delimited Format**   The resulting extract file will have one line for each occurrence of each block (a record), with each block reference and attribute specified in the template file normally separated by commas. String values will normally be enclosed by single quotes. The actual use of commas and/or single quotes can be altered within the template file.

```
Command: ATTEXT
CDF, SDF or DXF Attribute Extract? <C>: (enter C or just RETURN)
Template file <default>: (enter template file name)
Extract file name <drawing name>: (enter name)
```

Space Delimited Format   The resulting extract file will have one line for each occurrence of each block. The template file must specify the length of each field (block reference or attribute). The block references and attributes specified in the template file will appear in the resulting extract file but without the field separators or character string delimiters present in a CDF extract file.

```
Command: ATTEXT
CDF, SDF or DXF Attribute Extract? <C>: S
Template file <default>: (enter template file name)
Extract file name <drawing name>: (enter name)
```

Drawing Interchange File   A drawing interchange file follows Auto-CAD's DXF file format. Every piece of information associated with each block, including the insertion points and rotation angles, will be present in the file. You must select the number of decimal places used to calculate drawing coordinates and values.

```
Command: ATTEXT
CDF, SDF or DXF Attribute Extract? <C>: D
Enter decimal places of accuracy (0 to 16) <6>: (enter value)
```

Appendix D (see Figure D-3) shows a typical application of AutoCAD for geologic mapping. Each bore hole was inserted as a block along with associated attributes for the hole number, depth, and so on. This data was then output to analysis programs using the ATTEXT command.

Warnings   Both the template file and the extract file for CDF and SDF formats use the file extension .TXT. Be careful not to specify the same subdirectory and file name for the template and the extract file name. If you do, the resulting attribute extract file will overwrite the template file.

If you reuse an extract file name that was used previously, the new data will overwrite the older file.

When you use the SDF form of attribute extraction, if the template file specifies that a field will be a particular length but it is actually longer when extracted, AutoCAD will leave off any characters to the right of the specified length and report

```
** Field overflow in record <record number>
```

Tips   You can direct the extract file to the screen by specifying an extract file name *CON:*.

You can direct the extract file to a printer by specifying a printer port as the extract file name, for example, *PRN:* or *LPT2:*. Make sure the printer is connected and turned on; otherwise, you may cause your computer to lock up.

You can use the same template file for CDF and SDF formats if you plan ahead when designing the template. No significant differences in the template files are necessary for SDF and CDF extract files.

ATTMODE   [ADE-2] [all versions]

Overview   The ATTMODE system variable controls the current attribute display mode.

Procedure   The ATTMODE variable is normally controlled by the ATTDISP command, but it can also be changed with the SETVAR command and the AutoLISP (getvar) and (setvar) functions. The ATTMODE variable is saved with the drawing. A value of 0 causes all attributes to be invisible; a value of 1 causes attributes to display normally (visibility set individually, as determined by the ATTDEF command or the AFLAGS setting); a value of 2 causes all attributes to be visible regardless of their definitions. (See ATTDEF, ATT-DISP, and AFLAGS.)

Examples   ```
Command: SETVAR
Variable name or ?: ATTMODE
New value for ATTMODE <1>: 2
```
```

Warnings   Changing the ATTMODE value will not change the drawing until the drawing is regenerated. Regeneration is not automatic.

Tips   The initial ATTMODE value is determined from the prototype drawing (default value is 1, normal). The current value is saved with the drawing.

Remember, you can use the transparent 'SETVAR command to alter a system variable while another command is active.

ATTREQ [ADE-2] [r9]

Overview   The ATTREQ (Attribute Request) system variable determines whether AutoCAD will prompt for attribute values or use the default values during the insertion of blocks containing attributes.

Procedure   When the ATTREQ value is 1, attribute prompting occurs. An ATTREQ value of 0 disables the prompting for attribute values during the INSERT command. The default values, assigned during the ATTDEF command, will be automatically assigned during the insertion of the block. You can change these values later by using one of the attribute editing commands.

You can change the ATTREQ value with the SETVAR command or the (getvar) and (setvar) AutoLISP functions.

Examples   You can disable the prompting for attribute information during the INSERT command by changing the ATTREQ value to 0.

```
Command: SETVAR
Variable name or ?: ATTREQ
New value for ATTDIA <1>: 0
```

Tips   If ATTREQ has a value of 1, the method used to prompt for attribute information is determined by the ATTDIA system variable.

The initial ATTREQ value is determined from the prototype drawing (default value is 1, prompt for attribute information). The current value is saved with the drawing.

AUNITS

[ADE-1] [all versions]

Overview   The AUNITS (Angle Units) system variable determines the method or mode of angle representation.

Procedure   The AUNITS value is changed by the UNITS command or by the SETVAR command or AutoLISP's (getvar) and (setvar) functions. AUNITS is an integer with the following values representing the available settings:

VALUE	SETTING
0	Decimal degrees
1	Degrees/minutes/seconds
2	Grads
3	Radians
4	Surveyor's units

Examples   You can change the type of angle units used within a drawing, even while another command is active, by using the transparent 'SETVAR command to change the AUNITS value. In the following example, the AUNITS value is changed from decimal degrees to surveyor's units while the LINE command is active. The "rubberband" cursor disappears temporarily while the 'SETVAR command is being used but returns as soon as the LINE command is resumed. The representation of the angle unit also is changed.

```
To point: 'SETVAR
>>Variable name or ?: AUNITS
>>New value for AUNITS <0>: 4
Resuming LINE command.
To point:
```

Tips   The initial AUNITS value is determined from the prototype drawing (default value is 0, decimal degrees). The current value is saved with the drawing.

Use the transparent 'SETVAR command to change the AUNITS value whenever the need arises. If you need to do this often, build the feature into your menu.

AUPREC

[ADE-1] [all versions]

Overview   The AUPREC (Angle Unit Places Record) system variable controls the number of decimal places displayed when AutoCAD displays an angle measure.

Procedure   The AUPREC value is an integer. It may be changed with the UNITS command or with the SETVAR command or AutoLISP's (getvar) and (setvar) functions.

Examples   You can change the number of decimal places used to display angles, even while another command is active, by using the transparent 'SETVAR command to change the AUPREC value. In the following example, the AUPREC value is changed while the LINE command is active. The "rubberband" cursor disappears temporarily while the 'SETVAR command is being used but returns as soon as the LINE command is resumed. The representation of the angle unit also is changed.

```
To point: 'SETVAR
>>Variable name or ?: AUPREC
>>New value for AUPREC <4>: 6
Resuming LINE command.
To point:
```

Tips   The initial AUPREC value is determined from the prototype drawing (default value is 0 decimal places). The current value is saved with the drawing.

   The number of decimal places set by the AUPREC variable also determines the number of decimal places in the default value of the angular dimension for the dimension text (see DIM command).

AutoLISP

[ADE-3] [all versions]

Overview
AutoLISP is a programming language actually built into the AutoCAD program. With AutoLISP you can perform arithmetic operations, create variables, execute AutoCAD commands, assign values to system variables, and even create entirely new AutoCAD commands. It is the power of AutoLISP that enables many of the available third-party programs to provide enhancements to the basic AutoCAD program. An explanation of the programming and use of the AutoLISP language is beyond the scope of this book, but it is not necessary to learn AutoLISP to use AutoCAD.

Procedure
You can enter AutoLISP commands directly on the AutoCAD command line by enclosing them in parentheses. Most AutoLISP routines are saved in files with the file-name extension .LSP. Some routines are incorporated as part of AutoCAD menu files. Those that are part of menu files are run when the particular menu entry is selected. Routines contained in Auto-LISP files require that the files first be loaded.

To load an AutoLISP file, at the AutoCAD command prompt, enter the name of the AutoLISP file:

Command: **(load "d:/path/file name")**

where *d:* is the name of the drive on which the file is located, */path* is the path to the subdirectory in which the file is located, and *file name* is the actual name of the file. Do not include the .LSP file extension. The drive designation or path may be omitted if it is the current drive or subdirectory or is on the computer's path.

Examples
A series of AutoLISP routines are provided with the AutoCAD package. These routines are used to draw three-dimensional representations of cones, spheres, domes, and toruses. These routines are provided on the Support Files disk in the file 3D.LSP. Before you can use these routines, you must first load this file. Assuming that the file has been copied into your \ACAD subdirectory, load the file as follows:

Command: **(load "3D")**

AutoCAD will display the name of the first AutoLISP routine in the file. The other routines are DOME, DISH, SPHERE, and TORUS. To run any of these routines, at the AutoCAD command prompt simply enter the name of the routine as if it were an AutoCAD command.

Warnings   If AutoCAD displays the error message `** Insufficient node space **`, you need to increase the LISPHEAP environment variable (see Environment Variables).

To run an AutoLISP routine from the AutoCAD command prompt, you first must have defined the routine as a function by using the AutoLISP function *defun* and the format *defun C:XXXX*. Refer to the *AutoLISP Programmer's Reference* for a detailed description.

Tips   If you are interested in learning more about AutoLISP, all copies of AutoCAD that include the ADE-3 package also include Autodesk's *AutoLISP Programmer's Reference*. Other suggested books are *LISP*, by Winston and Horn (second edition), and *Looking at LISP*, by Tony Hasemer, both published by Addison-Wesley.

AutoCAD will automatically load the file ACAD.LSP if it exists and is saved in the same directory as the AutoCAD program files.

AXIS

[ADE-1] [all versions]

Options

On	Turns ruler lines on
Off	Turns ruler lines off
S	Locks ruler to snap spacing
A	Sets aspect [ADE-2]
number	Sets tick spacing
*number*X	Sets tick spacing as multiple of snap

Overview   The AXIS command lets you display ruler lines and change the spacing of the tick marks on the ruler. In addition, you can vary the spacing of the tick marks along the X-axis and the Y-axis.

Procedure

```
Command: AXIS
Tick spacing(X) or ON/OFF/Snap/Aspect <current value>:
```

Selecting ON turns the axis ruler lines on. Selecting OFF turns them off. Entering a value sets the tick spacing to a multiple of drawing units. For example, say you have selected architectural units. Entering 5 would place tick marks at every fifth drawing unit, in this case, every 5 inches. You can lock the tick marks along the ruler line to the current snap spacing by entering an *S* (for snap) or by entering a tick spacing value of zero. You can also set the axis to a multiple of the snap resolution by entering a value followed by an *X* (for example, *5X* would place axis ticks at every fifth snap point). Later, if the snap spacing is changed, the tick spacing will also change accordingly. (See SNAP.)

The Aspect option allows you to set different tick spacing for the X-axis and the Y-axis.

Examples

The following example sets the axis ruler lines to a multiple of the snap spacing. The X-axis is set to ten snap multiples, and the Y-axis is set to five snap multiples, as shown in Figure 10.

```
Command: AXIS
Tick spacing(X) or ON/OFF/Snap/Aspect <0'-0">: A
Horizontal spacing(X) <0'-0">: 10X
Vertical spacing(X) <0'-0">: 5X
Command: AXIS
Tick spacing(X) or ON/OFF/Snap/Aspect <A>: ON
```

Warnings

If the tick spacing is too small (or you use the ZOOM command to see more of your drawing) so that the axis display is not discernible, AutoCAD warns that

```
Axis ticks too close to display.
```

If the current snap style is Isometric, the Aspect option will not be usable.

Tips

When architectural or engineering units are in effect and the tick spacing is an exact fraction of an inch or a foot, some tick marks will be twice as long to indicate whole feet or inches.

The axis display can be turned on and off with the AXISMODE system variable. An AXISMODE value of 0 turns the axis display off; a value of 2

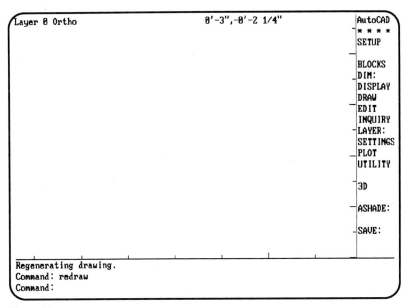

Layer 0 Ortho   0'-3",-0'-2 1/4"   AutoCAD
* * * *
SETUP

BLOCKS
DIM:
DISPLAY
DRAW
EDIT
INQUIRY
LAYER:
SETTINGS
PLOT
UTILITY

3D

ASHADE:

SAVE:

Regenerating drawing.
Command: redraw
Command:

Figure 10

turns the axis on; and a value of 1 returns the axis display to whatever the AXIS command was last set to. The spacing of the tick marks can be changed with the AXISUNIT system variable. Both the X and the Y can be specified, separated by a comma. You can change both system variables by using the SETVAR command or AutoLISP.

When you are using AutoCAD's three-dimensional visualization, the axis is displayed only when the VPOINT is set to 0,0,1 (see VPOINT).

You can also turn the AXIS on and off and change the axis spacing by using the dialog box provided by the DDRMODES command (available only in AutoCAD release 9; see DRRMODES).

AXISMODE   [ADE-1] [all versions]

Overview   The AXISMODE system variable controls whether AutoCAD's axis ruler line is turned on or off.

Procedure   An AXISMODE value of 0 means the ruler line is off; a value of 1 turns the ruler on. The AXIS command, the SETVAR command, and the AutoLISP functions (getvar) and (setvar) can all be used to alter the value.

Examples   The AXIS command will change the AXISMODE value. If the axis display is currently off, the AXISMODE value is 0. If you use the AXIS command to turn the axis on, the AXISMODE value has been changed to 1.

You can change the AXISMODE system variable by using the transparent 'SETVAR command even while another command is active. The following sequence will turn the axis on while the LINE command is being used.

```
To point: 'SETVAR
>>Variable name or ?: AXISMODE
>>New value for AXISMODE <0>: 1
Resuming LINE command.
To point:
```

Warnings   Changing the AXISMODE value will have no effect until a REDRAW command is executed. You can make the AXISMODE value change visible by adding a transparent 'REDRAW command after the sequence in the previous example.

Tips   The initial AXISMODE value is determined from the prototype drawing (default value is 0, axis is off). The current value is saved with the drawing.

If you use the screen axis often, you may want to use the transparent 'SETVAR command to build a transparent AXISMODE toggle into your AutoCAD menu system.

You can alter the AXISMODE value by selecting the axis check button in the dialog box presented by the DRRMODES command (available only in AutoCAD release 9; see DRRMODES).

AXISUNIT

[ADE-1] [all versions]

Overview   The AXISUNIT system variable controls the spacing of the tick marks along AutoCAD's axis ruler line.

Procedure   The AXISUNIT value is represented by a point (X,Y); thus, the variable can set different tick spacings along the X-axis and the Y-axis. The AXIS command, the SETVAR command, and the AutoLISP functions (getvar) and (setvar) can all be used to alter the AXISUNIT value.

Examples   The AXIS command will change the AXISUNIT value. If you use the AXIS command to change the spacing of axis ticks, the AXISUNIT value will be changed.

    You can use the transparent 'SETVAR command to change the AXIS-UNIT value even while another command is active. The following sequence will change the axis tick spacing from a current value of 1,1 to a new value of 5,5 while the LINE command is in use.

```
To point: 'SETVAR
>>Variable name or ?: AXISUNIT
>>New value for AXISUNIT <1,1>: 5,5
Resuming LINE command.
To point:
```

Warnings   Changing the AXISUNIT value will have no effect until a REDRAW command is executed. You can make the AXISUNIT value change visible by adding a transparent 'REDRAW command after the sequence in the previous example.

Tips   The initial AXISUNIT value is determined from the prototype drawing (default value is 0.0000,0.0000 — AutoCAD will match the AXISUNIT value to the current SNAP setting). The current value is saved with the drawing.

    You can alter the AXISUNIT value by entering new values from the dialog box presented by the DRRMODES command (available only in AutoCAD release 9; see DRRMODES).

Backup
[all versions]

Overview   AutoCAD saves a backup copy of your drawing file whenever you save or end the drawing.

Procedure Every time you save an AutoCAD drawing using the SAVE or the END command, AutoCAD always takes the previous version of the drawing file and makes it the backup file. What the program actually does is change the file extension from .DWG to .BAK. Thus, you are always assured of having the latest version of your drawing saved as a drawing file and the next most recent version as a backup, in case anything happens to your drawing. The previous version of a drawing's backup file is replaced by the newer version every time the drawing is saved.

Examples When you first save a drawing, AutoCAD adds the file extension .DWG. Subsequently, saving or ending the drawing will create a new file with the file extension .DWG and rename the old drawing file with the file extension .BAK. Thus, the first time you save a drawing with the filename EXAMPLE, AutoCAD names it EXAMPLE.DWG. If you continue to edit that drawing and save it again, the current drawing (the drawing as it appears on the screen) is saved as EXAMPLE.DWG and the old EXAMPLE.DWG file is renamed EXAMPLE.BAK. Any previous EXAMPLE.BAK file is discarded.

Tips To conserve disk space, some users erase all of the backup files each day when they start up their computer. This practice does free up space that would otherwise be taken up by duplicate copies of drawing files. It is not a good practice, however, unless you first establish an effective method of backing up your drawing files every day. The simplest way to do this is to always maintain a duplicate copy of the latest version of each drawing file on a separate floppy disk. This way, you will have every drawing stored on both your hard disk and on a floppy disk. Maintaining a .BAK file for every .DWG file would then be unnecessary.

BASE

[all versions]

Overview AutoCAD allows any existing drawing to be inserted into any other drawing as a block. The reference point within the inserted drawing (block) corresponding to the insertion point supplied during the INSERT command is the base point (see BLOCK and INSERT). By default, the base point is the 0,0 coordinate of the drawing to be inserted. The BASE command is used within the drawing to be inserted to set the base point to something other than 0,0.

Procedure Use the BASE command while you are in the drawing that will later be inserted.

```
Command: BASE
Base point: (specify coordinate or point on screen)
```

Examples Use the BASE command to determine the precise reference point for later insertions of the current drawing into another drawing. This is particularly useful when you are creating standard details or a symbol library for subsequent use. In the following drawing, the symbol was first drawn and then the BASE command was used, in conjunction with the center object snap override, to set the reference point to the center of the circle.

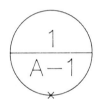

```
Command: BASE
Base point: CENTER
of: (select any point on circle)
```

Tips Use the BASE command so that you will know for sure where the inserted block drawing will occur in reference to the insertion point.

The base point is stored in the INSBASE system variable. You can change the base point by using the SETVAR command or AutoLISP.

You can use the BASE command to change the Z-coordinate of the insertion base point (only in version 2.6 and release 9 when ADE-3 is included).

BLIPMODE

[all versions]

Options

On   Enables temporary marker blips
Off   Disables temporary marker blips

Overview   The BLIPMODE command allows you to enable or suppress the generation of temporary marker blips on the graphics screen every time you select a point. The blips are not part of the drawing and are removed from the drawing when you execute a PAN, REDRAW, REGEN, or ZOOM command.

The BLIPMODE system variable stores the current status of the blip display.

Procedure   Command: **BLIPMODE**
ON/OFF <*current status*>:

Select ON to cause blips to appear. Select OFF to suppress the blips.

Using the BLIPMODE command to change the status of the blip display mode changes the BLIPMODE system variable setting. You can turn the blip display on and off with the SETVAR command or from within AutoLISP by changing the setting of the BLIPMODE system variable. The marker blips will display if BLIPMODE equals 1; they will be suppressed if BLIPMODE equals zero.

Examples   Although the enabling or disabling of blip display is often a personal preference, when you are writing an AutoLISP routine that will draw or select objects on the screen, it is generally wise to turn BLIPMODE off. You must, however, reset the BLIPMODE variable to its original condition when the AutoLISP routine has finished executing. The following example can be combined with any AutoLISP routine. It reads the current BLIPMODE value and saves it in the variable USRBLIP. It then resets the BLIPMODE value to its original condition at the end.

PROGRAM	EXPLANATION
`(setq USRBLIP (getvar "BLIPMODE"))`	Obtains current value
`(setvar "BLIPMODE" 0)`	Sets value to 0
.	Other parts of the routine would
.	occur here
`(setvar "BLIPMODE" USRBLIP)`	Restores original value

Tips   The initial BLIPMODE system variable value is determined from the prototype drawing (default value is 1, display blips). The current value is saved with the drawing.

The displaying of marker blips is handy in some instances. For example, when you are using object snaps to select the endpoint, midpoint, and so on, you can immediately determine if you have "snapped" to the right point if BLIPMODE is on.

When you are selecting many objects, the marker blips can clutter the screen. Turn BLIPMODE off in this instance.

You can turn BLIPMODE on and off by using the blips check button from within the DRRMODES dialog box (available only in release 9; see DRRMODES).

BLOCK

[all versions]

Options

?   Lists names of defined blocks

Overview   The BLOCK command lets you combine selected entities in a drawing into a single grouping, which you can then manipulate as a whole. Once combined into a grouping, the block can be inserted into the drawing in which it was created whenever you choose. Each insertion of the block can have different scale factors and rotation angles. The entities that make up the block are treated as a single object by all other AutoCAD commands. A special feature of blocks is that they can include attribute definitions. When you insert a block with attributes, you will be prompted to provide the attribute value (see ATTDEF).

71

Procedure Once you have drawn the objects, you can combine them into a block using the BLOCK command.

```
Command: BLOCK
Block name (or ?): (enter name) (up to 31 characters)
Insertion base point: (enter coordinate or point)
Select objects: (select by pointing, windowing, etc.)
```

The insertion base point you select will be the reference point that corresponds to the insertion point when you insert the block back into the drawing, so choose it wisely (see INSERT).

If you have forgotten the block names you have already used, enter a question mark (?) in response to `Block name (or ?):`. AutoCAD will list the names of all blocks in the drawing.

Examples The following example saves the illustrated symbol (a diaphragm valve) as a block with the name D-VALVE. The insertion point is the indicated intersection point, selected using the intersection object snap override. The symbol is then selected by using the windowing method of object selection.

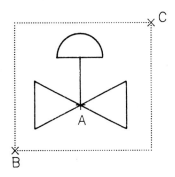

```
Command: BLOCK
Block name (or ?): D-VALVE
Insertion base point: INTERSECTION
of: (select point A)
Select objects: W
First corner: (select point B)
Second corner: (select point C) (objects will be highlighted)
Select objects: ( RETURN )
```

The selected entities that make up the block will be erased from the drawing. This helps to ensure that the block has been created properly. If you want to restore the individual entities to the screen, enter the OOPS command immediately after completing the BLOCK command. The block will have been defined. The entities restored to the screen will be individual entities, not part of the new block.

The airplane drawing (see Figure D-1 in Appendix D) is actually a two-dimensional drawing, with the appearance of perspective created by saving the internal structural pieces as blocks. These blocks were then inserted back into the drawing at successively larger scales.

Warnings   Each block name in a drawing must be unique. If you reuse a name, AutoCAD will warn you. For example, if you have a block named CHAIR and you use that name again in the block command, AutoCAD will prompt

```
Block CHAIR already exists.
Redefine it? <N>:
```

If you don't mean to change the definition of the block, press Return to indicate no. If you tell AutoCAD that you do want to redefine the block, the BLOCK command will proceed normally. After completion, all copies of the block in the drawing will be changed to the new description and the drawing will regenerate, unless REGENAUTO is off (see REGENAUTO). This is a very powerful feature.

All entities within blocks that were drawn on the layer named 0 (the default layer when a drawing is started, called the universal layer; see LAYER) will be drawn on whatever layer is current when the block is later inserted. All entities drawn on layers other than the universal layer will remain on those layers when the block is inserted.

Block definitions can contain other blocks. The individual block definitions are not revised when you redefine a block unless they are specifically revised. For example, if you have a block named CAR that contains a block named WHEEL, revising the CAR block does not automatically revise the WHEEL block. If you want to change the WHEEL block, you must do so explicitly.

If a block is defined with a color or linetype specified as BYLAYER or BYBLOCK, the block's color or linetype will be that of the layer on which it was created (blocks not on the universal layer) or on which it was inserted.

Tips   Give your blocks meaningful names and keep a record of them. If you don't like the name you have chosen for a block, you can change it using the RENAME command (see RENAME).

If you make a mistake when you create a block, simply redefine the block by creating a new block and reusing the same name. If the block definition was the last thing you did, you can remove the definition by typing *U* at the next command prompt (see UNDO).

Redefining blocks allows you to quickly update a change throughout a drawing. Every occurrence of a block will change when you redefine it.

A block is treated as an entity. The individual entities that make up a block cannot be edited individually. You can use the EXPLODE command to break a block back into its original entities (see EXPLODE), but then it is no longer a block.

Blocks that will be inserted onto several different layers, or that will need to be changed from one layer to another, must be created initially on the universal layer. If you want to be able to change the color or linetype of a block later, be sure color and linetype are specified initially as BYBLOCK. If you want the block to always assume the color and linetype of the layer on which it is inserted, specify the color and linetype as BYLAYER.

If a block always needs to be on a particular layer or have a particular color or linetype, assign it explicitly.

When entering an insertion point, you can specify the Z-coordinate. If you do, AutoCAD warns you (only in AutoCAD version 2.6 and release 9 when ADE-3 is included) by displaying the message

```
Beware, Z insertion base is not zero
```

BREAK   [ADE-1] [all versions]

Options

F   Respecify first point

Overview   Use the BREAK command to erase part of an object or to split it into two objects. The BREAK command operates on one object at a time and can be used only with arcs, circles, lines, polylines, and traces.

The BREAK command behaves differently, depending on which type of object is being broken. Arcs, lines, polylines, and traces are broken in two if the two points chosen fall along their length. If one point chosen for the break is off the end, the object is shortened. Traces and polylines with widths greater than zero are cut off square. Circles are broken counterclockwise from the first point to the second (subject to the ANGDIR system variable setting). Breaking polylines causes closed polylines to become open polylines. Breaking polylines that have been curve-fitted makes the curve-fit information permanent.

Procedure   Enter the command, select the object you want to break, select the first point, then select the second point. You can use any AutoCAD selection method (pointing, windowing, and so on) to select the object to be broken. You can select only one object for each execution of the command. If you select the object to be broken by windowing, AutoCAD uses the prompts

```
Command: BREAK
Select object: (select by windowing)
Select first point: (select)
Select second point: (select)
```

If you point to the object to be broken, AutoCAD issues a slightly different prompt. AutoCAD assumes that the point used to select the object to be broken is also the first break point.

```
Command: BREAK
Select object: (point to the object)
Enter second point (or F for first point):
```

If you selected only the object and now want to select the two break points, responding with an F causes AutoCAD to prompt

```
Enter first point:
Enter second point:
```

If you simply want to break an object in two without removing any part of the object, respond to Enter second point: by entering @. This tells Auto-CAD to use the last coordinate selected.

Examples   The following example is probably the most common usage of the BREAK command. The BREAK command is activated. AutoCAD's cross-hair cursor is replaced by the pick box, and you are prompted to select an object. In this case, the object is selected simply by pointing. In response to the prompt to select the second point, *F* is entered to allow independent selection of the two break points. When the second point is selected, the portion of the line between the two break points is immediately erased, and AutoCAD's command prompt reappears.

```
Command: BREAK
Select object: (point to the object)
Enter second point (or F for first point): F
Enter first point: (select first point A)
Enter second point: (select second point B)
```

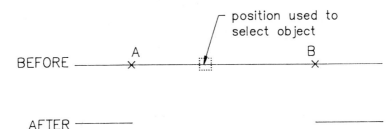

Warnings   When you use the BREAK command to shorten an object, if the second point selected is too far beyond the end of the object, the wrong part of the object may be deleted.

If you use object snap to select the intersection of two lines both to select the object to be broken and to select the first break point, Auto-CAD may choose the wrong object to break. AutoCAD will always choose the object most recently drawn. To select the correct object, first pick the object, then type *F* to go back and explicitly select the first and second break points.

Which portion of a circle will be erased depends on the direction in which AutoCAD evaluates angles (normally counterclockwise, determined by the setting of the ANGDIR system variable). The portion of the circle will be erased progressing counterclockwise from the first break point to the second break point.

Tips   If you make a mistake, you can restore the object just broken to its original unbroken condition by typing *U* at the next command prompt (see UNDO).

76

Build the F option into a menu macro so you can first pick the object to be broken and then select the break points. This macro and several other BREAK command macros are included as part of the standard AutoCAD menu.

Button Menu

[all versions]

Overview   If you are using a mouse or digitizer "puck" with more than one button, AutoCAD can use those additional buttons to enter commands.

Procedure   The first button on every mouse or puck is reserved as AutoCAD's "pick" button. This button is always used to select points on the screen or to select screen menu items. The extra buttons can be programmed within an Auto-CAD menu file (a text file with the file extension .MNU).

The button menu section of a menu file is simply individual lines that assign the functions of any available buttons. If the pointing device you are using has only one button, additional button assignments are ignored. Each button can be assigned any function, command, or series of commands (macro). You can even use other menu macros to call replaceable button menus (where the original button menu is replaced by a different set of button functions). For example, whenever you execute the LINE command, you might want your buttons reassigned so that each button functions as a different Object Snap mode. Menu swapping is accomplished by embedding the menu command $B= within a macro.

Examples   The AutoCAD program is supplied with a standard menu file that includes button definitions for the first nine buttons on a multibutton puck. You can change any of these command/function assignments, except the pick button. You can use any standard text editor to create or change a menu file. The following functions are assigned to any additional buttons in AutoCAD's standard menu:

BUTTON	KEY	FUNCTION
1	(Return)	
2		Displays "object snap overrides" submenu on screen menu area
3	(Ctrl-C)	Cancels
4	(Ctrl-B)	Toggles Snap mode on/off
5	(Ctrl-O)	Toggles Ortho mode on/off
6	(Ctrl-G)	Toggles Grid mode on/off
7	(Ctrl-D)	Toggles coordinate display
8	(Ctrl-E)	Toggles Isoplane left/top/right
9	(Ctrl-T)	Toggles Tablet mode on/off

These functions are assigned by the following button menu code within the standard ACAD.MNU file:

FUNCTION	EXPLANATION
***BUTTONS	Button menu section header
;	Semicolon is same as a Return
$S=osnapb	$S = calls a screen menu
^C^C	^C = cancels
^B	Toggles Snap mode
^O	Toggles Ortho mode
^G	Toggles Grid mode
^D	Toggles coordinate display
^E	Toggles Isoplane
^T	Toggles Tablet mode

Tips   If you intend to customize AutoCAD's ACAD.MNU standard menu, make a copy of it first. The button menu section of a menu file follows immediately after the line ***BUTTONS. Each button assignment occurs on a separate line.

You can include control codes in menus by preceding the character with a caret (^). Thus, ^C is the same as Ctrl-C and ^H is the same as a backspace.

Spread out long menu macros over more than one line by placing a plus sign as the last character of the line and continuing the macro on the next line.

A semicolon (;) has the same effect as a Return.

A backslash (\) within a menu macro causes the macro to pause for user input. Once the input is provided, the macro continues. In button menus, you can use the first backslash to return the coordinates of the screen cross-hair.

If your version of AutoCAD includes ADE-3, you can include AutoLISP macros within your menu files.

If you are using AutoCAD release 9 (with the standard AutoCAD menu ACAD.MNU), the button normally used to select object snap overrides causes the Tools pull-down menu to be displayed. This pull-down menu allows you to select an object snap. (Available only with AutoCAD release 9; see Advanced User Interface.)

CDATE

[v2.5, v2.6, r9]

Overview The current date and time in calendar and clock format are stored in the CDATE (Current Date) system variable. CDATE is a read-only variable. Its value cannot be altered from within AutoCAD.

Procedure The CDATE value is read from the computer's system clock and is reported as a real number. The format for the CDATE variable is

YYYYMMDD.HHMMSSmsec

where

VARIABLE	MEANING (VALUES)
YYYY	year
MM	month (01 – 12)
DD	day (01 – 31)
HH	hours (00 – 23)
MM	minutes (00 – 59)
SS	seconds (00 – 59)
msec	milliseconds (000 – 999)

Examples   If the current time is 340 milliseconds past the hour of 10:48:25 PM on July 13, 1987, the CDATE variable would contain the following value:

```
19870713.224825340
```

This value can be read using the SETVAR command.

```
Command: SETVAR
Variable name or ?: CDATE
CDATE = 19870713.224825340 (read only)
```

Warnings   CDATE values are easily compared to determine which one is earlier, but they do not easily relate mathematically. The other time system variables are mathematically related and can be more easily used in AutoLISP formulas that utilize time and date checks.

CDATE is a read-only variable. You can read it by using either the SETVAR command or the AutoLISP (getvar) function. CDATE cannot be changed from within AutoCAD. It is possible to change the CDATE value by using the DOS DATE and TIME commands to alter the system clock.

CECOLOR   [v2.5, v2.6, r9]

Overview   The CECOLOR (Current Entity Color) system variable stores the current entity color, which is controlled by the COLOR command.

Procedure   The color name or number stored in the CECOLOR variable is represented as a string and is stored with the drawing. CECOLOR is a read-only variable. You can read its value by using the SETVAR command or the AutoLISP (getvar) function, but you cannot change it directly. The current entity color is controlled by the COLOR command.

Examples   In most instances, you will assign a particular color to each layer within your drawing. Afterward, every object drawn on a layer will inherit the color associated with that layer. It is possible, however, to override the layer's color and to specify the color for all subsequent entities explicitly. This is accomplished with the COLOR command. When you are drawing entities whose color is determined by the color assigned to the current layer, the CECOLOR variable has the value BYLAYER. However, if you use the COLOR command to determine the current color explicitly, the CECOLOR variable takes on the color name established by the COLOR command.

This example first uses the SETVAR command to look at the value of CECOLOR and then the COLOR command to change the current color to RED. Afterward, use of the SETVAR command ascertains that the CECO-LOR value has indeed changed.

```
Command: SETVAR
Variable name or ?: CECOLOR
CECOLOR = "BYLAYER" (read only)
Command: COLOR
New entity color <BYLAYER>: RED
Command: SETVAR
Variable name or ? <CECOLOR>: ( RETURN )
CECOLOR = "RED" (read only)
```

Warnings   Mixing color definitions can become confusing (see COLOR).

Tips   The initial CECOLOR value is determined from the prototype drawing (default value is BYLAYER). The current value is saved with the drawing file.

Although you cannot alter the CECOLOR variable except by using the COLOR command, you can use it to quickly determine the current color setting. Color can be set as BYLAYER or BYBLOCK, or explicitly set to a color. (See COLOR.)

You can also alter the current CECOLOR setting by changing the color setting from within the Select Color menu accessed from the Entity Creation Modes dialog box. Access this dialog box with the DDEMODES command (available only in release 9; see DDEMODES).

CELTYPE   [v2.5, v2.6, r9]

Overview   CELTYPE (Current Entity Linetype) is the AutoCAD system variable that stores the current entity linetype set by the LINETYPE command.

Procedure   The linetype name stored in the CELTYPE variable is represented as a string and is stored with the drawing. CELTYPE is a read-only variable.

You can observe its value by using the SETVAR command or the AutoLISP (getvar) function, but you cannot change the CELTYPE variable directly. The current linetype is controlled by the LINETYPE command.

Examples

AutoCAD allows you to assign one specific linetype to each layer you create. Generally, all entities drawn on a layer will inherit the linetype assigned to that layer. If you use the LAYER command to change the linetype assigned to a layer, all entities on that layer would take on the new linetype. You can use the LINETYPE command to draw entities on the current layer with a specific overriding linetype. Once these entities are drawn, however, changing the linetype assigned to the layer has no effect. To alter a specifically assigned linetype, you must use the CHANGE command to explicitly alter the linetype.

Unless you have specified a linetype using the LINETYPE command, the CELTYPE variable will be assigned the value BYLAYER. If you have used the LINETYPE command to specify a linetype, that current linetype name will be stored in the CELTYPE variable. The following example first uses the SETVAR command to look at the value of CELTYPE and then the LINETYPE command to change the current linetype to DASHED. Afterward, use of the SETVAR command ascertains that the CELTYPE value has indeed changed.

```
Command: SETVAR
Variable name or ?: CELTYPE
CELTYPE = "BYLAYER" (read only)
Command: LINETYPE
?/Create/Load/Set: S
New entity linetype (or ?) <BYLAYER>: DASHED
?/Create/Load/Set: ( RETURN )
Command: SETVAR
Variable name or ? <CELTYPE>: ( RETURN )
CELTYPE = "DASHED" (read only)
```

Tips

The initial CELTYPE value is determined from the prototype drawing (default value is BYLAYER). The current value is saved with the drawing file.

Although you cannot alter the CELTYPE variable except by using the LINETYPE command, it can provide a quick check to determine the current linetype setting. You can set linetype as BYLAYER or explicitly. (See LINETYPE.)

You can also alter the current CELTYPE setting by changing the linetype setting from within the Select Linetype menu accessed from the Entity Creation Modes dialog box. Access this dialog with the DDEMODES command (available only in release 9; see DDEMODES). Linetypes must have been previously loaded using the LINETYPE Load option or the LAYER command (see LINETYPE and LAYER).

CHAMFER

[ADE-1][all versions]

Options

D Sets chamfer distances
P Chamfer an entire polyline [ADE-3]

Overview The CHAMFER command trims two intersecting lines a specified distance from their intersection and connects the trimmed ends with a new line. If the two lines do not intersect, AutoCAD extends them until they do intersect and then trims them back the specified distance. If the two lines intersect and extend beyond the intersection, AutoCAD trims them back to the intersection and then trims them back the specified distance. If the two lines are parallel, AutoCAD rejects the command. If both lines to be chamfered are on the same layer, the chamfer line is drawn on that layer. Otherwise, the chamfer line is drawn on the current layer. The same rule applies for color and linetype.

Procedure Enter the CHAMFER command, select the first line, and then select the second line.

```
Command: CHAMFER
Polyline/Distance/<select first line>: (point to line)
Select second line: (point to second line)
```

The chamfer distance is the distance from the point at which the lines intersect back to the point at which they are trimmed. Two chamfer distances are required: the distance to trim back the first line and the distance to trim back the second line. To set the chamfer distance, enter the CHAMFER command and then type a *D* in response to the first prompt.

AutoCAD then prompts for each chamfer distance. You may enter a number or indicate the distance by specifying two points on the screen (the distance set being the distance between those two points). The current chamfer distance is the initial default value. The first chamfer distance you enter becomes the default setting for the second chamfer distance. These values stay in effect until you change them. Changing the chamfer distance value does not affect any chamfers that have already been drawn. Once the chamfer distance is set, you will need to restart the CHAMFER command (by pressing Return) to actually chamfer two objects.

```
Command: CHAMFER
Polyline/Distance/<select first line>: D
Enter first chamfer distance <current value>: (enter distance)
Enter second chamfer distance <current value>: (enter distance)
```

The concept of chamfer distance is illustrated in the following drawing.

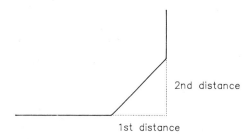

2nd distance

1st distance

You can chamfer an entire polyline at once by selecting the P option. A chamfer segment is added between each existing straight segment of the polyline. Arc segments separating two straight segments are removed. Segments too short to be chamfered are ignored.

```
Command: CHAMFER
Polyline/Distance/<select first line>: P
Select polyline: (select a single polyline)
```

The following drawing shows a typical polyline before and after it has been chamfered.

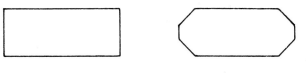

BEFORE   AFTER

When the polyline chamfer is complete, AutoCAD displays a report indicating any special conditions that prevented a chamfer from being added.

Examples

The CHAMFER command has many uses. You can quickly connect two lines with a third line segment. In the following example, a steel plate was drawn as a rectangle. The corners were then chamfered using the CHAMFER command, with both chamfer distances set to 0.25 inches.

```
Command: CHAMFER
Polyline/Distance/<select first line>: D
Enter first chamfer distance <0>: .25
Enter second chamfer distance <.25>: (RETURN)
Command: (RETURN) (restarts command)
Polyline/Distance/<select first line>: (touch first line)
Select second line: (point to second line)
```

The following drawing shows the steel plate after each corner has been chamfered.

Warnings

If the intersection point of the two lines chosen is outside the drawing limits and LIMITS checking is on, AutoCAD will not chamfer the lines (see LIMITS).

Once you chamfer a polyline, the chamfer segments that are added to the polyline become part of the polyline.

When you chamfer a spline curve polyline, only the spline curve is chamfered, not the frame (see PEDIT; release 9).

The shorter portion of the lines beyond their intersection point is erased during the CHAMFER command, as shown in the following drawing. The shorter lines (shown as dotted lines) will be erased.

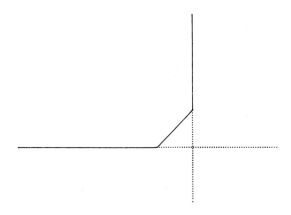

Tips   If you make a mistake, you can restore the last object chamfered to its previous condition by typing *U* at the next command prompt (see UNDO).

If you select a chamfer distance of 0, the two lines will be extended or trimmed to meet at their intersection.

The chamfer distances are stored in the system variables CHAMFERA and CHAMFERB and can be changed with the SETVAR command or AutoLISP.

CHAMFERA   [ADE-1] [all versions]

Overview   The first chamfer distance, set by the CHAMFER command, is stored in the system variable CHAMFERA. The value is a real number.

Procedure Changing the first chamfer distance from within the CHAMFER command changes the CHAMFERA value. Usually this will be the only method you will need to set the chamfer distance. You can use both the SETVAR command and AutoLISP, however, to read or alter the CHAMFERA value.

Examples The following steps change the first chamfer distance using the SETVAR command. In this example, the current value is 0, which is changed to 0.25.

```
Command: SETVAR
Variable name or ?: CHAMFERA
New value for CHAMFERA <0>: 0.25
```

Tips The initial CHAMFERA value is determined from the prototype drawing (default value is 0.0000). The current value is saved with the drawing.

CHAMFERA is provided for use within AutoLISP routines and as a variable in which AutoCAD can save the chamfer distance (see CHAMFER).

CHAMFERB

[ADE-1] [all versions]

Overview The second chamfer distance, set by the CHAMFER command, is stored in the CHAMFERB system variable. Like CHAMFERA, CHAMFERB is a real number that represents the chamfer distance in drawing units.

Procedure Changing the second chamfer distance from within the CHAMFER command changes the CHAMFERB value. Usually this will be the only method you will need to set the chamfer distance. You can use both the SETVAR command and AutoLISP, however, to read or alter the CHAMFERB value.

Examples The following steps change the second chamfer distance using the SETVAR command. In this example, the current value is 0, which is changed to 0.25.

```
Command: SETVAR
Variable name or ?: CHAMFERB
New value for CHAMFERB <0>: 0.25
```

Tips The initial CHAMFERB value is determined from the prototype drawing (default value is 0.0000). The current value is saved with the drawing.

CHAMFERB is provided for use within AutoLISP routines and as a variable in which AutoCAD can save the chamfer distance (see CHAMFER).

CHANGE

Options

P	Changes common properties
C	Color
E	Elevation [ADE-3]
LA	Layer
LT	Linetype
T	Thickness [ADE-3]

Overview

The CHANGE command modifies objects already in an AutoCAD drawing. It is the basic command for altering text in a drawing. The style, height, rotation angle, and actual wording of the text can be changed. You can rotate block using the CHANGE command. You can also use the CHANGE command to modify the properties of many existing objects such as their color, elevation, layer, linetype, and thickness. An attribute definition's tag, prompt string, and default value can be changed.

You can use the CHANGE command to specify a change point. Once the change point is selected, the endpoint of a line or lines can be changed to correspond to the change point. You can change the radius of a circle so that the circle passes through the change point. You can move the insertion point of a text string or a block to the change point.

Procedure

Activate the CHANGE command by entering the command at AutoCAD's command prompt or by selecting it from a menu. The command will prompt you to select objects. The screen cursor will be replaced by AutoCAD's pick box. Select the objects that you want to change (see ENTITY SELECTION). When you have finished selecting objects, press Return. If you want to change the properties of the objects selected, type *P*. If you are changing text entities, press Return again. To change the location of a selected entity, point to the new location. Each use of the CHANGE command is described in the examples below.

Examples   **Changing Properties**   Enter the CHANGE command, select the objects that you want to change, and respond to the `Properties/<Change point>:` prompt with a *P*. AutoCAD will ask for the property or properties that you want to change. Select any or all of the properties listed, one at a time, by entering the first one or two letters of the desired property. Only those letters that are capitalized are necessary, although you may spell out the entire word. You can change as many properties as you want. Remember that you are changing the properties of all the objects you selected globally.

```
Command: CHANGE
Select objects: (select objects you wish to change)
Select objects: ( RETURN )
Properties/<Change point>: P
Change what property (Color/Elev/LAyer/LType/Thickness)?:
```

Color: You can change the color of all the objects you selected in one step. When you type a *C* to the `Change what property` prompt, AutoCAD asks

```
New color <current>:
```

Enter any color number or specify the color as BYLAYER or BYBLOCK.
Elev: You can change the elevation at which an object occurs by entering a number in response to AutoCAD's prompt

```
New elevation <current>:
```

LAyer: You can move objects from one layer to another by entering the name of an existing layer in response to the prompt

```
New layer <current>:
```

LType: You can change the linetype of selected arcs, circles, lines, and polylines by entering the name of an existing linetype in response to the prompt

```
New linetype <current>:
```

Thickness: You can change the thickness of three-dimensional objects by entering a numeric value in response to the prompt

```
New thickness <current>:
```

After you have selected a property to change, entered the new property, and pressed Return, you will once again be presented with the prompt

```
Change what property (Color/Elev/LAyer/LType/Thickness)?:
```

You can continue to change other properties by entering another property to change or end the CHANGE command by pressing Return again.

Changing Text and Attribute Definitions   You can change individual text entries by pressing Return in response to the Properties/<Change point>: prompt. AutoCAD will then go through each text entry selected and prompt you for changes. You will be prompted for changes in the text start point, text style, text height, text rotation angle, and text string. Pressing Return in response to any or all of these prompts keeps the current value. If you select more than one text string to change, AutoCAD takes you through each text entry individually, prompting for changes to each text property. You cannot change the text alignment (fit, center, middle, and so on).

```
Command: CHANGE
Select objects: (select text)
Select objects: ( RETURN )
Properties/<Change point>: ( RETURN )
Enter text insertion point: ( RETURN )
Text style: STANDARD
New style or RETURN for no change: ( RETURN )
New height <0.2000>: ( RETURN )
New rotation angle <0>: ( RETURN )
New text <This is the text to change.>: This is the changed text.
```

If the entities you choose to change are attribute definitions, you will also be prompted for changes to the attribute definition tag, prompt string, and default value. The attribute definition cannot be changed if it is already part of a block.

```
Command: CHANGE
Select objects: (select attribute definition)
Select objects: ( RETURN )
Properties/<Change point>: ( RETURN )
Enter text insertion point: ( RETURN )
Text style: STANDARD
New style or RETURN for no change: ( RETURN )
New height <0.2000>: ( RETURN )
```

```
New rotation angle <0>: ( RETURN )
New tag <DETAIL>: ( RETURN )
New prompt <Enter detail>: Enter detail number
New default value <0>: ?
```

Changing Lines   You can use the CHANGE command to alter lines by selecting a change point in response to the `Properties/<Change point>:` prompt. Select the change point by pointing or entering coordinates. The end point closest to the change point will change to the location of the change point. If you have selected several entities, they will all converge at the change point, as shown in the following drawings.

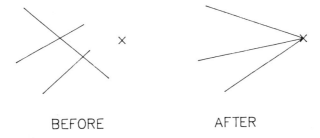

BEFORE   AFTER

Changing Circles   You can alter a circle selected within the CHANGE command so its circumference passes through the specified change point as shown in the following drawings.

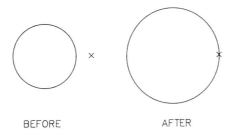

BEFORE   AFTER

Changing Blocks   By providing a change point to AutoCAD's prompt, you in effect change the point at which the block was inserted (similar to the MOVE command but not as flexible). You can change the block's rotation angle, the new angle being relative to the change point or the block's original insertion point if a change point was not selected. This feature is similar to the ROTATE command but is often easier to use.

Warnings   If ORTHO is on, the lines selected within the CHANGE command will all change to an orthagonal orientation. This can prove troublesome because AutoCAD will determine the orientation (0, 90, 180, or 270 degrees) by measuring the distance from the end point of the selected object to the change point. Thus, some lines that were vertical may turn to a horizontal orientation, and vice versa.

Tips   You can restore any entities altered by the CHANGE command to their condition prior to the command by entering *U* at the next command prompt (see UNDO).

You can change the layer of all selected entities at one time by entering *L* in response to the Properties/<Change point>: prompt. You are then prompted for the New layer <current>:. This is an undocumented holdover from an earlier version of AutoCAD. In a similar fashion, you can change the elevation and thickness of all selected entities at one time by entering *E* in response to the Properties/<Change point>: prompt. AutoCAD then prompts

```
Enter elevation:
Enter thickness:
```

You must enter a thickness value explicitly, since no default value is provided.

You can extend many orthogonal lines to the same length in one step by using the CHANGE command with ORTHO on. The response is similar to the EXTEND command and can often be used when EXTEND won't work (see EXTEND). Take care not to select lines whose end points are very far from the change point, since AutoCAD may change their orientation (see "Warnings").

If DRAGMODE is set to auto or if you enter *DRAG* in response to the Properties/<Change point>: prompt, you can visually drag individually selected objects into place. DRAG has no effect when you are altering properties.

Use the CHANGE command to lengthen lines rather than draw a new line segment. This helps keep down the size of your drawing file.

Use the CHANGE command to alter existing text rather than erase it and enter new text. It is faster and the only way to correct spelling mistakes other than erasing and starting again.

Use the CHANGE command to change the Z-coordinate of an insertion point and to change the X-, Y-, and Z-coordinates of 3DLines if you enter the Z-coordinates explicitly (only with version 2.6 and release 9 if ADE-3 is included).

CIRCLE

[all versions]

Options

2P	Specifies by two end points of diameter
3P	Specifies by three points on circumference
TTR	Specifies by two tangent points and radius [ADE-2]
D	Enters diameter instead of radius

Overview   Circles are one of AutoCAD's basic drawing entities. The CIRCLE command is used to draw new circle entities.

Procedure   AutoCAD provides five different methods for drawing a circle. The default method is to specify the center point and the radius. You can also draw circles by entering the center point and the diameter, by selecting three points on the circumference of the circle, by specifying two points that define end points of the circle's diameter, or by making AutoCAD generate the circle with a specified radius that is tangent to two objects already in the drawing. In each case, you can select the points by pointing on the screen, by entering actual or relative coordinates, or by a combination of the two.

Examples   **Center and Radius**   Enter the CIRCLE command, specify the center point, and specify the radius.

```
Command: CIRCLE
3P/2P/TTR/<Center point>: (select center point)
Diameter/<Radius>: (enter radius) (or point on screen)
```

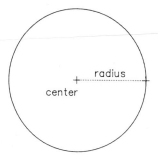

Center and Diameter   Enter the CIRCLE command, specify the center point, enter *D* to indicate that you will be specifying the diameter, then enter the diameter.

```
Command: CIRCLE
3P/2P/TTR/<Center point>: (select center point)
Diameter/<Radius>: D
Diameter: (enter diameter) (or point on screen)
```

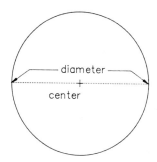

Three-Point Circles   Enter the CIRCLE command, type *3P* to indicate that you will be specifying three points on the circle's circumference, then select the three points in response to AutoCAD's prompts.

```
Command: CIRCLE
3P/2P/TTR/<Center point>: 3P
First point: (select point)
Second point: (select point)
Third point: (select point)
```

A typical three-point circle is shown in the following drawing.

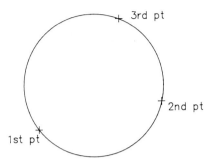

Two-Point Circles Enter the CIRCLE command, type *2P* to indicate that you will be specifying the end points of the circle's diameter, then select the two points.

```
Command: CIRCLE
3P/2P/TTR/<Center point>: 2P
First point on diameter: (select point)
Second point on diameter: (select point)
```

In the following drawing, the two points indicated specify the end points of a line passing through the center of the resulting circle.

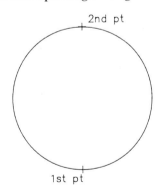

Tangent, Tangent, Radius Enter the CIRCLE command, type *TTR* to indicate that you will be specifying two lines and/or circles, which line or circle the circle will be tangent to, and the radius of the resulting circle.

```
Command: CIRCLE
3P/2P/TTR/<Center point>: TTR
Enter Tangent spec: (select point on first object)
```

```
Enter second Tangent spec: (select point on other object)
Radius: (enter radius)
```

In the following drawing, the two points shown on the small circles determine the tangent points. The radius of the resulting circle is indicated.

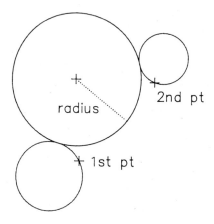

Warnings

When you are drawing TTR circles, there may be more than one circle that can be constructed. In such a case, AutoCAD will draw the circle that is tangent to the objects selected so that it is closest to the points selected.

Tips

Except for circles created with the center/diameter method or the tangent/tangent/radius method, you can use the DRAG command to visually drag the circles during their creation.

Use AutoCAD's Object Snap feature to make the program geometrically generate circles in specific relation to other objects. For example, you can create a circle tangent to three existing objects by using the 3P method and using object snap "tangent" when pointing to three objects in the drawing.

The two-point method is useful for drawing circles at the ends of lines, for example, for grid-line callouts on a floor plan.

If you are using release 9 (with the standard menu ACAD.MNU), you can select the CIRCLE command from the pull-down menu under the Draw menu bar. This selection displays an options menu in the regular screen menu area. The menu item repetition in release 9 will cause any of these options to repeat automatically until you cancel it by pressing Ctrl-C or by selecting another command from the menu (only with AutoCAD release 9; see Advanced User Interface and Repeating Commands).

CLAYER

[all versions]

Overview   The current layer name is stored in the CLAYER (Current Layer) system variable. The initial value is the current layer when the drawing is first created.

Procedure   The current layer name, determined by the Set option of the LAYER command, is stored in the CLAYER system variable as a string value. This variable is saved with the drawing. CLAYER is a read-only variable. You can use the AutoLISP (getvar) function and the SETVAR command to return its value, but you cannot change it. You can set the current layer only with the LAYER command.

Examples   Although you cannot change the CLAYER variable's value except by using the LAYER command, you can use the SETVAR command to retrieve the value. (This is only an example. In practice, the current layer name will be displayed on AutoCAD's status line. Retrieving its value in this fashion would therefore not normally be necessary.) In this example, the current layer is named 1FL-WALLS.

```
Command: SETVAR
Variable name or ?: CLAYER
CLAYER = "1FL-WALLS"  (read only)
```

Tips   The initial CLAYER value is determined from the prototype drawing (default value is 0, the universal layer). The current value is saved with the drawing.

The CLAYER variable is provided as a means to save the name of the current layer when the drawing is saved. Some AutoLISP routines may make use of the ability to read this variable.

You can also change the CLAYER value from the Select Layer menu accessed from the Entity Creation Modes dialog box. Access this dialog box by the DDEMODES command (available only in release 9; see DDEMODES). Layers must have been created previously with the LAYER command or the DDLMODES command dialog box (see DDLMODES and LAYER).

CMDECHO

[ADE-3] [v2.5, v2.6, r9]

Overview
The CDMECHO (Command Echo) system variable determines whether AutoCAD commands executed from within AutoLISP are echoed on the command line. In addition, the messages *n* selected, *n* found and Enter attribute values can be suppressed (in release 9 only) if input is coming from an AutoLISP function.

Procedure
If the value of CMDECHO is set to 1, any AutoCAD command called from within AutoLISP will be echoed on the screen. If the value is set to 0, the command will not be echoed.

Examples
Many third-party AutoLISP routines set this variable to 0 so that users cannot see what the routine is doing as it executes. In the following example, the same AutoLISP routine using the LINE command to draw a one-unit square is run first with CMDECHO set to 1 and then with CMDECHO set to 0.

```
Command: (Command "LINE" (list 1 2)(list 2 2)(list 2 3)(list 1 3) "C")
LINE from point:
To point:
To point:
To point:
To point: C
Command: nil

Command: SETVAR
Variable name or ?: CMDECHO
New value for CMDECHO <1>: 0

Command: (Command "LINE" (list 1 2)(list 2 2)(list 2 3)(list 1 3) "C")
nil
```

In both cases, the square is drawn properly. Notice that in the second case the To point prompts never appear.

Tips   The initial value of CMDECHO is always 1 when you enter the drawing editor. The current value is not saved when you leave the drawing editor.

When writing AutoLISP routines, you will usually set CMDECHO to 1 so you can follow the execution of the routine. Later, when you have completely debugged the routine, you may want to set the variable CMDECHO to 0 to keep the command line from scrolling off the screen as the routine executes.

COLOR [v2.5, v2.6, r9]

Options

number	Sets entity color by number
name	Sets entity color by standard name
BYBLOCK	Sets "floating" entity color
BYLAYER	Uses layer's color for entities

Overview   The COLOR command sets the color to be used for subsequently drawn entities. The initial setting is controlled by the prototype drawing. Specifying a color as BYLAYER means that entities will be drawn in the color chosen for the current layer within the LAYER command (see LAYER). You can specify the color to be used by entering a number (1 to 255) or a standard color name. Entities will then be drawn with the selected color regardless of the color of the current layer. You can use the COLOR command at any time to set a new color for subsequent use. You can also return to using the color defined for the current layer by entering *BYLAYER*.

As a special feature, you can specify the color to be selected as BYBLOCK. When you choose this option, entities are initially drawn with color number 7 (white), but once they are included as part of a block, they take on the color of the block when inserted.

COLOR

Colors are assigned differently on different types of computers. For consistency, however, the first seven colors have been standardized. They may be entered by their name or corresponding number.

NUMBER	COLOR
1	Red
2	Yellow
3	Green
4	Cyan
5	Blue
6	Magenta
7	White

Procedure Enter the COLOR command and respond to the prompt by entering a color name or number or by specifying that the color be chosen as BY-LAYER or BYBLOCK.

```
Command: COLOR
New entity color <current value>: (enter color option)
```

Release 9 If you are using AutoCAD release 9, the functions of the COLOR command are duplicated within the Select Color menu accessed from the Entity Creation Modes dialog box. This dialog box is accessed with the DDEMODES command (available only in release 9; see Advanced User Interface and DDRMODES).

Examples Defining a color as BYBLOCK is special case. The AutoCAD manual refers to this type of color specification as a "floating" color, since the actual color displayed depends on the color that is fixed for the block once it is inserted. If you insert the block without specifying a color, the entities remain white. In the case of blocks nested inside other blocks, the color "floats" up through the nested levels until a fixed color is encountered.

For example, if a block called DOOR, with entities defined as BYBLOCK, is specified as green, the DOOR will be green. Now, if the green DOOR block is included in the definition of the block HOUSE, and the HOUSE is inserted with the color blue, the HOUSE will be blue but the DOOR will remain green, since its color was set at a lower level. If the DOOR color hadn't been specified before it was included in the HOUSE block, it would also be blue, having floated up until it reached a fixed color.

Warnings   Mixing color definitions can become confusing. Blocks with entities drawn using different color specification methods may be difficult to deal with.

Blocks drawn with entities defined as BYLAYER always display with the color of the layer on which they were drawn or, when initially created on the universal (or zero) layer, on which they are inserted.

Pens often will run out of ink while you are plotting. When this happens, you can usually turn off the layers that have already plotted and replot only the affected layers. If you draw objects using specific colors, this may not be possible. AutoCAD maps colors to pens. If you have more than one color on a given layer, you may have to leave that layer on and replot some objects that have already successfully plotted in order to replot the objects that were drawn in the color matching the pen that ran out.

Tips   With the exception of blocks whose colors are fixed, you can use the CHANGE command to alter the color of existing objects.

You can read the current color value, stored in the system variable CECOLOR, using the SETVAR command or AutoLISP. You can change the value only by using the COLOR command.

Configuring AutoCAD   [all versions]

Overview   Before the AutoCAD program can be used, it must be configured to recognize the peripherals (graphics card, digitizer, plotter) connected to your computer.

Procedure   The first time you try to run AutoCAD, the program will display a message informing you that the program must first be configured. After you have set the initial configuration (by responding to a series of screen prompts), you can change AutoCAD's configuration at any time by selecting Task 5, Configure AutoCAD, from the program's main menu.

Examples   **First-Time Configuration**   When you first load AutoCAD (after you have copied the program files onto your computer's hard disk), you need to configure AutoCAD to recognize and use the exact peripherals installed on your system. AutoCAD will prompt you through the entire configuration routine, asking you to select the peripherals you are using from a list of available options.

If AutoCAD has not been configured, the screen shown in Figure 11 is displayed.

AutoCAD's Configuration Menu   When you select main menu task 5, AutoCAD displays its Configuration menu (see Figure 12).

To select any menu task, enter the task number and press Return. AutoCAD then presents a series of prompts.

Task 0. Exit to Main Menu   Selecting task 0 returns you to AutoCAD's main menu. Before doing so, AutoCAD asks if you really wish to change the old configuration.

Task 1. Show current configuration   Task 1 simply displays AutoCAD's current configuration.

```
            A U T O C A D
    Copyright (C) 1982,83,84,85,86,87 Autodesk, Inc.
    Version 2.6 (4/3/87) IBM PC
    Advanced Drafting Extensions 3
    Serial Number:  xx-xxxxxx

    AutoCAD is not yet configured.

    You must specify the devices to which AutoCAD will interface.

    In order to interface to a device, AutoCAD needs the
    control program for that device, called a device driver.
    The device drivers are files with a type of .DRV.
```

Figure 11

```
              A U T O C A D
Copyright (C) 1982,83,84,85,86,87 Autodesk, Inc.
Version 2.6 (4/3/87) IBM PC
Advanced Drafting Extensions 3
Serial Number:  xx-xxxxxx

Configuration menu

   0.   Exit to Main Menu
   1.   Show current configuration
   2.   Allow I/O port configuration

   3.   Configure video display
   4.   Configure digitizer
   5.   Configure plotter
   6.   Configure printer plotter
   7.   Configure system console
   8.   Configure operating parameters

Enter selection <0>:
```

Figure 12

Task 2. Allow I/O port configuration   If you have more than one serial and parallel port installed in your computer, you can use task 2 to tell AutoCAD which port to use for each installed device. AutoCAD asks

Do you wish to do I/O port configuration? <N>:

If you answer yes, the current configuration is displayed, showing the current I/O port assignments for each configured device. You may then subsequently select tasks 3, 4, and 5 and alter the I/O port assignments.

Task 3. Configure video display   Task 3 allows you to specify which graphics display card is installed in your computer. Additionally, you can turn the status line, screen menu, and command line areas on or off. If your computer's screen does not have a one-to-one aspect ratio (that is, if circles appear as ovals), you can adjust the aspect ratio of AutoCAD's graphics screen.

Task 4. Configure digitizer   Selecting task 4 will cause AutoCAD to display a list of available pointing devices. Select the installed device by entering its number.

Task 5. Configure plotter   Selecting task 5 will cause AutoCAD to display a list of available plotting devices. Select the installed device by entering its number. Once you have selected a device, AutoCAD will prompt you to set the intial plotting values. These prompts are similar to those issued by the PLOT command (see PLOT).

If you selected a single pen plotter, AutoCAD will ask

```
Do you want to change pens while plotting? <N>:
```

Answering yes will cause AutoCAD to pause during plotting to allow you to change pens (see PLOT).

Task 6. Configure printer plotter   Task 6 is similar to task 5. You can select from a list of supported devices and select the initial printer plotting values.

Task 7. Configure system console   On some computers, AutoCAD must be configured for the particular keyboard installed. Task 7 has no effect on PC-DOS and MS-DOS systems.

Task 8. Configure operating parameters   Task 8 displays the submenu shown in Figure 13.

Subtask 0 exits back to the Configuration menu.
Subtask 1 enables or disables an alarm tone that sounds whenever you enter an invalid entry. AutoCAD displays the prompt

```
Do you want the console alarm to sound when you make an input error? <N>
```

Subtask 2 allows you to specify the default prototype drawing (see Prototype Drawing for more information). AutoCAD prompts

```
Enter name of default prototype file for new drawing
or . for none <ACAD>:
```

Subtask 3 allows you to specify the name AutoCAD will use as the default value whenever you use the PLOT or PRPLOT command to plot to a file. AutoCAD displays the prompt

```
Enter default plot file name (for plot to file)
or . for none <AUTOSPOOL>:
```

```
                A U T O C A D
      Copyright (C) 1982,83,84,85,86,87 Autodesk, Inc.
      Version 2.6 (4/3/87) IBM PC
      Advanced Drafting Extensions 3
      Serial Number:  xx-xxxxxx

      Operating parameter menu

          0.   Exit to configuration menu
          1.   Alarm on error
          2.   Initial drawing setup
          3.   Default plot file name
          4.   Plot spooler directory
          5.   Placement of temporary files
          6.   Network node name
          7.   AutoLISP feature

      Enter selection <0>:
```

Figure 13

Subtask 4 allows you to specify the name of the subdirectory to be used for placement of plot files. If the default plot file name configured in subtask 3 is AUTOSPOOL and that default plot file name is accepted when you use the PLOT or PRPLOT command, AutoCAD places the plot file in the plot spooler directory. If you don't plot to a file or the plot file name is not AUTOSPOOL, the plot spooler directory name is ignored. The network node name configured in subtask 6 is used as the file extension for plot files sent to the plot spooler directory. AutoCAD prompts

```
Enter plot spooler directory name
<\SPFILES\>:
```

Subtask 5 allows you to specify the device and/or subdirectory that AutoCAD will use for the placement of its temporary files. These temporary files, used to "page out" portions of the drawing file and to save the list of commands for use with the UNDO command, are automatically created when you begin editing a drawing and are deleted when you QUIT or end the drawing.

On an IBM PC AT or compatible, you can use available memory above 640K as a RAM disk for storage of these temporary files. The RAM disk must first be configured within the DOS CONFIG.SYS file. For example,

```
device = VDISK.SYS 1536 /E
```

uses DOS's RAM disk VDISK, setting up 1536K bytes of extended memory for use as a RAM disk. If the last physical device installed in your computer was the hard disk (typically drive C:), you can respond to the subtask 5 prompt

```
Enter directory name for temporary files, or DRAWING to
place them in the same directory as the drawing being edited.
<DRAWING>: D:
```

Subtask 6 lets you specify a network node name, thus allowing AutoCAD to run in a local area network (LAN) by saving its temporary files with a different file extension for each configured copy of AutoCAD on the network. AutoCAD displays the prompt

```
Enter network node name (1 to 3 characters) <AC$>:
```

Subtask 7 allows you to turn the AutoLISP feature on or off. AutoCAD prompts

```
Do you want AutoLISP enabled? <current>:
```

If AutoLISP is enabled, AutoCAD requires more memory. If you do not have enough memory installed in your computer, AutoLISP will be disabled automatically every time you load AutoCAD. AutoLISP is available only if your copy of AutoCAD contains ADE-3.

Warnings
You cannot assign the video display and the digitizer or plotter to the same I/O port. AutoCAD will display an error message.

The name you specify for the plot spooler directory and for the placement of temporary files must already exist.

If you specify a drive for the placement of temporary files, be sure the drive has approximately one-and-a-half times the necessary free space to accommodate the drawing file. Running out of space on the temporary drive will cause AutoCAD to abort and return to the operating system prompt, after first allowing you to save any changes you have made. This can happen often if the temporary files are being placed on a RAM disk.

You can decrease the amount of temporary file storage required by turning off the UNDO feature, but at the expense of the availability of the UNDO command.

Prior to AutoCAD version 2.6, it was not possible to configure the default plot file name, the plot spooler directory, the placement of temporary files, or the network node name.

Tips

You can adjust the aspect ratio of the graphics screen display. Do this if circles appear as ovals. First, enter the drawing editor and draw a square. Use a ruler to accurately measure its sides. End the drawing and go into the configuration menu. Select the task to configure video display and answer no when asked if you want to select a different display. In this example, a 6×6 square was drawn that measured 3.5×3 on the screen. The following steps will adjust the display aspect ratio:

```
If you have previously measured the height and width of
a "square" on your graphics screen, you may use these
measurements to correct the aspect ratio.

Would you like to do so? <N>: Y

Width of square <1.0000>: 3.5
Height of square <1.0000>: 3.0
```

When you plot with a one-to-one plot scale, plotted lines should be exactly the same length as they were drawn. If this is not the case, you can calibrate your plotter or printer-plotter from within the tasks to configure the plotter or the printer-plotter. First, draw a square and plot it. Use a ruler to accurately measure the plot. Enter the configuration menu and select task 5 or 6, as appropriate. Answer no when asked if you want to select a different device. In this example, a 10×10 square was drawn that measured 10.25×10 when plotted. The following steps will calibrate the plotter:

```
If you previously measured the lengths of a horizontal
and a vertical line that were plotted to a specific scale,
you may use these measurements to calibrate your plotter.

Would you like to calibrate your plotter? <N>: Y

Enter measured length of horizontal line <1.0>: 10.25
Enter correct length of horizontal line <1.0>: 10.00
Enter measured length of vertical line <10.25>: 10.00
Enter correct length of vertical line <10.00>: 10.00
```

To conserve disk space, it is not necessary to copy all of the device drivers' files from the two disks labeled DRIVERS 1 and DRIVERS 2. Instead, when AutoCAD displays the screen shown in Figure 14, place the DRIVERS 1 disk into the floppy disk drive and enter the drive name (typically A:). Auto-CAD will read the disk and display a list of the graphics card drivers available. Select the desired device by entering its number. AutoCAD will copy the selected driver into its configuration file. To configure the digitizer, plotter, and printer-plotter options, repeat this procedure with the disk labeled DRIVERS 2. Only the selected device drivers will be copied into the configuration file. If you ever reconfigure AutoCAD, you will need to repeat this process.

Configuring AutoCAD to place its temporary files in a RAM disk can greatly speed up the editing of large drawings.

When you have configured AutoCAD to place its temporary files in a drive other than the one containing the drawing file you are editing, you can use the STATUS command to display the amount of available storage space remaining in both drives.

```
              A U T O C A D
Copyright (C) 1982,83,84,85,86,87 Autodesk, Inc.
Version 2.6 (4/3/87) IBM PC
Advanced Drafting Extensions 3
Serial Number:  xx-xxxxxx


AutoCAD is not yet configured.
You must specify the devices to which AutoCAD will interface.

In order to interface to a device, AutoCAD needs the
control program for that device, called a device driver.
The device drivers are files with a type of .DRV.

You must tell AutoCAD the disk drive or directory in
which the device drivers are located.  If you specify a
disk drive, you must include the colon, as in A:

Enter drive or directory containing the Display device drivers:
```

Figure 14

Besides configuring AutoCAD, you should alter your CONFIG.SYS file (located in the ROOT directory of the disk that your system boots from) to change the buffers and files settings. A buffer is an area in memory that holds data read from disk. A higher number means that required data may still be stored in memory. Too low a number results in more frequent disk reads. Too high a number results in system slowdown because DOS has to look through too many buffers. In addition, each configured buffer reduces the amount of RAM available. A setting of BUFFERS = 20 is recommended for PC XT machines. PC AT-class machines may respond a bit faster with a setting of BUFFERS = 32. The files setting is the limit on the number of files that can be open at one time. AutoCAD has several files open at a time and if that number exceeds the files limit, AutoCAD must close unneeded files when it needs to open another. This opening and closing of files results in slower operation. A setting of FILES = 15 should be sufficient for AutoCAD (although some other programs, database programs in particular, require higher settings). You can alter (or create) the CONFIG.SYS file using any ASCII word processor or EDLIN.

COORDINATES

[all versions]

Overview   AutoCAD permits you to locate points within a drawing using three methods: absolute coordinates (using their precise numeric value), relative coordinates (referencing a new coordinate to a previous coordinate), and polar coordinates (establishing a new coordinate by specifying a distance and an angle from an existing point).

Procedure   Every AutoCAD drawing contains an underlying Cartesian coordinate system. Any location in the drawing can be represented as a location relative to a 0,0,0 coordinate point. Within the normal two-dimensional drawing plane, positions horizontally to the right or left are measured along the X-axis and positions up and down on the plane are measured along the Y-axis. Thus, every point on the plane can be represented as a coordinate pair composed of an X- and a Y-coordinate. Positive coordinate points are located above and to the right of the 0,0 point, and negative points are located to the left and below the 0,0 coordinate. This concept is illustrated in Figure 15.

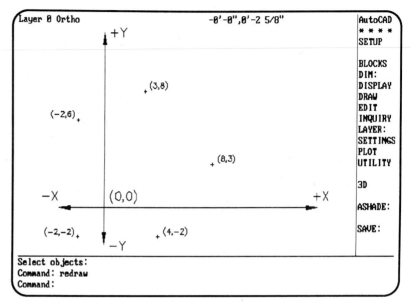

Figure 15

If AutoCAD's 3D feature is present, an additional coordinate axis, the Z-axis, is available. In this case coordinates would be represented as a coordinate triple of the form *X,Y,Z*. The Z-axis can be thought of as an axis extending straight up out of the screen at a ninety-degree angle to the normal two-dimensional plane. Positive Z-coordinate values are above the X-Y plane, and negative values are below the plane.

When creating a two-dimensional drawing, you do not have to deal with the Z-coordinate at all. For this reason, AutoCAD allows you simply to enter X- and Y-coordinates. You can use the ELEV command to set the current Z-coordinate and an extrusion thickness for objects drawn so that they can be represented in a three-dimensional projection. When you are creating two-dimensional drawings, the current elevation and thickness are normally set to zero. The only two AutoCAD commands that expect X-, Y-, and Z-coordinates are the 3DFACE and 3DLINE commands.

Examples   Because every location in a drawing is represented by an X- and a Y-coordinate (and a Z-coordinate when dealing in three-dimensional space), it is possible to select points by entering their absolute coordinates when you are responding to an AutoCAD prompt (such as From point:

when drawing a line). It is possible, but not necessary and often difficult, to select points by entering their absolute coordinates, since you would need to keep track of exactly where you are within the drawing at all times.

If you were always to use absolute coordinates when drawing, to draw a simple square on the screen, you would have to know the absolute coordinates of the four corners. For example, if the lower left corner was the point 4,5 and each side of the square was to be drawn 8½ units long, you would need to determine that the upper left corner was at coordinate 4,13.5. The upper right corner would be 12.5,13.5 and the lower right corner would be at 12.5,5. It quickly becomes apparent that using absolute coordinates is not always practical.

A simpler method is to use relative coordinates. AutoCAD lets you specify a location in the drawing by determining its position relative to the last coordinates specified. To use relative coordinates, precede the coordinate entry with the at-symbol, @. The coordinate pair that follows the at-symbol represents the distance along the X-axis and Y-axis to the next point "relative" to the last known point. The previous example could then be drawn using the following dialog:

```
Command: LINE
From point: 4,5
To point: @0,8.5 (offset 0 units along X-axis and 8.5 units along Y)
To point: @8.5,0 (offset 8.5 units along X and 0 units along Y)
To point: @0,-8.5 (offset 0 units along X and -8.5 units along Y)
To point: C (to close the square and return to the start point)
```

This example is illustrated in Figure 16.

This example would have been complicated even further had you wanted to draw the square tilted at, say, a forty-five-degree angle. Using absolute coordinates would have required trigonometric calculations on your part, something easily left to the computer if you use relative coordinates. Instead of entering relative coordinates as offsets along the X- and Y-axes, you can also enter them as an offset distance and an offset angle. To draw the square from the previous example, this time tilted at a forty-five-degree angle, the following dialog could be followed:

```
Command: LINE
From point: 4,5
To point: @8.5<45 (offset 8.5 units at 45 degrees)
To point: @8.5<315 (offset 8.5 units at 315 degrees)
To point: @8.5<225 (offset 8.5 units at 225 degrees)
To point: C (or you could have offset 8.5 units at 135 degrees)
```

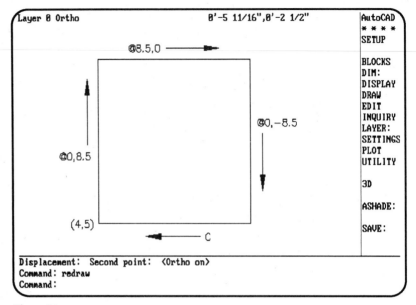

Figure 16

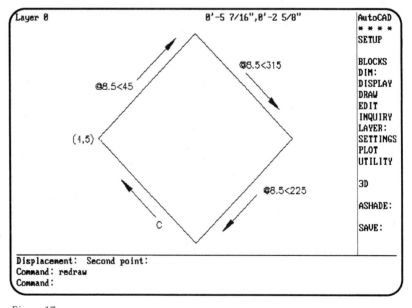

Figure 17

The resulting square is shown in Figure 17.

Clearly, using relative coordinates made this task much easier. There are times, however, when absolute coordinates are preferred. Since the geologic map in Appendix D (see Figure D-3) is drawn on a survey grid based on a real-world coordinate system, the location of the bore holes and the other geologic features are located from field survey information based on their absolute coordinates.

Tips   Become familiar with relative coordinates and use them often.

COORDS   [ADE-1] [v2.5, v2.6, r9]

Overview   The COORDS system variable stores a value that determines whether the coordinates displayed on the status line are continuously updated.

Procedure   The COORDS variable is always an integer. If the COORDS value is zero, the coordinate display is updated only when the pick button is pressed. If the value is 1, the display of absolute coordinates is updated constantly. If the value is 2, the distance and the angle from the last point are displayed when a distance or an angle is requested by an AutoCAD command. The absolute coordinates area is updated at other times.

Examples   You can read or change the COORDS variable using the SETVAR command or the (getvar) and (setvar) AutoLISP functions, respectively. In the following example, Figure 18 shows the status line display when COORDS is 1.

```
Command: LINE
From point: 1,2
To point: 'SETVAR
>>Variable name or ?: COORDS
>>New value for COORDS <0>: 1
Resuming LINE command.
To point:
```

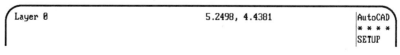

Figure 18

113

Layer 0	6.6763< 45	AutoCAD
		* * * *
		SETUP

Figure 19

Figure 19 shows the status line when the COORDS value is 2.

```
To point: 'SETVAR
>>Variable name or ?: COORDS
>>New value for COORDS <1>: 2
Resuming LINE command.
To point:
```

In both cases, the LINE command is active and the 'SETVAR command is used to change the COORDS value. The changed status reporting is seen immediately.

Tips

The Coordinate toggle key (F6; Ctrl-D on DOS systems) duplicates some of the features of the COORDS variable. In fact, using this key to toggle the coordinate display changes the COORDS value from 0 to 2. To always display the absolute coordinates, set the COORDS value to 1.

The initial COORDS value is determined from the prototype drawing (default value is 0, coordinate updating is off). The current value is saved with the drawing.

COPY
[all versions]

Overview

The COPY command is used to make a duplicate copy of an object or objects that already exist in a drawing. The original object is not affected.

Procedure

Select the objects to be copied and specify the displacement of the copy. The copy retains the original object's orientation and size. The COPY command can be used to make multiple copies of objects, each at a different specified displacement, by selecting the multiple option. In every case, the copy is composed of duplicates of the original's entities, which can be edited individually.

Examples

Single Copy   Enter the COPY command, select the objects to be copied, and define the displacement of the copy from the original.

```
Command: COPY
Select objects: (show what to copy)
<Base point or displacement>/Multiple: (first point A) (or x,y-distance)
Second point of displacement: (second point B) (or press Return)
```

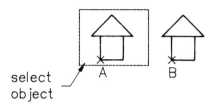

Multiple Copies   Enter the COPY command, select the objects to be copied, enter M to indicate that you are making multiple copies of the selected objects, then proceed to specify the displacement for each copy. Each time you specify a displacement, it is referenced to the base point. To end the command, press Return.

```
Command: COPY
Select objects: (show what to copy)
<Base point or displacement>/Multiple: M
Base point: (select a base point A)
Second point of displacement: (select point B) (or enter a displacement)
Second point of displacement: (select point C) (or enter a displacement)
Second point of displacement: (select point D) (or enter a displacement)
Second point of displacement: (press Return to end)
```

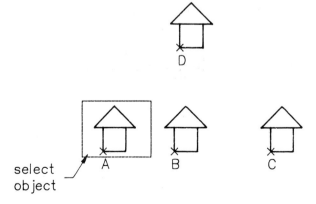

The intricate railings and moldings in the drawing of the Carson Mansion in Appendix D (see Figure D-4) were drawn once and then copied to different positions. AutoCAD eliminates the painstaking repetitous drawing that would have been required if this beautiful drawing had been done by hand.

Tips

If you make a mistake, you can restore the drawing to its condition prior to the COPY command by entering *U* at the next command prompt (see UNDO).

You can specify a displacement by simply entering the distance that you want to copy the object as an x-distance and y-distance in response to the Base point or displacement prompt and then pressing Return in response to the Second point prompt. Because a Return has a specific meaning when you are making multiple copies (it ends the command), you can't do the same thing. But you can accomplish the same result by entering a base point of 0,0 and then entering a relative x-distance and y-distance in response to the Second point prompt. To enter the relative displacement with a relative distance and angle, in either single- or multiple-copy procedures, first enter a base point of *0,0* and then enter the second point as a relative distance and angle, as in @6' < 45.

If you have DRAGMODE set to Auto or you enter DRAG in response to the Second point prompt, you can see the location of a copy before you actually copy an object.

If multiple copies need to be in a regular rectangular or circular pattern, use the ARRAY command instead of the COPY command.

When entering the displacement or coordinates, you can copy an object to a new Z-coordinate (only in version 2.6 and release 9 when ADE-3 is included).

If you are using AutoCAD release 9 (with the standard menu ACAD.MNU), you can select the COPY command from the Edit pull-down menu. In this situation, the Single and Auto selection modes are used. The menu item repetition of release 9 causes the command to repeat automatically until you cancel it by pressing Ctrl-C or by selecting another command from the menu (only with AutoCAD release 9; see Advanced User Interface and Entity Selection).

Crossing

[v2.5, v2.6, r9]

Overview

Crossing, an AutoCAD program feature, is one of the methods available for selecting objects.

Procedure   When an AutoCAD command prompts you to select objects, you can enter C to indicate that you will select objects using the crossing method. When you use the crossing method, AutoCAD prompts you to select

```
First corner:
Second corner:
```

Select two points. Once you have selected the first point, AutoCAD will display a "rubberband" box, which changes size as you move the cursor. Use this box to surround the object or objects you want to select. When you pick the second corner, AutoCAD selects all the objects that fall entirely within this box or that cross the boundary of the box. These objects are added to the current selection set (see Entity Selection).

Release 9 Only   If you are using AutoCAD release 9, there are three additional object selection options that you can use to effect a crossing selection. If you enter *Box* in reply to the prompt to select objects, you are immediately prompted to select the first and second corners as usual. After you have selected the first corner, if you move the cursor to the left, AutoCAD immediately assumes you will make a crossing selection. Moving the cursor toward the right indicates a window selection method. If your graphics card supports the Advanced User Interface (see Advanced User Interface), AutoCAD further indicates that you are making the selection by the crossing method by displaying the rubberband box with a dashed linetype. The window method of object selection is indicated by a rubberband box with a solid line.

In Figure 20, the crossing method of object selection is indicated by the dashed linetype of the rubberband box.

Entering *Auto* or *AU* in response to the prompt to select objects causes AutoCAD to prompt again for object selection. If you select a blank area of the screen, AutoCAD immediately goes into Box selection mode, as described previously.

Normally, AutoCAD continues prompting you to select objects until you press Return a final time. If you respond to the prompt to select objects by entering *Single* or *SI*, AutoCAD prompts again for object selection. You can then use any valid selection method, such as crossing, box, or auto. As soon as an object is selected, object selection is complete.

Examples   In this example, the crossing method is used with the ERASE command to erase the objects shown in the diagram.

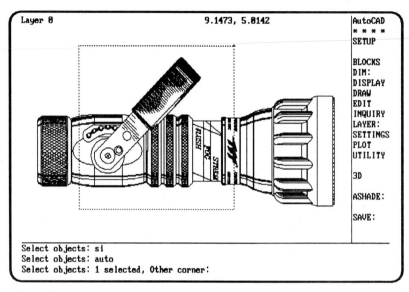

```
Layer 0                          9.1473, 5.8142        AutoCAD
                                                       * * * *
                                                       SETUP

                                                       BLOCKS
                                                       DIM:
                                                       DISPLAY
                                                       DRAW
                                                       EDIT
                                                       INQUIRY
                                                       LAYER:
                                                       SETTINGS
                                                       PLOT
                                                       UTILITY

                                                       3D

                                                       ASHADE:

                                                       SAVE:


Select objects: si
Select objects: auto
Select objects: 1 selected, Other corner:
```

Figure 20

```
Command: ERASE
Select objects: C
First corner: (select first point)
Second corner: (select second point)
```

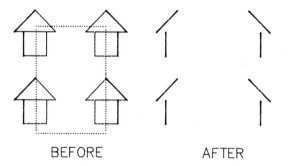

BEFORE   AFTER

Tips   Some commands, such as the TRIM and EXTEND commands, cannot be
           used with the crossing method to select objects. On the other hand, the
           STRETCH command requires the use of either the crossing or window
           method of selecting objects.

DATE

Overview   The current system date and time are stored in the system variable DATE.

Procedure   The DATE value is a real number in the form *<Julian date> · <Fraction>*. The Julian date is in actual days, and the fraction is represented in seconds. The number after the decimal point, multiplied by 86400, is the number of seconds past midnight.

Examples   The date 01/01/1980 is the earliest date that DOS allows. This date is represented in AutoCAD's DATE format as 2444240. The date 01/01/1981 is represented as 2444606, exactly 366 higher (remember to count leap years).

If the DATE variable contains the value 2447038.603877364, the current date and time is August 30, 1987, at 2:29:35 PM. This date is arrived at as follows:

CALCULATION	RESULT
(2447038 − 2444240)	= 2798, subtracting the known date 1980
(2798 − 2556)	= 242, subtracting 7 years
(242 − 212)	= 30, subtracting the first 7 months of the year

Thus, the result is August 30, 1987.
The time is calculated as follows:

CALCULATION	RESULT
(0.603877364*86400)	= 52175
(52175/3600)	= 14.49305, which truncates to 14 hours, or 2 PM
(0.49305*60)	= 29.58333, which truncates to 29 minutes
(0.58333*60)	= 35 seconds

DATE can be returned while you are in AutoCAD by using the SETVAR command or AutoLISP. In this example, the SETVAR command is used from AutoCAD's command prompt to return the previous value:

```
Command: SETVAR
Variable name or ?: DATE
DATE = 2447038.603877364 (read only)
```

119

Note that the variable is a read-only value. You cannot change it except by actually changing the computer's internal time and date values set by DOS.

Tips

The number returned by the DATE variable is represented as a fraction of a day. Times returned by the DATE variable can be subtracted directly to determine differences in time using AutoLISP. The previous exercise for determining the date and time is given as an example:

```
(defun C:GETDATE (/ S YEAR YEARS LEAP LEAPS DAY DAYS MONTH HOUR MINS SECONDS
                 SECS TIME)
  (setq S (getvar "date"))
  (setq YEARS (- (fix S) 2444240.0))
  (setq YEARS (fix YEARS))
  (setq LEAPS (fix (/ YEARS 1461)))
  (setq YEAR (- YEARS (* LEAPS 1461)))
  (setq YEAR (fix (/ YEAR 365)))
  (if (= YEAR 0) (setq LEAP 1) (setq LEAP 0))
  (setq DAYS (- YEARS (* LEAPS 1461) (* YEAR 365)))
  (setq DAYS (+ LEAP DAYS))
  (setq YEAR (+ 1980 (* LEAPS 4) YEAR))
  (if (= DAYS 0) (setq YEAR (- YEAR 1)))
  (if (= DAYS 0) (setq DAYS (+ LEAP 365)))
  (cond ((<= DAYS 31) (setq MONTH "January"))
        ((<= DAYS (+ LEAP 59)) (setq MONTH "February"))
        ((<= DAYS (+ LEAP 90)) (setq MONTH "March"))
        ((<= DAYS (+ LEAP 120)) (setq MONTH "April"))
        ((<= DAYS (+ LEAP 151)) (setq MONTH "May"))
        ((<= DAYS (+ LEAP 181)) (setq MONTH "June"))
        ((<= DAYS (+ LEAP 212)) (setq MONTH "July"))
        ((<= DAYS (+ LEAP 243)) (setq MONTH "August"))
        ((<= DAYS (+ LEAP 273)) (setq MONTH "September"))
        ((<= DAYS (+ LEAP 304)) (setq MONTH "October"))
        ((<= DAYS (+ LEAP 334)) (setq MONTH "November"))
        ((<= DAYS (+ LEAP 365)) (setq MONTH "December")))
  (cond ((= MONTH "January") (setq DAY DAYS))
        ((= MONTH "February") (setq DAY (- DAYS 31)))
        ((= MONTH "March") (setq DAY (- DAYS (+ LEAP 59))))
        ((= MONTH "April") (setq DAY (- DAYS (+ LEAP 90))))
        ((= MONTH "May") (setq DAY (- DAYS (+ LEAP 120))))
        ((= MONTH "June") (setq DAY (- DAYS (+ LEAP 151))))
        ((= MONTH "July") (setq DAY (- DAYS (+ LEAP 181))))
        ((= MONTH "August") (setq DAY (- DAYS (+ LEAP 212))))
```

```
((= MONTH "September") (setq DAY (- DAYS (+ LEAP 243))))
((= MONTH "October") (setq DAY (- DAYS (+ LEAP 273))))
((= MONTH "November") (setq DAY (- DAYS (+ LEAP 304))))
((= MONTH "December") (setq DAY (- DAYS (+ LEAP 334)))))
(setq SECONDS (* 86400.0 (- S (fix S))))
(setq HOUR (fix (/ SECONDS 3600)))
(setq MINS (fix (* 60 (- (/ SECONDS 3600) HOUR))))
(setq SECS (* 60 (- (* 60 (- (/ SECONDS 3600) HOUR)) MINS)))
(if (>= HOUR 13) (setq TIME " PM") (setq TIME " AM"))
(if (= TIME " PM") (setq HOUR (- HOUR 12)))
(if (< MINS 10) (setq MINS (strcat "0" (itoa MINS))) (setq MINS(itoa MINS)))
(if (< SECS 10.0) (setq SECS (strcat "0" (rtos SECS)))
    (setq SECS (rtos SECS)))
(strcat MONTH " " (itoa (fix DAY)) ", " (itoa YEAR) "  " (itoa HOUR)
    ":" MINS ":" SECS TIME)
)
```

DBLIST

Overview   The DBLIST (Database List) command causes an informational listing of every entity in the active drawing to be listed on the screen.

Procedure   Type *DBLIST* in response to AutoCAD's command prompt. You will not use this command often while drawing. It is provided mainly for training and debugging purposes. If printer echoing is turned on, the list will also be printed. The information displayed is in the same format as displayed with the LIST command.

Examples   To use this command, simply enter the command DBLIST at the command prompt.

Command: **DBLIST**

Tips   The DBLIST can take a long time to display all of its information on the screen and even longer to direct it to a printer. The length of time depends on the size of the drawing. Pressing Ctrl-C will cancel the command at any time and return you to the command prompt.

DDATTE

[ADE-3] [r9]

Overview   The DDATTE (Dynamic Dialog Attribute Edit) command causes the Edit Attributes dialog box to be displayed on the screen. This dialog box allows interactive editing of existing attributes.

Procedure   You can enter the DDATTE command at the command prompt or select it from a menu. The command then prompts for the selection of a block. The block selected must have attributes associated with it.

As soon as you have selected an acceptable block, the DDATTE command displays a dialog box showing all of the block's variable attributes. You can change any attribute value by highlighting its value button. Once the value button is highlighted, simply type in the new value. The new value automatically replaces the existing value. Complete the editing of the selected attribute by clicking on the OK button displayed alongside the selected attribute value or by pressing Return. If you decide not to keep the value just entered, you can restore the original value by selecting the Cancel button displayed alongside the selected attribute or by pressing Ctrl-C or Escape. The Edit Attributes dialog box will remain displayed. You can continue to edit other attributes associated with the selected block. When you have finished editing attributes, complete the DDATTE command by clicking on the OK button at the bottom of the dialog box. If you decide not to keep any of the new values, click on the Cancel button or press Ctrl-C or Escape. The original values (before you activated the DDATTE command) will be restored.

The DDATTE dialog box will display up to ten attributes at a time. If the block selected has more than ten associated attributes, you can edit additional attributes by scrolling through them. Page-up and Page-down scroll buttons allow you to scroll through the attributes a full page at a time. The Up and Down scroll buttons move through the attributes individually. To view other attributes, simply click on the appropriate scroll button.

Examples   In the following example, the DDATTE command is used to edit the Room Name value in the selected block. The DDATTE command is entered at the command prompt. The command prompts for the selection of a block and the Roomtag block is selected by being pointed to on the graphics screen.

Figure 21

The Edit Attributes dialog box, shown in Figure 21, is immediately displayed, overlaying the drawing. The Room Name value is edited by simply moving the pointing device (thus moving the arrow cursor) until the Room Name value Living Room is highlighted. The new value Great Room is then typed in at the keyboard. The entering of the new value is completed by moving the arrow cursor to the right so that the OK button is highlighted and then pressing the pick button. The DDATTE command is completed by moving the arrow cursor so that the bottom OK button is highlighted and then pressing the pick button again.

Command: **DDATTE**
Select block: (*select block using any valid selection method*)

Warnings The DDATTE command is available only with AutoCAD release 9.

If your graphics display does not support the Advanced User Interface, attempting to execute the DDATTE command will result in AutoCAD displaying the message

** Command not allowed with this display configuration **

and the command will be ignored.

123

If the entity you select is not a block, AutoCAD displays the dialog alert box shown in Figure 22. You must click on the OK button to continue.

If the block you select has no attributes, AutoCAD displays the dialog alert box shown in Figure 23. You must click on the OK button to continue.

Tips

The dialog box displayed by the DDATTE command is similar to the dialog box displayed by the INSERT command, except that the INSERT command's dialog box is titled Enter Attributes (see INSERT).

You can use the DDATTE command for multiple continuous editing of several different blocks by using the MULTIPLE command modifier of release 9. Entering the DDATTE command in this way causes the command to repeat until you cancel it by pressing Ctrl-C. (See MULTIPLE.)

Command: **MULTIPLE DDATTE**

Because of width limitations when dialog boxes are used to edit attributes, only the first twenty-four characters of an attribute prompt and the first

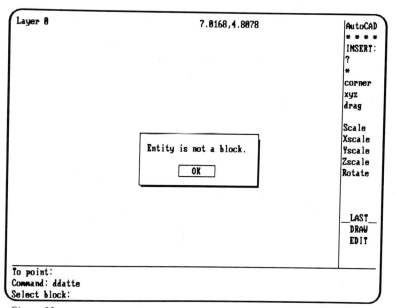

Figure 22

Block has no attributes.

OK

Figure 23

thirty-four characters of an attribute value appear in the dialog box. If you use the DDATTE command to edit an attribute value, you will be able to enter only thirty-four characters, since that is all that will fit into the dialog box.

The ATTEDIT command may prove more useful when you are editing attributes globally. The DDATTE can be used to edit attributes only one block at a time. (See ATTEDIT.)

You can use either the keyboard arrow keys or a pointing device to move the arrow cursor within a dialog box.

For more information, see Advanced User Interface, Dialog Box, and INSERT.

'DDEMODES
[ADE-3] [r9]

Overview   The DDEMODES (Dynamic Dialog Entity Modes) command displays the Entity Creation Modes dialog box. This dialog box allows you to control the current color, linetype, layer, elevation, and thickness used for subsequent entities. The DDEMODES command can be used transparently, but some of the settings changed may not take effect until after the active command has been completed.

Procedure   You can enter the DDEMODES command at the command prompt or select it from a menu. If you are using the standard ACAD.MNU provided with AutoCAD release 9, you can select the transparent 'DDEMODES command by picking Entity Creation from the Modes pull-down menu, shown in Figure 24.

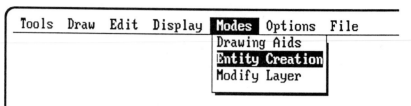

Figure 24

The Entity Creation Modes dialog box duplicates the functions of the LAYER and ELEV commands, as well as the LINETYPE set and the LAYER set of the LINETYPE and LAYER commands, respectively. The Elevation and Thickness functions allow you to enter a new value directly into the input button. Selecting the Color, Layer name, or Linetype requestor button causes a secondary dialog box to be displayed.

Command: **DDEMODES**

Figure 25 illustrates the Entity Creation Modes dialog box.

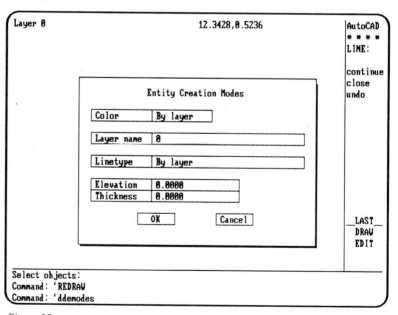

Figure 25

Color   To change the current color, select the Color requestor button (move the arrow cursor until that button is highlighted and then press the pick button). A Select Color dialog box will overlay the previous dialog box, as shown in Figure 26. The first eight colors are listed, as well as the selections By layer and By block. Each is displayed with a check box alongside the selection. To change the current color to one of those listed, simply click on the appropriate button. Notice that the corresponding color number is displayed in the Color Number input button. To change to a color number above 7, move the arrow cursor until the Color Number input button is highlighted, then type in the color number. You can also select one of the standard colors in this fashion. To complete the selection of the current color, you must click on the OK button. This will return you to the Entity Creation Modes dialog box. To disregard any changes you have just made, press Ctrl-C or Escape or click on the Cancel button.

Layer   To change the current layer, select the Layer requestor button by clicking on it. A Select Layer dialog box will overlay the Entity Creation Modes dialog box, as shown in Figure 27. Only those layers

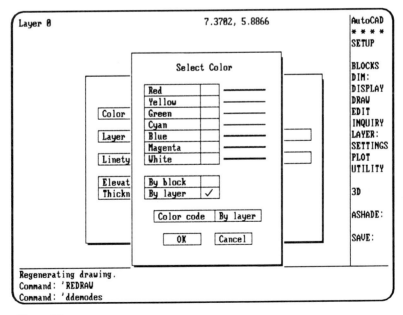

Figure 26

already created by the LAYER command or the DDLMODES command will appear in the dialog box. A check button is provided alongside the name of each layer. To set a different current layer, click on the desired check button. If more than five layers exist, use the Page-up and Page-down buttons to move through the list one page at a time, or use the Up and Down buttons to move through the list one line at a time. The current layer name also appears in the Layer input button. You can highlight this button and select a new current layer by typing the name of an existing layer; when you are finished, select the OK button immediately alongside the Layer input button or press Return. To disregard the name you just entered, select the Cancel button or press Ctrl-C or Escape. To finalize selection of the current layer, click on the OK button at the bottom of the dialog box. This will return you to the Entity Creation Modes dialog box. To disregard any changes you just made and to return to the previous dialog box, press Ctrl-C or Escape or click on the Cancel button.

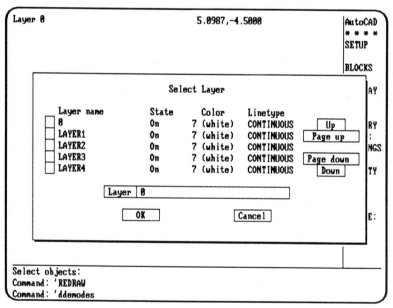

Figure 27

Linetype   To change the current linetype, select the Linetype requestor button by clicking on it. A Select Linetype dialog box will overlay the previous dialog box. The Select Linetype dialog box is shown in Figure 28. Any linetypes already loaded into the drawing will be displayed, as well as the selection By layer and By block. A check button is provided alongside each available selection. Only five linetypes will display in the dialog box at a time. If more than five linetypes are loaded, use the Page-up and Page-down buttons to move through them one page at a time or the Up and Down buttons to move through the list one line at a time. To change the current linetype, click on the check button of one of the linetypes. The new current linetype will appear in the Linetype input button. Alternatively, you can highlight the Linetype input button and type the name of a linetype. To finalize the selection of the current linetype, click on the OK button. This will return you to the Entity Creation Modes dialog box. To disregard any changes you just made and to return to the previous dialog box, press Ctrl-C or Escape or click on the Cancel button.

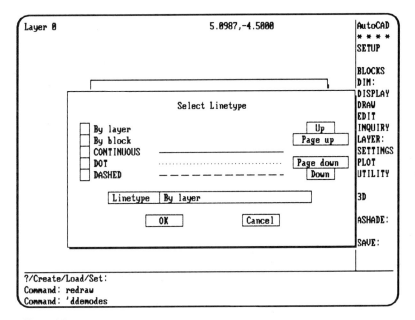

Figure 28

Elevation   To change the current elevation, move the arrow cursor within the Entity Creation Modes dialog box until the Elevation input button is highlighted as shown in Figure 29. Type a new elevation. Indicate when you have finished entering the new elevation by clicking on the OK button or by pressing Return. To cancel the elevation you just entered, click on the Cancel button or press Ctrl-C or Escape.

Thickness   To change the current thickness, move the arrow cursor within the Entity Creation Modes dialog box until the Thickness input button is highlighted as shown in Figure 30. Type a new thickness. Indicate when you have finished entering the new thickness by clicking on the OK button or by pressing Return. To cancel the thickness value you just entered, click on the Cancel button or press Ctrl-C or Escape.

Examples   You can activate the DDEMODES command transparently by preceding the command with an apostrophe, as in 'DDEMODES. In the following example, the current layer is changed in the middle of the LINE command, which was impossible to do in earlier versions of AutoCAD.

When the DDEMODES command is activated, the line command is temporarily suspended. The Entity Creation Modes dialog box appears on the screen. To change the current layer, you click on the Layer name requestor button, which causes the Select Layer dialog box to be displayed. The new current layer is selected by clicking on the check button adjacent to the desired layer name. After you click on the OK button to return to the Entity Creation Modes dialog box, you return to the LINE command by clicking on the OK button at the bottom of the DDEMODES dialog box.

```
To point: 'DDEMODES
Resuming LINE command.
To point:
```

Figure 29

Figure 30

Warnings   The DDEMODES command is available only with AutoCAD release 9.

If your graphics display does not support the Advanced User Interface, attempting to execute the DDEMODES command will result in AutoCAD displaying the message

```
** Command not allowed with this display configuration **
```

and the command will be ignored.

Only previously defined layer names (those appearing in the dialog box) can be entered into the Layer input button. If you enter a layer name that has not been defined, AutoCAD will display the dialog alert box shown in Figure 31. You must click on the OK button to continue.

Only previously loaded linetypes (those appearing in the dialog box) can be entered into the Linetype input button. If you enter a linetype that has not been loaded using either the LINETYPE Load option or by defining a layer with that linetype, AutoCAD will display the dialog alert box shown in Figure 32. You must click on the "OK" button to continue.

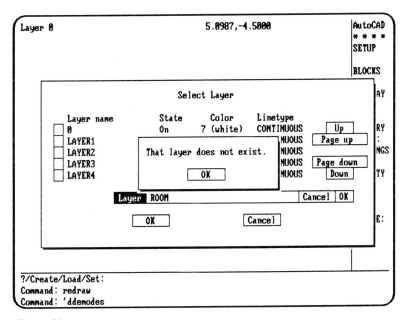

Figure 31

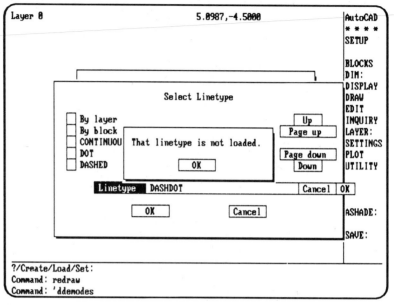

Figure 32

Tips   If you are using the standard menu "ACAD.MNU" provided with release 9, you can activate the DDEMODES command transparently by selecting Entity Creation from the Modes pull-down menu.

The DDEMODES command changes the values of the CECOLOR, CELTYPE, CLAYER, ELEVATION, and THICKNESS system variables, similar to using the SETVAR command. (For more information, see SETVAR and the individual system variables.)

You can use either the keyboard arrow keys or a pointing device to move the arrow cursor within a dialog box.

For more information, see Advanced User Interface and Dialog Box.

'DDLMODES

[ADE-3] [r9]

Overview   The DDLMODES (Dynamic Dialog Layer Modes) command displays the Modify Layer dialog box. This dialog box gives you complete control over the functions of the LAYER command. You can create new layers, set a new current layer, turn layers on and off, freeze and thaw layers, rename layers, and change the linetype and color associated with any layer. You can use the DDLMODES command transparently, but some of the changes may not take effect until after the active command has been completed and the drawing regenerated.

Procedure   You can enter the DDLMODES command at the command prompt or select it from a menu.

Command: **DDLMODES**

If you are using the standard ACAD.MNU provided with AutoCAD release 9, you can select the transparent 'DDLMODES command by picking Layer Control from the Modes pull-down menu shown in Figure 33.

The Layer Control dialog box duplicates the functions of the LAYER command. All existing layer names appear in the dialog box. If there are more than five layer names, you can view additional names by scrolling through the layer names. Scroll through the names a page at a time by clicking on the Page-up or Page-down button. Move through the layers one at a time by clicking on the Up or Down button.

You can alter any of the accessible Layer options before completing the command. To remove the dialog box and return to the command prompt, select the OK button (to accept all changes) or the Cancel button (to disregard all changes). In addition, you can cancel all the changes by pressing Ctrl-C or Escape. The Modify Layer dialog box is shown in Figure 34.

Current Layer   You can select a new current layer by clicking on the check button alongside the appropriate layer name. A small check mark will appear in the button adjacent to the current layer. This duplicates the capabilities of the Set option of the LAYER command.

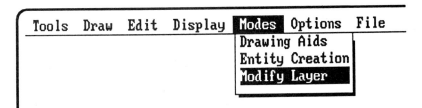

Figure 33

133

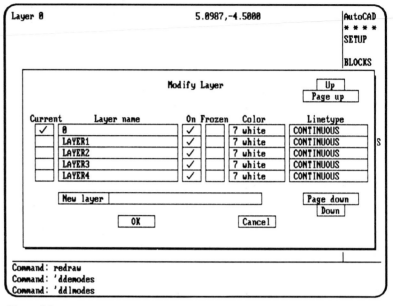

Figure 34

Layer Name   You can change any layer name by moving the arrow cursor so that the layer name is highlighted. Type in the new layer name. Indicate that you are finished and verify your entry by clicking on the OK button that appears alongside the name you just entered or by pressing Return. If you made a mistake, backspace over the name you entered. If you decide not to rename the layer, cancel the change by clicking on the Cancel button adjacent to the layer name (see Figure 35) or press Ctrl-C or Escape.

On/Off   You can turn layers on and off by clicking on the On check button adjacent to the appropriate layer name. If a check mark appears in the button, the layer is currently on. The absence of a check mark indicates that the layer is off.

Figure 35

134

Freeze/Thaw   You can "freeze" or "thaw" layers by clicking on the Frozen check button adjacent to the appropriate layer name. If a check mark appears in the button, the layer is currently frozen. The absence of a check mark indicates that the layer is currently thawed. (For an explanation of frozen layers, see LAYER.)

Color   You can change the color of any layer by clicking on the Color requestor button adjacent to the appropriate layer name. This causes a Select Color dialog box to overlay the previous dialog box. This dialog box is shown in Figure 36. The first eight colors are listed with a check button alongside each selection. To change the color of a layer, simply click on the appropriate button. Notice that the corresponding color number is displayed in the Color Number input button. To change to a color number above 7, move the arrow cursor until the Color Number input button is highlighted and type in the color number. You can also select one of the standard colors in this fashion. To complete the selection of the desired color, click on the OK button. This will return you to the Modify Layer dialog box. To disregard any changes you just made, press Ctrl-C or Escape or click on the Cancel button.

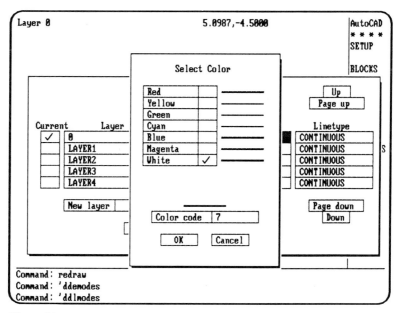

Figure 36

Linetype You can change the linetype of any layer by clicking on the Linetype requestor button adjacent to the appropriate layer name. This causes a Select Linetype dialog box to overlay the previous dialog box. This dialog box is shown in Figure 37. Only linetypes already loaded into the drawing will be displayed. A check button is provided alongside each available selection. Only five linetypes display in the dialog box at a time. If more than five linetypes are loaded, use the Page-up and Page-down buttons to move through them one page at a time or the Up and Down buttons to move through the list one line at a time. To change the selected linetype, click on the check button of one of the linetypes. The new current linetype appears in the Linetype input button. Alternatively, you can highlight the Linetype input button and type the name of a linetype. To finalize the selection of the linetype, click on the OK button. This returns you to the Modify Layer dialog box. To disregard any changes you just made and return to the previous dialog box, press Ctrl-C or Escape or click on the Cancel button.

New Layers You can create new layer names very quickly using the New Layer input button. Move the arrow cursor so that the New Layer

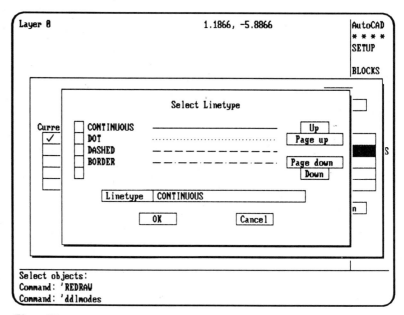

Figure 37

input button is highlighted. Then type in the new layer names. The new layer name appears in the input button as you type each character. If you make a mistake while typing, simply backspace over the mistake and retype it. Indicate when you have completed entering the layer name. To accept the new layer name, click on the OK button that appears to the right of the input button or press Return. To disregard the layer name you just entered, click on the Cancel button or press Ctrl-C or Escape.

As you add layer names, the new names appear in the list in the dialog box. The dialog list scrolls upward as each new layer is added at the bottom of the list. New layers are initially created with the color white (color number 7) and the CONTINUOUS linetype. You can subsequently use the Color and Linetype selections to change the color and the linetype, as well as turn the newly created layers on or off or make them frozen or thawed.

Examples   You can use the DDLMODES command to quickly and interactively create new layers. By placing the arrow cursor so that the New Layer input button remains highlighted, you can enter new layer names from the keyboard. Press Return after you type each layer name. The new names are added to the list of layer names. When you have completed entering the new names, you can use the other facilities to select the appropriate color and linetype for the new layers.

The dialog box in Figure 38 is activated by the DDLMODES command. The key strokes were then entered while the New Layer input button remained highlighted. Each line of key strokes was completed by pressing Return. When the four new layer names were entered, the DDLMODES command was completed by clicking on the OK button.

```
Command: DDLMODES  (Modify Layer dialog box appears)

1FL-WALLS
1FL-DOORS
1FL-WINDOWS
1FL-PLUMBING
```

Warnings   The DDLMODES command is available only with AutoCAD release 9.

If your graphics display does not support the Advanced User Interface, attempting to execute the DDLMODES command will result in AutoCAD

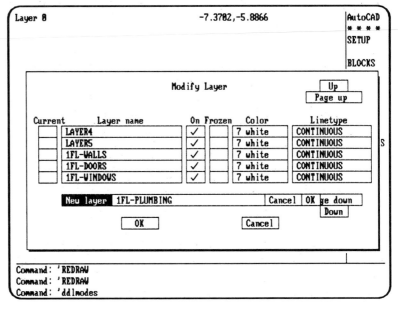

Figure 38

displaying the message

`** Command not allowed with this display configuration **`

and the command will be ignored.

If you attempt to rename the 0 layer, AutoCAD will display the dialog alert box shown in Figure 39. You must click on the OK button to continue.

If you turn off the current layer, AutoCAD will display the dialog alert box shown in Figure 40. You must click on the OK button to continue.

It is possible to freeze the current layer. If you do so, AutoCAD will display the dialog alert box shown in Figure 41. You must click on the OK button to continue. Even if the current layer is frozen, new entities can be drawn on it. They won't appear, however, until the current layer is thawed.

If you thaw a layer, AutoCAD displays the dialog alert box shown in Figure 42. You must click on the OK button to continue.

Only previously loaded linetypes (those appearing in the dialog box) can be entered into the Linetype input button. If you enter a linetype that has not been loaded using either the LINETYPE—Load option or by defining a layer with that linetype using the regular LAYER command,

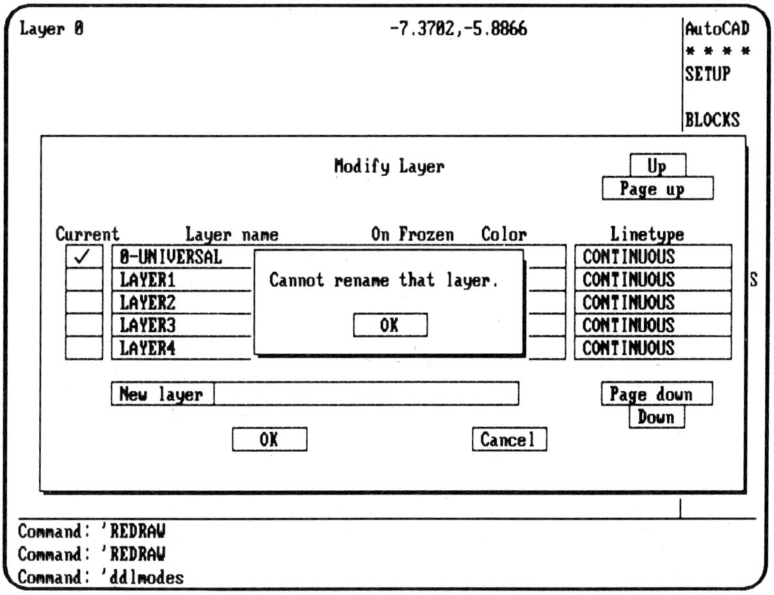

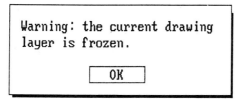

Figure 39

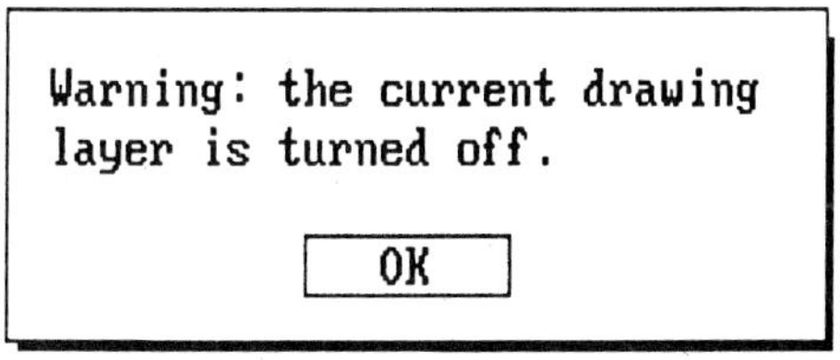

Warning: the current drawing
layer is turned off.

OK

Figure 40

Warning: the current drawing
layer is frozen.

OK

Figure 41

AutoCAD will display the dialog alert box shown in Figure 43. You must click on the OK button to continue.

> Changes to frozen status will take effect on the next REGEN.
>
> OK

Figure 42

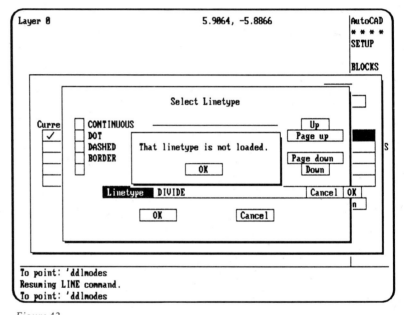

Figure 43

If you enter any other form of nonvalid information, such as a layer name with spaces between words, AutoCAD will not take any action. You must clear the entry by clicking on the Cancel button adjacent to the input button or by pressing Ctrl-C or Escape.

Tips   If you are using the standard menu ACAD.MNU provided with release 9, you can activate the DDLMODES command transparently by selecting Layer Control from the Modes pull-down menu.

You can enter layer names of up to thirty-one characters, but only the first twenty-four characters will be displayed in the dialog box.

You can use either the keyboard arrow keys or a pointing device to move the arrow cursor within a dialog box.

For more information, see Advanced User Interface and Dialog Box.

'DDRMODES

[ADE-3] [r9]

Overview   The DDRMODES (Dynamic Drawing Modes) command displays a dialog box that permits interactive control over most of AutoCAD's on-screen drawing aids. All of the functions of the AXIS, GRID, ISOPLANE, ORTHO, and SNAP commands are duplicated. The current Blip-mode setting can also be controlled.

Procedure   You can enter the DDRMODES command at the command prompt or select it from a menu.

Command: **DDRMODES**

If you are using the standard ACAD.MNU provided with AutoCAD release 9, you can select the transparent 'DDRMODES command by picking Drawing Aids from the Modes pull-down menu shown in Figure 44.

Mode Toggles   You can turn AutoCAD's Snap, Grid, Axis, Ortho, and Blip modes on and off simply by moving the arrow cursor until the appropriate check button adjacent to the desired mode is highlighted and then pressing the pick button (clicking). If the selected mode is currently on, a check mark appears in the check button immediately to the right of the

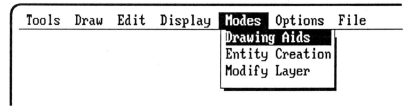

Figure 44

141

mode. If no check mark appears, the mode is currently off. Clicking on the appropriate check button toggles the selected mode on and off.

Changing any one of these modes has the same effect as changing the SNAPMODE, GRIDMODE, AXISMODE, ORTHOMODE, or BLIPMODE system variable (see the individual system variables).

Snap Spacing   You can change the current X and Y snap spacing. Move the arrow cursor until the current setting is highlighted. Type in the new snap spacing. If you make a mistake, use the Backspace key to delete the incorrect spacing value. Accept the new value by clicking on the OK button adjacent to the highlighted setting or by pressing Return. To disregard your new input, click on the Cancel button or press Ctrl-C or Escape.

Changing the snap spacing from this dialog box has the same effect as changing the SNAPUNIT system variable (see SNAPUNIT).

Snap Angle   You can change the current snap angle. Move the arrow cursor so that the Snap Angle input button is highlighted. Type in the new value. The new angle is interpreted in the current angle units. Delete mistakes by backspacing. Accept your new value by clicking on the OK button or by pressing Return. Disregard your new value by clicking on Cancel or by pressing Ctrl-C or Escape.

Changing the snap angle in this way is identical to using the SETVAR command to alter the SNAPANG system variable (see SNAPANG).

Snap Base   You can change the current snap base by moving the arrow cursor until the X-base or Y-base input button is highlighted. Then type in the new value and accept it as you did with Snap Spacing.

This is analogous to changing the SNAPBASE system variable (see SNAPBASE).

Grid Spacing   You can change the grid spacing in a fashion similar to changing the snap spacing. Only numeric values are accepted. A value of 0.0000 indicates that the grid spacing is the same as the snap spacing. This is identical to changing the GRIDUNIT system variable (see GRIDUNIT).

Axis Spacing   The spacing of the axis markers are changed in a fashion identical to changing the grid spacing. Again, a value of 0.0000 indicates that the axis spacing is the same as the snap spacing. This is identical to changing the AXISUNIT system variable (see AXISUNIT).

Isometric Mode   You can place AutoCAD into its isometric drawing mode, similar to selecting the Isometric style from within the SNAP command, by clicking on the Isometric check button. This method is identical to using the SETVAR command to change the SNAPSTYL system variable (see SNAPSTYL).

Isoplane Toggle   You can toggle between the left, top, and right isometric planes by selecting one of the Isoplane check buttons. The selected isoplane will have a check mark in its adjacent button. This method is identical to using the SETVAR command to change the SNAPISOPAIR system variable (see SNAPISOPAIR).

Examples   You can activate the DDRMODES command while another command is active by preceding the command with an apostrophe, as in 'DDRMODES. In the following example, the DDRMODES command is activated while the LINE command is active. You could change any of the available mode settings and then return to the LINE command by selecting the OK button at the bottom of the dialog box.

　　　While the dialog box is on the screen, no keyboard input appears on the command line (the command line appears exactly as follows regardless of the changes implemented from the dialog box). The only way to leave the dialog box and return to the LINE command is to click on the OK button or Cancel button or to press Ctrl-C or Escape.

```
To point: 'DDRMODES
Resuming LINE command.
To point:
```

Warnings   The DDRMODES command is available only with AutoCAD release 9.
　　　If your graphics display does not support the Advanced User Interface, attempting to execute the DDRMODES command will result in AutoCAD displaying the message

```
** Command not allowed with this display configuration **
```

and the command will be ignored.

Although the capabilities of this dialog box duplicate the functions of the SNAP, GRID, AXIS, and ISOPLANE commands, not all of the options normally associated with these commands are available. For example, you can enter only number values for the Grid and Axis spacing. Values with an X suffix are not accepted.

Tips Changing a mode setting while in the DDRMODES dialog box is similar to changing that same setting using the SETVAR command. Changing the snap, grid, or axis spacing or the snap base or snap angle does not turn the Snap, Grid, or Axis mode on or off.

When you change the X-axis spacing value for the snap, grid, or axis spacing, the new value you enter is automatically copied to the Y-axis. If you need to set different X and Y values, set the X-axis first, then Y.

The effect of changing a setting from this dialog box is the same as changing the setting using the SETVAR command to change the associated system variable. (For more information, see SETVAR and the individual system variables.)

If you are using the standard menu ACAD.MNU provided with release 9, you can activate the DDRMODES command transparently by selecting Entity Creation from the Modes pull-down menu.

You can use either the keyboard arrow keys or a pointing device to move the arrow cursor within a dialog box.

For more information, see Advanced User Interface and Dialog Box.

DELAY

Overview The DELAY command causes AutoCAD to pause for a specified period of time before continuing with the execution of the next command.

Procedure You can type the DELAY command in response to the command prompt, but it is really meant to be used as part of a script file. The format of the DELAY command is *DELAY* followed by a number. The number represents the approximate time in milliseconds that the computer will pause before executing the next command in the script file. If typed at the AutoCAD command prompt, the command would appear:

```
Command: DELAY
Delay time in milliseconds: (enter a number)
```

Examples

The following simple script file draws a line of text on the screen, lets it stay there for 2 seconds so that it can be read, and then erases it.

```
TEXT 0,0 1/8 0 AutoCAD Version 2.6
DELAY 2000
ERASE L
```

Make sure you enter the blank spaces between each word as well as two blank spaces after the L in the last line. Save the file with the file extension .SCR so that AutoCAD knows it is a script file. You can run the file later using the SCRIPT command (see SCRIPT).

Tips

One of AutoCAD's powerful features is that you can preprogram any number of command sequences. One possible way to program these sequences is through the use of scripts, using the SCRIPT command. A script is nothing more than a text file that contains a series of AutoCAD commands. Normally, when a script file is run, the commands execute in rapid succession. The speed with which the commands execute depends solely on the speed of the computer. One of the major reasons for using a script is to put on some sort of presentation; however, the script will often cause things to happen too fast for viewers to comprehend. The DELAY command allows you to build pauses into your script files.

The number that you enter depends on the speed of your computer (PC XTs are faster than PCs, PC ATs are faster than PC XTs, and the new PS/2s are faster than PC ATs). The exact delay that you will need to build into your scripts will depend on what type of computer you will be running your script on and how long the information needs to stay on the screen before the next sequence starts. As a general rule of thumb, a delay of 1000 is approximately one second. The highest number you can enter as a delay is 32767, which is about 32 seconds.

Dialog Box

[ADE-3] [r9]

Overview

A feature of the Advanced User Interface, dialog boxes are rectangular displays that pop-up over AutoCAD's graphics screen in response to certain commands. Specifically, a dialog box will be displayed in response to

145

the DDATTE, DDEMODES, DDLMODES, and DDRMODES commands and when using the INSERT command to insert a block containing variable attributes (subject to the setting of the ATTDIA system variable).

Dialog boxes are available only when you are using AutoCAD release 9 with a graphic display device that supports the features of the Advanced User Interface.

Procedure   When a dialog box appears, the normal cross-hair screen cursor is replaced by an arrow cursor. You move the cursor within the dialog box by moving the pointing device.

Dialog boxes have several selections available, each within a box, called a *button*. All dialog boxes have an OK selection. Most also have a Cancel selection. Dialog boxes may present different types of selections.

- *Check buttons* are small rectangles that are either blank or contain a check mark. This type of selection is used for selecting the current isoplane or for turning AutoCAD's Snap, Grid, Axis, Ortho, or Blips modes on and off.
- *Action buttons* simply perform an operation, such as selecting OK or Cancel.
- *Input buttons* allow you to type in a value. When you move the arrow cursor over an input button, the input button will be highlighted. You can then type in a value and press the Return key or select the OK button. If the value you enter is invalid, no action will be taken. You must either backspace and enter the value again or cancel the input.
- *Requestor buttons* are selections within dialog boxes that cause a second dialog box to overlay the first. You must enter a response or cancel the top dialog box before you will be returned to the underlying box. At other times, a dialog box selection may cause a warning to be displayed, such as when you freeze the current layer. This warning message must be acknowledged by clicking on the OK button displayed within the alert box before AutoCAD will continue.

The dialog box shown in Figure 45 has both check buttons and input buttons.

A dialog box will remain on the screen until an acceptable selection is made or the dialog box is canceled. Pressing the Return key has the same effect as pressing the pick button when the OK selection is highlighted (also referred to as clicking on a selection) except that you must actually click on the OK button to complete the command that called the

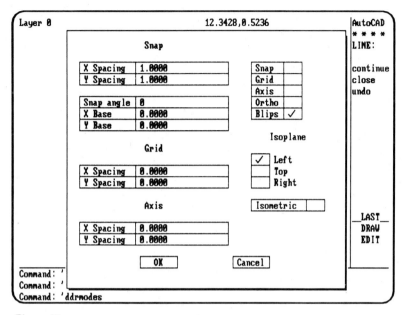

Figure 45

dialog box onto the screen. Ctrl-C or Escape have the same effect as selecting Cancel.

Examples For specific examples of commands using dialog boxes, see DDATTE, DDEMODES, DDLMODES, DDRMODES, and INSERT.

Warnings Dialog boxes are available only with AutoCAD release 9 when used with graphics devices that support the Advanced User Interface. Dialog boxes are not user-customizable.

   If your graphics display does not support the Advanced User Interface, attempting to execute a dialog-oriented command will result in AutoCAD displaying the message

`** Command not allowed with this display configuration **`

and the command will be ignored.

   If one dialog box overlays a lower box, you must complete the response to the upper box before you can access the lower dialog box.

Tips
You can execute the DDRMODES, DDEMODES, and DDLMODES commands transparently by preceding the command with an apostrophe. If you are using the standard menu ACAD.MNU supplied with release 9, these transparent commands can be selected by clicking on Drawing Aids, Entity Creation, or Layer Control, respectively, from the Modes pull-down menu.

You can use either the keyboard arrow keys or a pointing device to move the arrow cursor within a dialog box.

All of the actions accomplished from a dialog box will be treated as a single command by the UNDO command.

DIM

Overview
The DIM command places AutoCAD into Dimensioning mode. When in this mode, a new set of commands control the dimensioning features of the program. Most of the regular commands have no effect. While in Dimensioning mode, the normal command prompt is replaced with a Dim: prompt to indicate that you are in the Dimensioning mode.

Many drawings that you create with AutoCAD will require the inclusion of dimensions and notations. The Dimensioning mode provides all the tools necessary to accomplish this.

For more information, see also the individual subcommand entries (for example, see DIM—Horizontal for more information on AutoCAD's horizontal dimensioning subcommand). See Associative Dimensions and Dimensioning for general information of AutoCAD's dimensioning features.

Subcommands:

ALIGNED	HOMETEXT [v2.6, r9]	STATUS
ANGULAR	HORIZONTAL	STYLE
BASELINE	LEADER	UNDO
CENTER	NEWTEXT [v2.6, r9]	UPDATE [v2.6, r9]
CONTINUE	RADIUS	VERTICAL
DIAMETER	REDRAW	
EXIT	ROTATED	

Procedure
To enter the Dimensioning mode, enter the DIM command at the command prompt. After the completion of each dimensioning command, the

dim prompt reappears. To leave Dimensioning mode, type *EXIT* in response to the dim prompt.

Example

Enter Dimensioning mode by typing *DIM* in response to AutoCAD's command prompt. In this example, the horizontal dimension of the rectangle in the drawings that follow is annotated using the HORIZONTAL dimensioning subcommand. Return to AutoCAD's command prompt afterward by typing *EXIT* in response to the dim prompt.

```
Command: DIM
Dim: HORIZONTAL
First extension line or RETURN to select: ( RETURN )
Select line, arc, or circle: (select point A)
Dimension line location: (select point B)
Dimension text <6.0>: ( RETURN )
Dim: EXIT
Command:
```

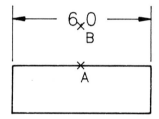

Warnings

While you are in Dimensioning mode, most AutoCAD commands have no meaning. Other commands, such as STATUS, have different meanings. The exceptions are the REDRAW command (which redraws the screen normally) and the mode toggle keys (such as ORTHO, COORDS, SNAP), which behave normally.

AutoCAD's dimensioning system variables control such dimensioning factors as the appearance of arrowheads, the height of the dimensioning text, and whether a plus-or-minus symbol is appended to the dimension text string. You should edit your prototype drawing to preset as many of these variables as practical and then control the others individually as necessary. (For more information on the dimension system variables, see the individual variables by name. They are listed alphabetically immediately after the individual dimensioning subcommands.)

The dimensioning system variable DIMASO controls whether Auto-CAD's associative dimensioning feature is enabled or disabled (see DIMASO and Associative Dimensions).

The U or UNDO command, if executed immediately after you exit from the Dimensioning mode, will undo all of the dimensions and annotations added during the entire period of time you were at the dim prompt. If you accidentally execute the UNDO command, the REDO command will immediately restore the dimensions and annotations. To remove individual dimensions or annotations, use the UNDO dimensioning subcommand (the U or UNDO command executed while at the dim prompt) or erase the individual dimensions when you are once again at the command prompt.

Tips

You can leave the Dimensioning mode by typing *EXIT* at the dim prompt or by pressing Ctrl-C. You can cancel any dimensioning command by pressing Ctrl-C.

REDRAW redraws the entire screen, removing any marker blips placed on the screen when BLIPMODE is on.

STATUS displays the settings of all of the dimensioning system variables.

STYLE allows you to change the current text style while in Dimensioning mode.

UNDO (or U), issued at the dim prompt, removes just the last dimension command.

Often the annotations AutoCAD places in the drawing are not quite right. The dimension text is in the wrong place or overwrites other drawing information. AutoCAD places the annotations into the drawing based on constraints internal to the program. There isn't much you can do to change the initial placement. After you have annotated your drawing, you will need to leave the Dimensioning mode and use the other Auto-CAD commands, such as MOVE and CHANGE, to correct some of the annotations. Autodesk calls the Dimensioning mode "semi-automatic dimensioning," meaning it won't do everything for you. This is true and you must simply be content to make whatever changes are necessary.

When dimensioning polylines, each segment of a polyline is treated as a separate entity.

DIM1

[ADE-1] [v2.5, v2.6, r9]

Overview   The DIM1 command places AutoCAD into Dimensioning mode and allows the execution of one of the special dimensioning subcommands. As soon as the dimensioning subcommand has been completed, you are returned to the normal AutoCAD command prompt. While at the dim prompt, you can use any of AutoCAD's dimensioning commands, but the completion of any chosen command returns you to the command prompt.

Procedure   The use of the DIM1 command is essentially identical to the DIM command. The only difference is that you do not need to type *EXIT* to leave the Dimensioning mode.

Examples   The DIM command example is repeated here. Notice that after completion of the HORIZONTAL dimensioning subcommand, AutoCAD's command prompt returns automatically. You don't have to type *EXIT* to leave the Dimensioning mode.

```
Command: DIM1
Dim: HORIZONTAL
First extension line or RETURN to select: ( RETURN )
Select line, arc, or circle: (select point A)
Dimension line location: (select point B)
Dimension text <6.0>: ( RETURN )
Command:
```

Tips   Use the DIM1 command instead of DIM to place just one annotation into a drawing. It is particularly useful with the CENTER (which places a center mark into the drawing) or LEADER (which places an arrow, leader line, and line of text into the drawing) subcommands, which often are needed only one at a time.

DIM — ALIGNED

[ADE-1] [all versions]

Overview   The ALIGNED dimensioning subcommand measures, draws, and annotates an object within a drawing parallel to the object or to the chosen extension line origin points.

151

Procedure

Use the ALIGNED dimensioning subcommand to dimension objects at a nonorthagonal orientation, that is, when the objects do not lie at ninety-degree angles. Use the DIM or DIM1 command to enter Dimensioning mode. At the dim prompt, type *ALIGNED*.

AutoCAD prompts you to select the

`First extension line origin or RETURN to select:`

If you select a point, you are prompted for the

`Second extension line origin:`

and the dimension will be calculated as the distance between these two points at the angle from the first point to the second point.

If you press Return when prompted to select the first extension point, the screen cross-hairs will change to a pick box and AutoCAD will prompt you to

`Select line, arc, or circle:`

Point to the object (other selection methods are not valid). If the object is a line or an arc, the endpoints are assumed to be the extension-line origins and the dimension is measured accordingly. If the object chosen is a circle, the point used to select the circle is assumed to be one end of its diameter.

You are then prompted to select the location of the dimension line. This is a point through which the dimension line will pass. It will be drawn parallel with the alignment of the chosen object. Finally, you are prompted to accept or change the proposed dimension text. If it is acceptable, simply press Return.

Examples

In the first example, the rectangle in the following drawing is not positioned at a ninety-degree angle in the drawing. The two extension origins are chosen individually. Object snap overrides are used to select the two extension-line origins. (Remember that object snap overrides and dimensioning subcommands can be abbreviated down to their first three letters, for example, ALI for ALIGNED).

```
Dim: ALIGNED
First extension line origin or RETURN to select: ENDPOINT
of: (select point A)
Second extension line origin: ENDPOINT
of: (select point B)
```

```
Dimension line location: (select point C)
Dimension text <6.0>: ( RETURN )
```

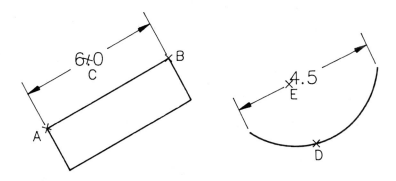

In the second example, the arc in the drawing is dimensioned using the automatic extension-line method.

```
Dim: ALIGNED
First extension line origin or RETURN to select: ( RETURN )
Select line, arc, or circle: (select point D)
Dimension line location: (select point E)
Dimension text <4.5>: ( RETURN )
```

Warnings   When selecting extension-line origins, use object snap to ensure that you select the correct points. Remember, the dimension will be measured solely between the two origin points you select.

Tips   Remember, factors such as the height of the dimension text, the size and appearance of the dimensioning arrowhead, and the location of the extension line in relation to the dimension line and extension origins are determined by AutoCAD's dimensioning system variables. There are twenty-nine dimensioning variables. (For more information, see the individual dimensioning system variables. For a complete list of all twenty-nine dimensioning system variables, see DIM—STATUS.)

If DIMASO is on (associative dimensioning enabled), definition points will be placed at the extension-line origin points and at the intersection of the dimension line and the first extension line.

The ALIGNED subcommand can be abbreviated as ALI.

DIM—ANGULAR

[ADE-1] [all versions]

Overview The ANGULAR dimensioning subcommand measures the angle between any two nonparallel lines. The two lines do not need to intersect.

Procedure Enter the ANGULAR dimensioning subcommand in response to the dim prompt. AutoCAD prompts you to select the two lines whose angle you want to measure. The screen cross-hair is replaced by the pick box. You can use any selection method to choose the two lines, as long as only two lines are selected (pointing is the most acceptable method).

Once the two lines have been selected, AutoCAD prompts you to enter the location of the dimension-line arc. The dimension line for angular measurement is an arc drawn between the two selected lines that spans the measured angle and passes through the point you select. Select a point between the two lines.

You will next be prompted for the dimension text (for more information, see Dimensioning). Finally, AutoCAD prompts you to select the text location. If you select a point, the text will be placed there. If you press Return, Auto-CAD will break the arc and center the text inside the break. If there is not enough room to fit the text inside the broken portion of the arc, you will be prompted to select a new text location. In this case, you must select a point. The dimension text will be placed there, and the arc will not be broken.

Examples In the following examples, the angle between the two lines is dimensioned using the ANGULAR subcommand and allowing AutoCAD to place the dimension text. In the second example, the text does not fit within the arc and AutoCAD prompts for a new text location.

```
Dim: ANGULAR
Select first line: (select point A)
Second line: (select point B)
Enter dimension line arc location: (select point C)
Dimension text <73>: ( RETURN )
Enter text location: ( RETURN )   (accept AutoCAD's location)

Dim: ANGULAR
Select first line: (select point D)
Second line: (select point E)
```

```
Enter dimension line arc location: (select point F)
Dimension text <15>: [ RETURN ]
Enter text location: [ RETURN ]   (accept AutoCAD's location)
Text does not fit.  Enter new text location: (select point G)
```

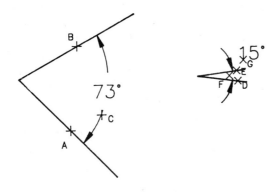

Warnings   Only nonparallel straight lines can be dimensioned using the ANGULAR subcommand. Only two lines can be selected at a time.

Tips   Extension lines will be added to the selected lines if they do not intersect with the placement of the dimension line, as shown in the following drawing.

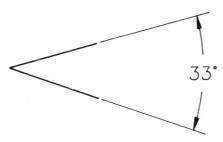

Remember, the dimensioning system variables control the appearance of dimensioning annotations (for more information, see the individual dimensioning system variables).

If DIMASO is on (associative dimensioning is enabled), definition points will be placed at the end points of the lines and at the point provided in response to the prompt

```
Enter dimension line arc location
```

The ANGULAR subcommand can be abbreviated as ANG.

DIM — BASELINE
[ADE-1] [all versions]

Overview   The BASELINE dimensioning subcommand causes AutoCAD to measure additional linear dimensions (aligned, horizontal, rotated, or vertical) from a common first extension-line origin.

Procedure   Use the BASELINE subcommand to reference multiple linear dimensions to a common datum point. Draw the first dimension using the ALIGNED, HORIZONTAL, ROTATED, or VERTICAL dimensioning subcommand. When AutoCAD reissues the dim prompt, enter the BASELINE subcommand. You will be prompted to select another second extension-line origin. Each baseline dimension will be drawn using the same orientation as the previous dimension command.

After you select the next extension-line origin, the dimension line will be placed automatically. You will be prompted to verify the dimension text.

Examples   In the following drawing, each hole in the baseplate is dimensioned back to a common datum point (the first extension-line origin) using the BASE-LINE subcommand. Object snap overrides are used to select the extension-line origins.

```
Dim: HORIZONTAL
First extension line origin or RETURN to select: ENDPOINT
of: (select point A)
Second extension line origin: CENTER
of: (select point B)
Dimension line location: (select point C)
Dimension text <0.37>: ( RETURN )
Dim: BASELINE
Second extension line origin: CENTER
of: (select point D)
Dimension text <1.50>: ( RETURN )
Dim: BASELINE
Second extension line origin: CENTER
of: (select point E)
Dimension text <2.62>: ( RETURN )
```

Tips   Each successive dimension line is offset from the previous line by a distance determined by the DIMDLI dimensioning system variable (see

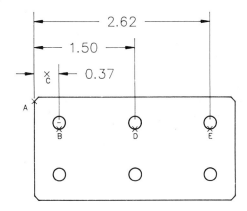

DIMDLI). This prevents each line from being drawn over the previous line. Extension lines are lengthened as necessary.

If you select the wrong point for one of your second extension-line origin points, you can enter a *U* to undo the last dimension added and then reenter the BASELINE subcommand to correctly place the next dimension.

If DIMASO is on (associative dimensioning enabled), definition points will be placed at the extension-line origin points and at the intersection of the dimension line and the first extension line.

The BASELINE subcommand can be abbreviated as BAS.

DIM — CENTER

[ADE-1] [all versions]

Overview   The CENTER dimensioning subcommand draws a center-line mark at the centerpoint of a selected arc or circle.

Procedure   If the CENTER subcommand is entered in response to the dim prompt, AutoCAD asks for the selection of an arc or circle. The pick box replaces the screen cross-hairs. Select a circle or arc by pointing. A center-line mark is added to the drawing.

Examples   In the following drawing, a center-line mark is placed at the center of a circle.

```
Dim: CENTER
Select arc or circle: (select point on circle)
```

157

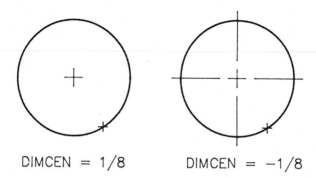

DIMCEN = 1/8   DIMCEN = −1/8

Tips   The size of the center-line mark is determined by the DIMCEN dimensioning system variable.

The CENTER subcommand can be abbreviated as CEN.

DIM–CONTINUE

[ADE-1] [all versions]

Overview   The CONTINUE dimensioning subcommand causes AutoCAD to measure additional linear dimensions (aligned, horizontal, rotated, or vertical) from the previous second extension-line origin.

Procedure   Use the BASELINE subcommand to draw continuous strings of multiple linear dimensions. Draw the first dimension using the ALIGNED, HORIZONTAL, ROTATED, or VERTICAL dimensioning subcommand. When AutoCAD reissues the dim prompt, enter the CONTINUE subcommand. You are prompted to select another second extension-line origin. Each continued dimension is drawn using the same orientation as the previous dimension command.

After you have selected the next extension-line origin, the dimension line is placed automatically. You are prompted to verify the dimension text.

Examples   In the following drawing, each hole in the baseplate is dimensioned along a continuous line using the CONTINUE subcommand. Object snap overrides are used to select the extension-line origins.

158

```
Dim: HORIZONTAL
First extension line origin or RETURN to select: ENDPOINT
of: (select point A)
Second extension line origin: CENTER
of: (select point B)
Dimension line location: (select point C)
Dimension text <0.37>: ( RETURN )
Dim: CONTINUE
Second extension line origin: CENTER
of: (select point D)
Dimension text <1.12>: ( RETURN )
Dim: CONTINUE
Second extension line origin: CENTER
of: (select point E)
Dimension text <1.12>: ( RETURN )
```

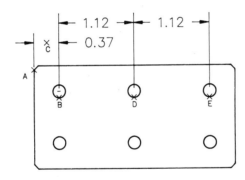

Warnings   If the previous dimension or the new dimension added by the CONTINUE subcommand has its arrows outside the extension lines, the dimension will be offset a distance determined by the DIMDLI dimensioning system variable.

Tips   If you select the wrong point for one of your second extension-line origin points, you can enter *U* to undo the last dimension added and then reenter the CONTINUE subcommand to correctly place the next dimension.

If DIMASO is on (associative dimensioning enabled), definition points will be placed at the extension-line origin points and at the intersection of the dimension line and the first extension line.

The CONTINUE subcommand can be abbreviated as CON.

DIM – DIAMETER

Overview   The DIAMETER dimensioning subcommand measures, draws, and annotates a diameter dimension of a selected arc or circle.

Procedure   Enter the DIAMETER subcommand in response to the dim prompt. Auto-
CAD prompts you to select an arc or a circle. The pick box replaces the
cross-hair screen cursor. Select the arc or the circle by pointing.
   Once the arc or circle has been selected, you will be prompted to verify
the dimension text. The default text will always begin with a diameter
symbol ($\varnothing$). The diameter dimension line is always drawn through the
center of the arc or circle. The angle at which the line is drawn depends on
the point used to select the object.
   If AutoCAD determines that the dimension text will fit inside the circle
or arc, the dimension is drawn and the subcommand completed. If, however, the text will not fit, you will be prompted to enter the length of a
leader line extending from the diameter dimension line. A rubberband
cursor will appear, and you can use your pointing device to indicate the
length of the leader line. If you press Return, the default leader-line length
(twice the length of the arrowheads) will be used. You can specify only the
length. The leader line will always be drawn as an extension of the diameter dimension line.

Examples   In the following drawing, the diameters of the two circles are dimensioned using the DIAMETER subcommand. In the second circle, the text
will not fit within the circle and AutoCAD prompts for a new text location.

```
Dim: DIAMETER
Select arc or circle: (select point)
Dimension text <>: ( RETURN )

Dim: DIAMETER
Select arc or circle: (select point)
Dimension text <>: ( RETURN )
Text does not fit.  Enter leader length for text: ( RETURN )
```

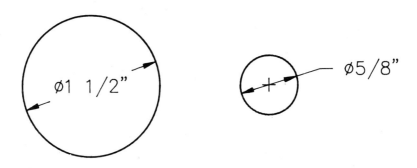

Warnings   If the leader-line length specified is less than the default value, the default value will be used.

Tips   If the value of the dimensioning system variable DIMCEN is not zero, the DIAMETER command will automatically add a center mark at the center point of the arc or circle.

   If the diameter dimension line is drawn at an angle of greater than fifteen degrees, the leader line will be drawn with an additional horizontal line attached.

   If the length of the diameter dimension line is less than four arrowhead lengths, no diameter dimension line is drawn. Instead, the dimension text is placed outside the arc or circle and an arrow is added to the end of the leader line. In addition, a center mark is automatically placed at the center point of the arc or circle as shown in the following drawing.

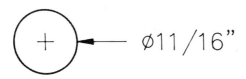

If DIMASO is on (associative dimensioning enabled), definition points will be placed at the end points of the diameter dimension line.

   The DIAMETER subcommand can be abbreviated as DIA.

DIM – EXIT

[ADE-1] [all versions]

Overview   The EXIT dimensioning subcommand causes AutoCAD to leave Dimensioning mode and return to the normal AutoCAD command prompt.

Procedure   Typing *EXIT* in response to the dim prompt exits from Dimensioning mode.

Examples   To return to AutoCAD's command prompt, type *EXIT*.

Dim: **EXIT** Command:

Tips   It is not necessary to use the EXIT subcommand when you are using the DIM1 command.

    You can also leave Dimensioning mode by typing Ctrl-C in response to the dim prompt.

DIM – HOMETEXT

[ADE-1] [v2.6, r9]

Overview   The HOMETEXT dimensioning subcommand relocates associative-dimension text back to its original default position.

Procedure   Enter the HOMETEXT subcommand in response to the dim prompt. The subcommand causes AutoCAD to prompt you to select objects. Valid objects are any associative dimension. If the text has been moved from its default position, it will be returned to that location. If the object selected is not an associative-dimension text or the text has not been moved, no change will be made. The subcommand continues prompting until you press Return a final time. No action is taken until you press Return.

D

Examples   In the following drawing, the dimension text had been moved previously to one side using the STRETCH command. The HOMETEXT subcommand causes the text to be returned to its "home" position. Notice that the dimension line is broken automatically to accept the dimension text (the STRETCH command had caused it to be healed).

```
Dim: HOMETEXT
Select objects: (select the dimension)
Select objects: (RETURN)
```

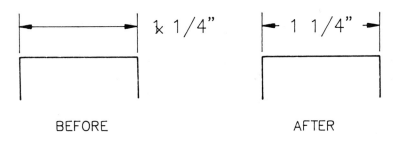

BEFORE   AFTER

Warnings   HOMETEXT is not available in AutoCAD version 2.5.

Tips   Associative-dimension text altered using the HOMETEXT subcommand reflects any changes to system variables (such as changes in dimension text height and arrow size), changes in the current text style, and changes to the current UNITS made since the last time the dimension was altered.
    The HOMETEXT command can be abbreviated as HOM.

DIM—HORIZONTAL

[ADE-1] [all versions]

Overview   The HORIZONTAL dimensioning subcommand measures, draws, and annotates an object within a drawing with a dimension line drawn horizontally across the drawing.

Procedure   Use the HORIZONTAL subcommand to dimension objects with a horizontal orientation or when the dimension line needs to be drawn horizontally regardless of the actual orientation of the object being measured.

At the dim prompt, activate the HORIZONTAL subcommand. AutoCAD will prompt you to select the

```
First extension line origin or RETURN to select:
```

If you select a point, you will be prompted for the

```
Second extension line origin:
```

and the dimension will be calculated as the distance between these two points measured horizontally regardless of the angle between the two points.

If you press Return when prompted to select the first extension point, the screen cross-hairs will change to a pick box, and AutoCAD will prompt you to

```
Select line, arc, or circle:
```

Point to the object (other selection methods are not valid). If the object is a line or an arc, the end points are assumed to be the extension-line origins and the dimension is measured accordingly.

You will then be prompted to select the location of the dimension line. This is a point through which the dimension line will pass. It is drawn horizontally regardless of the orientation of the object being measured. Finally, you will be prompted to accept or change the proposed dimension text. If it is acceptable, simply press Return.

Examples   In the following drawing, the rectangle is not positioned at a ninety-degree angle in the drawing. The two extension origins are chosen individually. Object snap overrides are used to select the two extension-line origins. (Remember that object snap overrides and dimensioning subcommands can be abbreviated down to their first three letters; for example, HOR for HORIZONTAL.)

```
Dim: HORIZONTAL
First extension line origin or RETURN to select: ENDPOINT
of: (select point A)
Second extension line origin: ENDPOINT
of: (select point B)
Dimension line location: (select point C)
Dimension text <6.0>: ( RETURN )
```

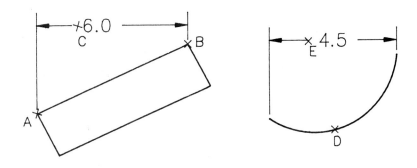

In the same drawing, the arc is dimensioned using the automatic extension-line method.

```
Dim: HORIZONTAL
First extension line origin or RETURN to select: (RETURN)
Select line, arc, or circle: (select point D)
Dimension line location: (select point E)
Dimension text <4.5>: (RETURN)
```

Warnings   When selecting extension-line origins, use object snap to ensure that you select the correct points. Remember, the dimension will be measured solely between the two origin points that you select.

Horizontal dimensions are always drawn horizontally regardless of the angle between the two extension-line origin points.

Tips   Remember, factors such as the height of the dimension text, the size and appearance of the dimensioning arrowhead, and the location of the extension line in relation to the dimension line and extension origins are determined by AutoCAD's dimensioning system variables. There are twenty-nine dimensioning variables. (For more information, see the individual dimensioning system variables. For a complete list of all twenty-nine dimensioning system variables, see DIM – STATUS.)

If DIMASO is on (associative dimensioning enabled), definition points will be placed at the extension-line origin points and at the intersection of the dimension line and the first extension line.

The HORIZONTAL subcommand can be abbreviated as HOR.

DIM—LEADER

[ADE-1] [all versions]

Overview The LEADER dimensioning subcommand allows you to draw a line or series of lines. An arrowhead is placed at the first point. Dimension text is annotated after the final line segment is drawn.

Procedure Enter the LEADER subcommand in response to the dim prompt. AutoCAD prompts for points in a fashion similar to the LINE command. You may enter as many line segments as you want. When you are finished drawing the line segments, press Return in response to the To point: prompt. AutoCAD then prompts for dimension text, giving the most recent measurement as the default. You may use any of AutoCAD's dimension text prefixes or suffixes. (For more information on dimension text, see Dimension Text.)

Examples Use the LEADER subcommand for placing notes into a drawing with a leader line and arrow pointing to the position in the drawing being noted.

```
Dim: LEADER
Leader start point: (select point A)
To point: (select point B)
To point: (select point C)
To point: (RETURN)
Dimension text <>: (type in your note here)
```

This example is illustrated in the following drawing.

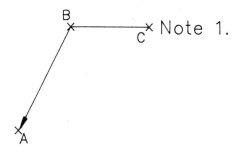

You can also use the LEADER subcommand for placing dimensioning text at a location other than where AutoCAD would normally place it. In

D

the following drawing, a horizontal dimension is measured but no dimension text placed. The LEADER command is then utilized to draw a leader line pointing to a dimension line and then to place the text at a position above the dimension line.

```
Dim: HORIZONTAL
First extension line origin or RETURN to select: ENDPOINT
of: (select point A)
Second extension line origin: ENDPOINT
of: (select point B)
Dimension line location: (select point C)
Dimension text <1.0>: (type a single space and press  RETURN  )
Dim: LEADER
Leader start: (select point D)
To point: (select point E)
To point:  RETURN
Dimension text <1.0>:  RETURN
```

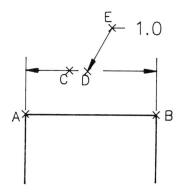

Warnings   If the length of the first leader-line segment is less than the length of two arrows, AutoCAD draws only lines. If the distance is greater, AutoCAD automatically places an arrow at the first leader-line point.

Tips   You can type *U* in response to any of the To point: prompts to undo the preceding line segments, even all the way back to the first segment drawn.

To draw a leader line without annotating any text, enter a single blank space and press Return when prompted for the dimension text.

The LEADER subcommand can be abbreviated as LEA.

The LEADER subcommand allows only single lines of text to be entered. The following AutoLISP routine utilizes the LEADER command as its basis and permits multiple lines of text to be entered.

```
(defun c:leader (/ pt1 pt2 pt3 pt a b t d)
  (setq t (getvar "textsize"))
  (setq b (getvar "orthomode"))
  (setvar "orthomode" 0)
  (setq pt1 (getpoint "\nLeader begin: "))
  (command "dim" "leader" pt1)
  (command (setq pt2 (getpoint pt1 "\nNext point:")))
  (setvar "orthomode" 1)
  (command (setq pt3 (getpoint pt2 "\nNext point:")))
  (command "")
  (command (setq a (getstring T "\nEnter text:")))
  (command "exit")
  (setq pt (cdr (assoc 10 (entget (entlast)))))
  (setq d (cdr (assoc 40 (entget (entlast)))))
  (if (< (car pt3) (car pt2))
    (progn
      (setq pt (list (- (car pt3) (* d 0.66666)) (- (cadr pt3) (* d 2.20833))))
      (command "dtext" "r" pt d""))
    (progn
      (command "text" pt d"""")
      (command "dtext" "")))
  (setvar "orthomode" b)
  (setvar "textsize" t)
  (command "text" PT T"""".")
  (command "erase" "l" "")
)
```

DIM — NEWTEXT

[ADE-1] [v2.6, r9]

Overview   The NEWTEXT dimensioning subcommand allows you to change an existing associative-dimension text string.

Procedure   Enter the NEWTEXT subcommand in response to the dim prompt. The subcommand causes AutoCAD to prompt for new dimension text. After entering the new text, you are prompted to select objects. Valid objects are any associative dimension. If the object selected is not an associative dimension, no change will be made. The subcommand continues prompting until you press Return a final time. Once you press Return, the dimension text of all the associative dimensions you selected will be changed to the new dimension text you entered.

Examples   In this example, two associative dimensions are selected. Notice that the text of both dimensions are changed but not their locations. Notice also that the size of the break in the dimension line is automatically adjusted to the different lengths of the new text.

```
Dim: NEWTEXT
Enter new dimension text: 4
Select objects: (select one dimension)
Select objects: (select other dimension)
Select objects: ( RETURN )
```

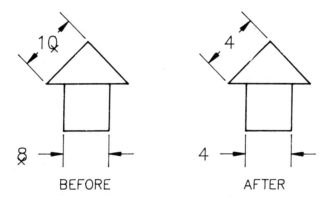

BEFORE   AFTER

Warnings   NEWTEXT is not available in AutoCAD version 2.5.

Tips   You can use the NEWTEXT command to restore an associative-dimension text string to its actual value by entering < > in response to the prompt

```
Enter new dimension text
```

In this case, the dimension text value of each associative dimension selected will be changed to its actual value.

169

Associative-dimension text altered using the NEWTEXT subcommand reflects any changes to system variables (such as changes in dimension text height and arrow size), changes in the current text style, and changes to the current UNITS made since the last time the dimension was altered. The NEWTEXT command can be abbreviated as NEW.

DIM—RADIUS

[ADE-1] [all versions]

Overview   The RADIUS dimensioning subcommand measures, draws, and annotates a radius dimension of a selected arc or circle. It is similar to the DIAMETER subcommand.

Procedure   Enter the RADIUS subcommand in response to the dim prompt. AutoCAD then prompts you to select an arc or a circle. The pick box replaces the cross-hair screen cursor. Select the arc or circle by pointing.

Once the arc or circle has been selected, you will be prompted to verify the dimension text. The default text will always begin with the letter *R*, to indicate a radius. The radius dimension line is always drawn from the center of the arc or circle to the point used to select it.

If AutoCAD determines that the dimension text will fit inside the circle or arc, the dimension is centered along the radius dimension line and the subcommand is completed. If, however, the text will not fit, you will be prompted to enter the length of a leader line extending from the radius dimension line. A rubberband cursor will appear, and you can use your pointing device to indicate the length of the leader line. If you press Return, the default leader-line length (twice the length of the arrowheads) will be used, and the leader will extend out from the object along the radius line. If you specify the leader-line length by pointing to a location on the opposite side of the circle or arc from the point used to select it, the leader line will be drawn back through the center point.

Examples   In the following examples, the radii of the two circles are dimensioned using the RADIUS subcommand.

```
Dim: RADIUS
Select arc or circle: (select point A)
Dimension text <1">: RETURN
```

In the second circle, the text will not fit within the circle, so AutoCAD prompts for a new text location.

```
Dim: RADIUS
Select arc or circle: (select point B)
Dimension text <1/2">: RETURN
Text does not fit.  Enter leader length for text: RETURN
```

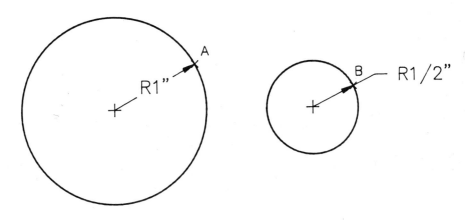

Warnings   If the leader-line length specified is less than the default value, the default value will be used.

Tips   If the value of the dimensioning system variable DIMCEN is not zero, the RADIUS command automatically adds a center mark at the center point of the arc or circle.

If the radius dimension line is drawn at an angle of greater than fifteen degrees, the leader line will be drawn with an additional horizontal line attached.

If DIMASO is on (associative dimensioning enabled), definition points will be placed at the point used to select the arc or circle and at its center point.

The RADIUS subcommand can be abbreviated as RAD.

DIM—REDRAW

Overview   The REDRAW dimensioning subcommand causes the screen to be refreshed. It is identical to the regular REDRAW command (see REDRAW).

Procedure   Enter the REDRAW subcommand in response to any dim prompt.

Examples   To refresh the screen while in Dimensioning mode (to remove blip marks, for example), type *REDRAW* in response to the dim prompt and press Return.

Dim: **REDRAW**

Tips   The REDRAW subcommand is one of the few commands that behaves the same way in Dimensioning mode as it does from the command prompt.

The REDRAW subcommand can be abbreviated as RED.

DIM—ROTATED

Overview   The ROTATED dimensioning subcommand measures, draws, and annotates an object within a drawing at a user-specified orientation angle.

Procedure   Use the ROTATED dimensioning subcommand to dimension objects when their dimensions will be measured at a user-specified angle. Use either the DIM or DIM1 command to enter Dimensioning mode. At the dim prompt, type *ROTATED*.

You will first be prompted to enter the dimension-line angle. You can enter a number or indicate the angle by pointing. If you use the pointing method, once you select the first point, AutoCAD will prompt you to select the second point.

Once you have provided an angle, AutoCAD will prompt you to select the

```
First extension line origin or RETURN to select:
```

If you select a point, you will be prompted for the

```
Second extension line origin:
```

and the dimension will be calculated as the distance between those two points at the specified angle.

If you press Return when prompted to select the first extension point, the screen cross-hairs will change to a pick box and AutoCAD will prompt you to

```
Select line, arc, or circle:
```

Point to the object (other selection methods are not valid). If the object is a line or an arc, the end points are assumed to be the extension-line origins and the dimension is measured accordingly.

You will then be prompted to select the location of the dimension line. This is a point through which the dimension line will pass. It will be drawn parallel to the specified angle regardless of the orientation of the selected object. Finally, you will be prompted to accept or change the proposed dimension text. If it is acceptable, simply press Return.

Examples In the first example that follows, the rectangle is not positioned at a thirty-degree angle in the drawing. The two extension origins are chosen individually. Object snap overrides are used to select the two extension-line origins. (Remember that object snap overrides and dimensioning subcommands can be abbreviated down to their first three letters, for example, ROT for ROTATED.)

```
Dim: ROTATED
Dimension line angle <0>: 30
First extension line origin or RETURN to select: ENDPOINT
of: (select point A)
Second extension line origin: ENDPOINT
of: (select point B)
Dimension line location: (select point C)
Dimension text <6.0>: RETURN
```

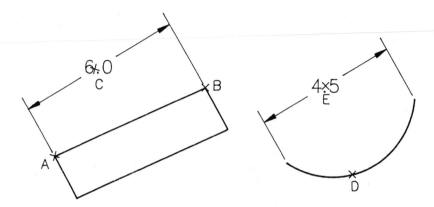

In the second example, the arc is dimensioned using the automatic extension-line method. An angle of thirty degrees is used as the rotation angle.

```
Dim: ROTATED
Dimension line angle <0>: 30
First extension line origin or RETURN to select: ( RETURN )
Select line, arc, or circle: (select point D)
Dimension line location: (select point E)
Dimension text <4.5>: ( RETURN )
```

Warnings   When selecting extension-line origins, use object snap to ensure that you
select the correct points. Remember, the dimension will be measured
solely between the two origin points that you select.

Tips   Remember, factors such as the height of the dimension text, the size and
appearance of the dimensioning arrowhead, and the location of the extension line in relation to the dimension line and extension origins are
determined by AutoCAD's dimensioning system variables. There are
twenty-nine dimensioning variables. (For more information, see the individual dimensioning system variables. For a complete listing of all twenty-
nine dimensioning system variables, see DIM – STATUS.)

The ROTATED subcommand can be made to behave exactly like any
other linear dimensioning subcommand. If an angle of zero degrees is
entered, ROTATED will yield the same result as HORIZONTAL. If an angle
of ninety degrees is entered, the VERTICAL subcommand will be mim-
icked. By using object snaps to "point" to the rotation angle, the subcommand can yield the same results as the ALIGNED subcommand.

If DIMASO is on (associative dimensioning enabled), definition points will be placed at the extension-line origin points and at the intersection of the dimension line and the first extension line.

The ROTATED subcommand can be abbreviated as ROT.

DIM—STATUS

[ADE-1] [all versions]

Overview   The STATUS dimensioning subcommand causes all of the dimensioning system variables and their values to be displayed.

Procedure   To view all the dimensioning system variables and their values, enter the STATUS command in response to the dim prompt. The list is too long to fit on the screen, so AutoCAD displays the message

```
-- Press RETURN for more --
```

whenever more of the list remains to be viewed.

Examples   In the following example, the STATUS subcommand is entered at the dim prompt. The resulting list of dimensioning system variables is provided as it would appear on the screen.

```
Dim: STATUS
DIMSCALE    1.0000  Overall scale factor
DIMASZ      0.1800  Arrow size
DIMCEN      0.0900  Center mark size
DIMEXO      0.0625  Extension line origin offset
DIMDLI      0.3800  Dimension line increment for continuation
DIMEXE      0.1800  Extension above dimension line
DIMTP       0.0000  Plus tolerance
DIMTM       0.0000  Minus tolerance
DIMTXT      0.1800  Text height
DIMTSZ      0.0000  Tick size
DIMRND      0.0000  Rounding value
DIMDLE      0.0000  Dimension line extension
DIMTOL      Off     Generate dimension tolerances
DIMLIM      Off     Generate dimension limits
DIMTIH      On      Text inside extensions is horizontal
DIMTOH      On      Text outside extensions is horizontal
DIMSE1      Off     Suppress the first extension line
DIMSE2      Off     Suppress the second extension line
```

DIMTAD	Off	Place text above the dimension line
DIMZIN	Off	Edit zero inches
DIMALT	Off	Alternate units selected
DIMALTF	25.40	Alternate units scale factor
DIMALTD	2	Alternate units decimal places
DIMLFAC	1.000	Linear unit scale factor
DIMBLK	DOT	Arrow block name
DIMASO	On	Create associative dimensions
DIMSHO	Off	Update dimensions while dragging
DIMPOST		Default suffix for dimension text
DIMAPOST		Default suffix for alternate text

Warnings   The STATUS subcommand only displays the list. The individual values can be changed only by entering the name of the variable in response to the Dimensioning mode prompt or by using the SETVAR command or Auto-LISP. (For more information, see the individual dimensioning system variables.)

The STATUS command behaves differently when entered from the command prompt (see STATUS).

Tips   The Dimensioning mode STATUS subcommand is a handy command to build into a menu. In fact, it is part of AutoCAD's standard menu.

The STATUS subcommand can be abbreviated as STA.

DIM—STYLE   [ADE-1] [v2.5, v2.6, r9]

Overview   The STYLE dimensioning subcommand enables you to change the current text style while in AutoCAD's Dimensioning mode.

Procedure   The STYLE subcommand is similar to the regular AutoCAD STYLE command. Entering STYLE in response to the dim prompt causes AutoCAD to prompt you for a new text style. The current style is displayed as the default. Simply press Return to retain the current style.

Examples   Enter the STYLE subcommand at the dim prompt. In this example, the current text style, STANDARD is changed to SMOOTH.

176

```
Dim: STYLE
New text style <STANDARD>: SMOOTH
SMOOTH is now the current text style.
```

Warnings   The text style that you enter when prompted for a new text style must already exist in your drawing. The STYLE subcommand does not allow you to create a new text style from within the Dimensioning mode.

Tips   Use the regular STYLE command to create a text style for dimensioning prior to starting the Dimensioning mode. Use a width factor less than 1 for your dimensioning text style. This will help ensure that AutoCAD is able to place the dimensioning text between the arrows. A width factor of 0.85 gives an acceptable appearance. A good style name for your dimensioning text is DIME.

The STYLE subcommand can be abbreviated as STY.

DIM—UNDO (*or* U)

[ADE-1] [all versions]

Overview   The UNDO (or U) dimensioning subcommand removes all of the annotations produced by the most recent dimensioning command.

Procedure   The UNDO subcommand behaves in a fashion similar to AutoCAD's regular UNDO (or U) command. Entering *UNDO* or *U* in response to the dim prompt causes the last dimension placed in the drawing to be removed. All extension lines, arrows, dimension lines, and dimension text added by the most recent dimensioning command are removed.

Examples   Entering *UNDO* or *U* in response to the dim prompt causes the latest dimensioning command to be reversed. All of its annotations are removed. In the following example, the horizontal dimension is first added using the HORIZONTAL subcommand. Then the entire dimension is removed in one step using the UNDO subcommand. The before and after drawings illustrate this procedure.

```
Dim: HORIZONTAL
First extension line origin or RETURN to select: ( RETURN )
Select line, arc, or circle: (select point A)
```

Dimension line location: **(select point B)**
Dimension text <6'-8">: (RETURN)

Dim: **UNDO**

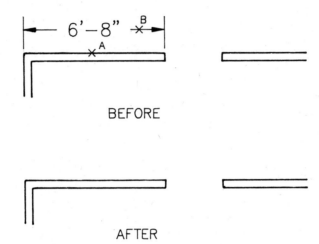

BEFORE

AFTER

Warnings   Remember that in Dimensioning mode, the UNDO subcommand reverses one dimensioning command at a time. The UNDO (or U) command issued from AutoCAD's command prompt reverses an entire Dimensioning mode editing session. A REDO command, issued at the command prompt, reverses the action of an UNDO issued at the command prompt. There is no REDO command for the UNDO dimensioning subcommand.

Tips   The UNDO subcommand, issued from the dim prompt, can be used to step back through the Dimensioning mode editing session, reversing one dimension command at a time.

The UNDO dimensioning subcommand can be abbreviated as U.

DIM–UPDATE

[ADE-1] [v2.6, r9]

Overview   The UPDATE dimensioning subcommand allows you to change an existing associative dimension so that it reflects the current settings of the dimensioning system variables, the current text style, and the current units.

D

Procedure   Enter the UPDATE subcommand in response to the dim prompt. The sub-command causes AutoCAD to prompt you to select objects. Valid objects are any associative dimension. If the object selected is not an associative dimension, no change will be made. The subcommand continues prompting until you press Return a final time. Once you press Return, all the associative dimensions you selected will be changed to reflect the current settings of the dimensioning system variables, the current text style, and the current units.

Examples   In this example, many of the dimensioning system variables controlling the text height and arrow size in the before drawing have been changed. In addition, the UNITS command has been used to change from decimal to architectural units. Notice the changes in the after drawing.

```
Dim: UPDATE
Select objects: (select one dimension)
Select objects: (select other dimension)
Select objects: ( RETURN )
```

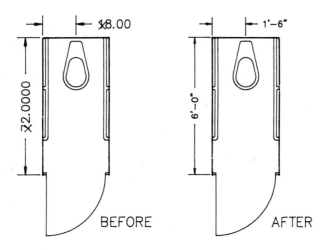

BEFORE   AFTER

Warnings   UPDATE is not available in AutoCAD version 2.5. Only associative dimensions can be changed.

Tips Associative-dimension text altered using the UPDATE subcommand reflects any changes to system variables (such as changes in dimension text height and arrow size), changes in the current text style, and changes to the current UNITS made since the last time the dimension was altered.

The UPDATE subcommand can be abbreviated as UPD.

DIM — VERTICAL

[ADE-1] [all versions]

Overview The VERTICAL dimensioning subcommand measures, draws, and annotates an object within a drawing. The dimension line is always drawn at a vertical orientation.

Procedure Use the VERTICAL subcommand to dimension objects with a vertical orientation or when the dimension line needs to be drawn vertically regardless of the actual orientation of the object being measured.

At the dim prompt, activate the VERTICAL subcommand. AutoCAD will prompt you to select the

```
First extension line origin or RETURN to select:
```

If you select a point, you will be prompted for the

```
Second extension line origin:
```

and the dimension will be calculated as the distance between these two points measured vertically regardless of the angle between the two points.

If you press Return when prompted to select the first extension point, the screen cross-hairs will change to a pick box, and AutoCAD will prompt you to

```
Select line, arc, or circle:
```

Point to the object (other selection methods are not valid). If the object is a line or an arc, the end points are assumed to be the extension-line origins, and the dimension is measured accordingly.

You will then be prompted to select the location of the dimension line. This is a point through which the dimension line will pass. It will be drawn vertically regardless of the orientation of the object being measured. Finally, you will be prompted to accept or change the proposed dimension text. If it is acceptable, simply press Return.

Examples
In the first example, the rectangle is not positioned at a ninety-degree angle in the drawing. The two extension origins are chosen individually. Object snap overrides are used to select the two extension-line origins. (Remember that object snap overrides and dimensioning subcommands can be abbreviated down to their first three letters, for example, VER for VERTICAL.)

```
Dim: VERTICAL
First extension line origin or RETURN to select: ENDPOINT
of: (select point A)
Second extension line origin: ENDPOINT
of: (select point B)
Dimension line location: (select point C)
Dimension text <6.0>: ( RETURN )
```

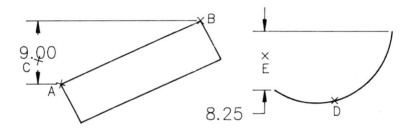

In the second example, the arc is dimensioned using the automatic extension-line method.

```
Dim: VERTICAL
First extension line origin or RETURN to select: ( RETURN )
Select line, arc, or circle: (select point D)
Dimension line location: (select point E)
Dimension text <4.5>: ( RETURN )
```

Warnings
When selecting extension-line origins, use object snap to ensure that you select the correct points. Remember, the dimension will be measured solely between the two origin points that you select.

Vertical dimensions will always be drawn vertically regardless of the angle between the two extension-line origin points.

Tips   Remember, factors such as the height of the dimension text, the size and appearance of the dimensioning arrowhead, and the location of the extension line in relation to the dimension line and extension origins are determined by AutoCAD's dimensioning system variables. There are twenty-nine dimensioning variables. (For more information, see the individual dimensioning system variables. For a complete listing of all twenty-nine dimensioning system variables, see DIM—STATUS.)

          If DIMASO is on (associative dimensioning enabled), definition points will be placed at the extension-line origin points and at the intersection of the dimension line and the first extension line.

          The VERTICAL subcommand can be abbreviated as VER.

DIMALT
[ADE-1] [v2.5, v2.6, r9]

Overview   The DIMALT (Alternate Dimensions) dimensioning system variable controls whether alternate dimension are annotated during Dimensioning mode when linear dimensions are measured. The value can be either on or off.

Procedure   If DIMALT is on, the measured distance is adjusted by the DIMLFAC global scale factor and multiplied by the DIMALTF value to determine an alternate dimension, which is then appended to the dimension text. The alternate text is placed inside square brackets [ ] and displayed to the number of decimal places determined by the DIMALTD variable. If DIMALT is off, no alternate dimensions are added. (See DIMALTD, DIMALTF, and DIMLFAC.)

          You can manipulate the DIMALT variable directly from the dim prompt or by using the SETVAR command or AutoLISP.

Examples   The most common use of this feature is to add dimensions to a drawing using both inches and metric dimensions. To do so, set DIMALT on and DIMALTF to its default value of 25.4 (the number of millimeters per inch). If DIMALTD is set to its default value of two decimal places, the dimension 1'-0" will appear, as in the following drawing.

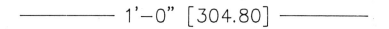

Warnings All measured distances, including limits (if DIMLIM is on) and tolerances (if DIMTOL is on), will have alternate units appended (see DIMLIM and DIMTOL).

Tips Prefixes and suffixes can be added to dimensions (for more information, see DIMAPOST and Dimension Text).

The initial setting of the DIMALT dimensioning system variable is determined by the prototype drawing (default value is 0, off). The current value is saved with the drawing file. If you use alternate dimensions often, edit the prototype drawing and set the DIMALT value to on.

DIMALTD

Overview The DIMALTD (Alternate Dimensions Decimal) dimensioning system variable controls the number of decimal places displayed in alternate unit dimensions.

Procedure The DIMALTD variable has no bearing on dimensioning a drawing unless the DIMALT variable is on. You can manipulate the DIMALTD variable directly from the dim prompt or by using the SETVAR command or AutoLISP. DIMALTD can be only a decimal number.

Examples You can change the DIMALTD variable from the dim prompt. If in the example under DIMALT the value were changed from its default of 2 to 4, it would appear as follows:

```
Dim: DIMALTD
Current value <2>  New value: 4
```

————— 1'–0" [304.8000] —————

Tips The DIMALTD variable controls only the number of decimal places shown for alternate dimensions. The number of decimal places in regular dimension text is determined by the number of decimal places set by the UNITS command.

The initial DIMALTD value is determined by the prototype drawing (default is 2). The current value is saved with the drawing.

DIMALTF

[ADE-1] [v2.5, v2.6, r9]

Overview   The DIMALTF (Alternate Dimension Factor) dimensioning system variable is multiplied by measured linear dimensions to yield an alternate system of measurement if the DIMALT variable is on. DIMALTF can be only a decimal number.

Procedure   The DIMALTF variable has no bearing on dimensioning a drawing unless the DIMALT variable is on. You can manipulate the DIMALTF variable directly from the dim prompt or by using the SETVAR command or AutoLISP.

    If DIMALT is on, whenever you use one of the linear dimensioning subcommands (ALIGNED, BASELINE, CONTINUE, HORIZONTAL, RO-TATED, or VERTICAL), AutoCAD automatically multiplies the measured dimension (subject to the DIMLFAC global scale factor) by the DIMALTF value to arrive at an alternate measurement. That measurement is annotated to the drawing inside square brackets, with the number of decimal places determined by the DIMALTD variable. (See DIMALT and DIMLFAC.)

Examples   The value of DIMALTF is usually set to 25.4 (the number of millimeters per inch), which causes alternate dimensions to be displayed in metric units. This is demonstrated in the examples under DIMALT and DIMALTD. In the following example, normal units are displayed in feet and inches. The DIMALTF value is set to 1, so the alternate units are displayed in decimal inches. The DIMALTD value of 2 causes two decimal places to be shown.

$$\text{———— } 1'-0'' \ [12.00] \ \text{————}$$

Tips   The initial value for DIMALTF is determined by the prototype drawing (default is 25.4000). The current value is saved with the drawing. If you use a different alternate scale factor often, edit the prototype drawing and replace the default DIMALTF with your own.

DIMAPOST

[ADE-1] [v2.6, r9]

Overview   The DIMAPOST (Alternate Dimension Postscript) dimensioning system variable determines a text string suffix to be added to alternate dimensions when the DIMALT variable is on. DIMAPOST is assumed to be a text string.

Procedure   The DIMAPOST variable has no bearing on dimensioning a drawing unless the DIMALT variable is on. You can manipulate the DIMAPOST variable directly from the dim prompt or by using the SETVAR command or AutoLISP.

When DIMALT is on and a linear dimension subcommand is used, an alternate dimension is added to the drawing inside square brackets. If DIMAPOST has a value, the text string contained in the variable is added to the alternate dimension.

To disable the DIMAPOST variable, enter a single period (.).

Examples   You can change the DIMAPOST variable from the dim prompt. If the value were set to *mm*, the example under DIMALT would appear as follows:

```
Dim: DIMAPOST
Current value <>  New value: mm
```

$$\text{---------}\quad 1'-0''\ [304.80\text{mm}]\ \text{-------}$$

Warnings   DIMAPOST is not available in AutoCAD version 2.5.

Tips   You can add individual prefixes and suffixes to dimensions by using the < > method (see Dimension Text).

If limits or tolerance dimensioning is enabled (see DIMLIM and DIMTOL), the alternate dimension suffix is automatically applied to them also.

The initial value for DIMAPOST is determined by the prototype drawing (default is " "—blank, no suffix). The current value is saved with the drawing. If you often use a different alternate dimension suffix, edit the DIMAPOST value in prototype drawing accordingly, so your suffix is always preset when you start a new drawing.

DIMASO

Overview   The DIMASO (Associative Dimension) dimensioning system variable controls whether dimensions subsequently added to a drawing will be drawn as associative or nonassociative dimensions. DIMASO is either on or off.

Procedure   When DIMASO is on, all subsequent dimensions (linear, angular, diameter, and radius) are drawn as associative dimensions and are subject to the rules that govern associative dimensions.

    You can manipulate the DIMASO variable directly from the Dimensioning mode prompt or by using the SETVAR command or AutoLISP. Its value can be either on or off.

Examples   AutoCAD's associative dimensioning feature is described under Associative Dimensions.

Warnings   DIMASO is not available in AutoCAD version 2.5.

Tips   The intitial value of DIMASO is determined by the prototype drawing (default value is 1, on). The current value is saved with the drawing. If you never use this feature, you can edit the prototype drawing to set DIMASO to off.

DIMASZ

Overview   The DIMASZ (Dimension Arrow Size) dimensioning system variable controls the size of the arrows drawn at the end of the dimension line.

Procedure   The value contained in the DIMASZ system variable specifies the size, in drawing units, of the dimensioning arrow. You can manipulate the DIMASZ variable directly from the dim prompt or by using the SETVAR command or AutoLISP. You can enter the DIMASZ value in any format acceptable by the current UNITS. You can also indicate the size of the arrows by selecting two points in the drawing when prompted for a new value. The DIMASZ value will be the distance between the two points.

Examples   When you use AutoCAD's standard arrow, the DIMASZ value represents the actual size of the arrows, as illustrated in the following example.

```
Dim: DIMASZ
Current value <0.18>  New value: 1
```

Warnings   If the variable DIMBLK has a user-assigned arrow block name, the user-defined arrow block drawing will be drawn at a scale factor determined by the DIMASZ value (see DIMBLK).

Tips   AutoCAD determines whether dimension lines and text will fit between the extension lines by multiplying the arrow-size variable DIMASZ by a factor of 4.1 (multiplied by the DIMSCALE variable). This result is subtracted from the distance between the extension lines. If the resulting distance allows enough room for the dimension, it is placed between the extension lines. If not, the text is placed beyond the second extension line.

The initial value of DIMASZ is determined by the prototype drawing (default is 0.1800). The current value is saved with the drawing. If most of your drawings are plotted at a particular scale, edit the prototype drawing to set the DIMASZ variable to a more appropriate value. (You can keep AutoCAD's default setting and simply alter the scale factor applied to all variables that control the sizes and distances of dimensioning features; see DIMSCALE).

DIMBLK

[ADE-1] [v2.5, v2.6, r9]

Overview   The DIMBLK (Dimension Block) dimensioning system variable allows you to instruct AutoCAD to insert a user-defined arrowhead block in place of AutoCAD's standard dimensioning arrow. DIMBLK is assumed to be a text string representing the name of the arrow block.

Procedure   If the DIMBLK value equals the name of an existing block, that block will be drawn in place of AutoCAD's standard dimension arrow. The block will be inserted with X- and Y- scale factors equal to the product of DIMASZ * DIMSCALE.

You can change the DIMBLK value at any time directly from the dim prompt or by using the SETVAR command or AutoLISP.

Examples   Before you can use DIMBLK, you must first draw your alternate arrow block. Create the block representing the right arrow of a horizontal dimension. Draw the arrow block so it has an arrowhead and a short line extending from it to the left. The total length of the arrowhead and the tail should be exactly one drawing unit. Use AutoCAD's BLOCK command to save the arrow block drawing. When prompted for the insertion point, select the point where the arrowhead would touch the extension line (the tip of the arrowhead). When prompted to select objects, select the entire arrow block. This procedure is illustrated in the following example, along with several typical arrow blocks.

```
Command: BLOCK
Block name (or ?): (enter name)
Insertion base point: (select a point at the tip of the arrow)
Select objects: (each individual block you use is created separately)
```

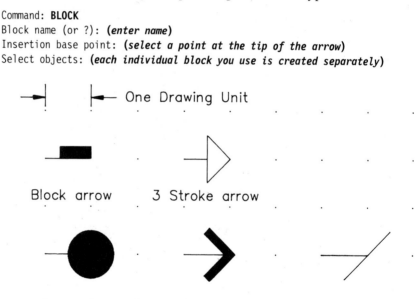

Warnings   AutoCAD shortens the dimension line by a distance equal to DIMASZ* DIMSCALE so be sure to include a short tail on your arrow block.

You need to draw only the right arrow. The left arrow will be inserted with a rotation angle of 180 degrees when you insert the arrow block.

If the block name you specify in the DIMBLK variable does not exist in the drawing, AutoCAD will ignore it and use its standard arrow.

Tips   If you want to use a dot as your arrow, you do not need to define an arrow block. AutoCAD will automatically draw a dot arrowhead if you use the DIMBLK value DOT.

If you have set a value for DIMBLK and later want to use AutoCAD's standard arrow, change the DIMBLK value to a single period (.):

```
Dim: DIMBLK
Current value <myarrow>  New value: .
```

The arrow block must exist as a block in the current drawing. AutoCAD will not look for arrowhead blocks as drawing files. If you have saved your block as a drawing file (using the WBLOCK command, for example), use the INSERT command to temporarily place the arrow into your drawing. Then erase the arrow. Although the arrow is not in the drawing, its block definition is now part of the current drawing, and it can be used as the DIMBLK value.

The initial DIMBLK value is determined by the prototype drawing (default value is a blank: use AutoCAD's standard arrow). The current value is saved with the drawing file. If you have a custom block you use often, save that block drawing in the prototype drawing and change the DIMBLK value to your block name. In that way, these values will be preset for you whenever you begin a new drawing.

DIMCEN   [ADE-1] [all versions]

Overview   The DIMCEN (Dimension Center) dimensioning system variable controls the drawing of centerline marks by the CENTER, DIAMETER, and RADIUS dimensioning subcommands.

Procedure   The DIMCEN variable accepts a numeric value. If the value is zero, no centerline marks are drawn. If the value is a positive number, a center mark is drawn. The value determines the size of the center mark in drawing units. If DIMCEN is a negative value, a centerline mark is drawn instead of a center mark. The absolute value of DIMCEN determines the size of the center mark portion of the centerline mark.

You can change the DIMCEN variable directly from the dim prompt or with the SETVAR command or AutoLISP. You can use any format acceptable by the current UNITS to specify the DIMCEN value, or you can indicate the value as the distance between two points in the drawing.

Examples   The following drawings illustrate the difference between a center mark and a centerline mark. In the first example, the DIMCEN value is a positive number; in the second example, it is negative.

```
Dim: DIMCEN
Current value <0.09>  New value: 1
```

```
Dim: DIMCEN
Current value <0.09>  New value: -1
```

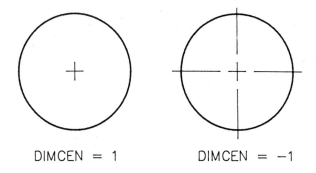

DIMCEN = 1   DIMCEN = −1

Tips   When you use the DIAMETER dimensioning subcommand, center marks are drawn only when the dimension text is placed outside the arc or the circle.

The initial value of DIMCEN is determined from the prototype drawing (default value is 0.0900). The current value is saved with the drawing. If you often require a different size center mark, edit the prototype drawing so that whenever you begin a new drawing the proper value is predetermined. (You can keep AutoCAD's default setting and simply alter the scale factor applied to all variables that control the sizes and distances of dimensioning features. See DIMSCALE.)

DIMDLE

[ADE-1] [v2.5, v2.6, r9]

Overview   The DIMDLE (Dimension Line Extension) dimensioning system variable determines the distance, in drawing units, that the dimension line will extend beyond the extension line when you use ticks instead of arrows.

Procedure   When dimensioning, AutoCAD will draw ticks instead of arrows if the variable DIMTSZ is not zero. In this situation, the variable DIMDLE determines how far to extend the dimension line beyond the extension line. This value is expressed in drawing units.

You can set the DIMDLE variable directly from the dim prompt or by using the SETVAR command or the (setvar) AutoLISP function. You can use any format accepted by the current UNITS to specify the DIMDLE value. As an alternative method, if you respond to the prompt for a new value by selecting a point in the drawing, you will then be prompted to select a second point. The DIMDLE distance will be the distance between those two points.

Examples   In this example, the DIMTSZ variable value is 1 and the DIMDLE variable is set to 0.5. The resulting dimension line extension is shown in the following drawing.

```
Dim: DIMDLE
Current value <0>  New value: 0.5
```

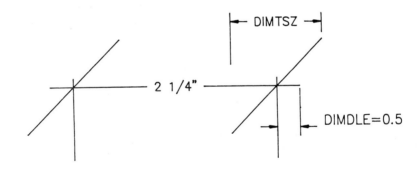

Warnings   The DIMDLE variable has meaning only when DIMTSZ is a value other than zero.

Tips   The initial value of DIMDLE is determined from the prototype drawing (default value is 0.0000). The current value is saved with the drawing.

DIMDLI

[ADE-1] [all versions]

Overview The DIMDLI (Dimension Line Increment) dimensioning system variable determines the distance to offset successive dimension lines drawn using the BASELINE and CONTINUE dimensioning subcommands (see DIM—BASELINE and DIM—CONTINUE).

Procedure When you use the BASELINE dimensioning subcommand, each successive dimension line is offset so that it is not drawn over a previous dimension line. The same thing occurs when you use the CONTINUE subcommand if the command causes the arrows to be placed outside the extension lines. The distance with which these subsequent dimension lines is offset is determined by the DIMDLI variable. The distance is expressed as a number representing drawing units.

You can set the DIMDLI value directly from the dim prompt or by using the SETVAR command or AutoLISP. You can use any format accepted by the current UNITS to specify the DIMDLI value. As an alternative method, if you respond to the prompt for a new value by selecting a point in the drawing, you will be prompted to select a second point. The DIMDLI distance will be the distance between those two points.

Examples The following drawing illustrates how AutoCAD measures the DIMDLI distance. The value is set to 0.5.

```
Dim: DIMDLI
Current value <0.38>  New value: 0.5
```

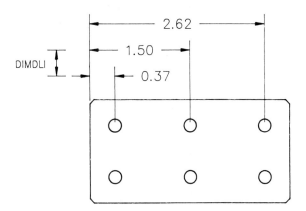

D

The initial DIMDLI value is determined from the prototype drawing (default value is 0.3800 unit). The current value is saved with the drawing. Edit the prototype drawing to adjust this value to suit the scale at which most of your drawings are plotted. (You can keep AutoCAD's default setting and simply alter the scale factor applied to all variables that control the sizes and distances of dimensioning features. See DIMSCALE).

Dimension Text

[ADE-1] [v2.5, v2.6, r9]

Overview

When using AutoCAD's dimensioning subcommands, you can have the program automatically place dimensions into your drawing. The way in which these dimensions are displayed is controlled by the type of dimensioning subcommand used and the settings of various dimensioning system variables. In addition, you can override AutoCAD's default text or append a prefix or suffix to it.

Procedure

As you use the various dimensioning subcommands, AutoCAD will eventually prompt for the dimension text and will present a default value. This default value will be the measured dimension. The units used and the number of decimal places will depend on the current UNITS.

At this point, you have the choice of accepting AutoCAD's default dimension text (press Return) or altering it manually. You can simply enter your own text at this point. Alternatively, you can tell AutoCAD to use the default dimension text but to append a prefix or suffix. You do this using the angle bracket < > method.

To append a prefix to a default dimension, simply type in the desired prefix followed by left and right angle brackets (the less-than symbol and the greater-than symbol). To add a suffix, type in < > and then the desired suffix. AutoCAD adds any text (including spaces) that comes before the angle brackets as a prefix and any text that comes after the angle brackets as a suffix. The default dimension text occurs in place of the angle brackets.

Examples

The following examples illustrate the use of the angle bracket method. Alternate dimensions (see DIMALT) and suffixes set using the DIMAPOST

193

are included along with any user-supplied prefix and suffix. In each case, the following text is added to the dimension text string:

Dimension text <1'-0">: **About <> or so**

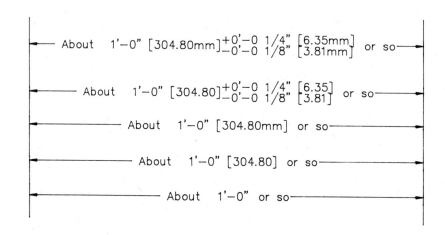

Tips

The angle bracket method can be used with any AutoCAD dimensioning subcommand that provides a default dimension text string (ALIGNED, ANGULAR, BASELINE, CONTINUE, DIAMETER, HORIZONTAL, RADIUS, ROTATED, and VERTICAL).

DIMEXE

[ADE-1] [all versions]

Overview

The DIMEXE (Dimension Extension-line Extension) dimensioning system variable determines how far, in drawing units, to extend the extension lines beyond the dimension line.

Procedure

It is a typical drawing convention to draw extension lines slightly beyond the dimension line. The DIMEXE variable determines the distance to extend them. The distance is given in drawing units.

You can enter the DIMEXE value directly at the dim prompt or by using the SETVAR command or AutoLISP. You can use any format accepted by the current UNITS to specify the DIMEXE value. As an alternative method, if you respond to the prompt for a new value by selecting a point in the drawing, you will be prompted to select a second point. The DIMEXE distance will be the distance between those two points.

Examples   The following example illustrates how AutoCAD determines the distance to extend extension lines. The DIMEXE value is set to 0.5.

```
Dim: DIMEXE
Current value <0.18>  New value: 0.5
```

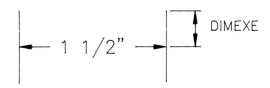

Tips   The initial value of DIMEXE is determined from the prototype drawing (default value is 0.1800 unit). The current value is saved with the drawing. You should change this value, either through a setup macro or by editing the prototype drawing, to change DIMEXE to suit the scale at which you will plot the drawing. (You can keep AutoCAD's default setting and simply alter the scale factor applied to all variables that control the sizes and distances of dimensioning features. See DIMSCALE.)

DIMEXO

[ADE-1] [all versions]

Overview   The DIMEXO (Dimension Extension-line Offset) dimensioning system variable determines the offset distance from the points chosen as the extension origins. The extension lines will actually be offset this distance away from those origin points.

Procedure   It is a standard drawing convention to stop extension lines just short of the object you are dimensioning to. The DIMEXO value specifies the distance from the end of the extension line to the extension-line origin point. The distance is measured in drawing units.

You can set the DIMEXO value directly at the dim prompt or by using the SETVAR command or AutoLISP. You can use any format accepted by the current UNITS to specify the DIMEXO value. As an alternative method, if you respond to the prompt for a new value by selecting a point in the drawing, you will be prompted to select a second point. The DIMEXO distance will be the distance between those two points.

Examples   The following drawing illustrates how the DIMEXO distance is measured. The value is set to 0.25 directly at the Dimensioning mode prompt.

```
Dim: DIMEXO
Current value <0.0625>  New value: 0.25
```

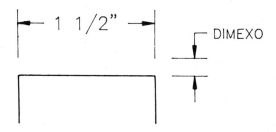

Tips   The initial DIMEXO value is determined from the prototype drawing (default value is 0.0625). The current value is saved with the drawing file. Edit the prototype drawing to set a value for your typical drawings or use a setup macro when you begin a drawing to change DIMEXO (and other dimensioning variables) to suit the scale at which the drawing will be plotted. (You can keep AutoCAD's default setting and simply alter the scale factor applied to all variables that control the sizes and distances of dimensioning features. See DIMSCALE.)

DIMLFAC

[ADE-1] [v2.5, v2.6, r9]

Overview   The DIMLFAC (Dimension Linear Factor) dimensioning system variable is a global scale factor that is applied to any length measured using a dimensioning subcommand.

Procedure   All dimensions measured by a dimensioning subcommand (except AN-GULAR) are first multiplied by the DIMLFAC scale factor. The dimension text displayed as the default is the product of the actual measured dimension multiplied by the DIMLFAC value.

You can change the DIMLFAC variable directly from the dim prompt or by using the SETVAR command or AutoLISP. DIMLFAC must be a decimal value.

Examples   For most drawings, the DIMLFAC value remains set to 1.0 (one measured unit equals one dimensioned unit). If you are drawing to a different scale, you can change the DIMLFAC value so that the dimensions are annotated to that scale factor.

In the following drawing, the length of the line is first dimensioned using a DIMLFAC of 1 and then again using a value of 0.5. Notice that the dimension text is different even though the actual size of the object is the same.

```
Dim: DIMLFAC
Current value <1>  New value: 0.5
```

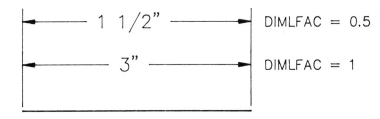

Tips   The initial value of DIMLFAC is determined from the prototype drawing (default value is 1.0000; one drawing unit equals one dimensioning unit). The current value is saved with the drawing file.

DIMLIM
[ADE-1] [all versions]

Overview   The DIMLIM (Dimension Limits) dimensioning system variable enables or disables the generation of the default dimension text as upper and lower dimension limits.

197

Procedure   The DIMLIM variable can be either on or off. If the value is on, dimension text is generated as an upper and lower dimension limit. The upper-limit dimension is the actual measured dimension plus the value of the variable DIMTP. The lower-limit dimension is the actual measured dimension minus the value of the variable DIMTM. The measured dimension is also subject to adjustment by the DIMLFAC global scale factor.

You can change the DIMLIM variable directly from the dim prompt or by using the SETVAR command or AutoLISP.

Examples   In the following drawing, DIMLIM has been set to on. The DIMTM value is 0.02. The DIMTP value is 0.03. The actual measured dimension is 4.75 inches.

Dimension text <4.75>: (RETURN)

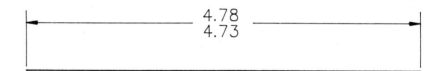

Warnings   If you set the DIMLIM variable on, the DIMTOL variable (which controls the inclusion of tolerances) is set to off. DIMTOL and DIMLIM cannot be on at the same time.

Tips   The number of decimal places displayed in the dimension is determined by the number of decimal places specified in the UNITS command and saved in the LUPREC system variable.

The displaying of limit dimensions is independent of the angle bracket method. Any prefix or suffix added using the angle bracket feature will be added before or after the limit dimensions. (For more information on the angle bracket feature (< >), see Dimension Text.)

If alternate dimensions are enabled (that is, DIMALT is on), the alternate dimensions will also be displayed as upper and lower limits. If a suffix is specified in the DIMPOST (or DIMAPOST) variable, the suffix will be appended to the upper and lower limits.

The initial setting of DIMLIM is determined from the prototype drawing (default value is 0, off). The current value is saved with the drawing.

DIMPOST

[ADE-1] [v2.6, r9]

Overview   The DIMPOST (Dimension Postscript) dimensioning system variable defines a text string suffix to be added to dimensions.

Procedure   The DIMPOST variable contains the dimensioning suffix that will be appended to any dimension text string placed into the drawing by a LINEAR, DIAMETER, or RADIUS dimensioning command. The suffix will be added automatically if DIMPOST has a value.

You can manipulate the DIMAPOST variable directly from the dim prompt or by using the SETVAR command or AutoLISP. To disable the DIMAPOST variable, enter a single period (.).

Examples   You can use the DIMPOST variable to generate dimensions for use in working in metric dimensions. If the value is set to *mm*, that suffix will be appended to all measured dimensions, as shown in the following example.

```
Dim: DIMPOST
Current value <> New value: mm
```

4.75mm

Warnings   DIMPOST is not available in AutoCAD version 2.5.

Tips   The suffix applied by the DIMPOST value is always placed before any suffix appended using the angle bracket method. (For more information about the angle bracket method, see Dimension Text.)

If limits or tolerance dimensioning is enabled (see DIMLIM and DIMTOL), the dimension suffix automatically will be applied to them also.

The initial value for DIMPOST is determined by the prototype drawing (default is " " —blank, no suffix). The current value is saved with the drawing. If you often use a different dimension suffix, edit the DIMPOST value in your prototype drawing accordingly so that your suffix is always preset when you start a new drawing.

DIMRND

Overview The DIMRND (Dimension Rounding) dimensioning system variable sets the rounding-off value for linear, radius, and diameter dimensions.

Procedure You can set the DIMRND number value directly from the dim prompt or by using the SETVAR command or AutoLISP. DIMRND determines the nearest unit value to round dimensions to. The value may be expressed in any form acceptable by the UNITS in effect. You can also indicate a distance by selecting two points in the drawing.

Examples If the DIMRND value is 1.0, all dimensions will be rounded to the nearest drawing unit. If drawing units represent inches, dimensions will be rounded to the nearest inch. Similarly, if DIMRND equals 0.25, dimensions will be rounded to the nearest quarter-inch (0.25 unit).

Warnings The DIMRND value has no bearing on the number of decimal places displayed for a dimension. The number of decimal places is determined by the UNITS command and is stored in the system variable LUPREC.

Tips The initial DIMRND value is determined from the prototype drawing (default value is 0.0000 unit, no rounding). The current value is saved with the drawing file.

DIMSCALE

Overview The DIMSCALE (Dimension Scale) dimensioning system variable is a general scale factor that is applied to all dimensioning variables that define a size, distance, or offset.

Procedure The DIMSCALE value is applied automatically to the size and the distance determined by the following dimensioning system variables: DIMASZ, DIMCEN, DIMDLE, DIMDLI, DIMEXE, DIMEXO, DIMTSZ, and DIMTXT.

You can change the DIMSCALE value directly from the dim prompt or by using the SETVAR command or AutoLISP. DIMSCALE must be a decimal value.

Examples   AutoCAD's default values for the dimensioning system variables listed under Procedure establish a good working ratio of dimensioning factors. For instance, the relationship of the size of dimension text to the size of arrows is already established. Rather than change each of the aforementioned variables individually (and thus be required to reestablish the proper ratios), you can use the DIMSCALE factor variable to simply change the size or distance of all of these variables without disturbing the ratio.

The following table lists appropriate DIMSCALE values for drawings that will ultimately be plotted at the specified scales.

DIMSCALE Values (to achieve a standard height at a plotted scale)

FINISHED SHEET SCALE	DIMSCALE value for default test height (0.1800 unit)	DIMSCALE value for 1/8" dimension text height	DIMSCALE value for 1/10" dimension text height
1/32" = 1'	384	266.66666	213.33333
1/16" = 1'	192	133.33333	106.66666
1/8" = 1'	96	66.66666	53.33333
3/16" = 1'	64	44.44444	35.55555
1/4" = 1'	48	33.33333	26.66666
3/8" = 1'	32	22.22222	17.77777
1/2" = 1'	24	16.66666	13.33333
3/4" = 1'	16	11.11111	8.88888
1" = 1'	12	8.33333	6.66666
1 1/2" = 1'	8	5.55555	4.44444
3" = 1'	4	2.77777	2.22222
1" = 10'	120	83.33333	66.66666
1" = 20'	240	166.66666	133.33333
1" = 30'	360	250	200
1" = 40'	480	333.33333	266.66666
1" = 50'	600	416.66666	333.33333

Table continues

FINISHED SHEET SCALE	DIMSCALE value for default test height (0.1800 unit)	DIMSCALE value for 1/8″ dimension text height	DIMSCALE value for 1/10″ dimension text height
1″ = 60′	720	500	400
1″ = 70′	840	583.33333	466.66666
1″ = 80′	960	666.66666	533.33333
1′ = 90′	1080	750	600
1″ = 100′	1200	833.33333	666.66666
1/100	100	69.44444	55.55555
1/80	80	55.55555	44.44444
1/64	64	44.44444	35.55555
1/40	40	27.77777	22.22222
1/32	32	22.22222	17.77777
1/20	20	13.88888	11.11111
1/16	16	11.11111	8.88888
1/10	10	6.94444	5.55555
1/8	8	5.55555	4.44444
1/4	4	2.77777	2.22222
3/8	2.6666666	1.85185	1.48148
1/2	2	1.38888	1.11111
5/8	1.6	1.11111	0.88888
3/4	1.3333333	0.92592	0.74074
7/8	1.1428671	0.79366	0.63495
1	1	0.69444	0.55555
2	0.5	0.34722	0.27777
4	0.25	0.17361	0.13888
10	0.1	0.06944	0.05555
20	0.05	0.03472	0.02777

Tips

Write a setup macro using screen menus to allow you to automatically set the DIMSCALE value and the text height, based on the scale that you will ultimately plot the drawing.

The initial DIMSCALE value is determined from the prototype drawing (default value is 1.0000). The current value is saved with the drawing file.

DIMSE1

[ADE-1] [all versions]

Overview

The DIMSE1 (Dimension Suppress Extension-line 1) dimensioning system variable controls whether AutoCAD draws the first extension line when adding dimensions to a drawing. DIMSE1 may be on or off.

Procedure

As you use the dimensioning subcommands to add dimensions to a drawing, AutoCAD prompts you for the first and second extension-line origins and then the dimension-line location. You can enter DIMSE1 in response to any of these prompts. AutoCAD will display the current DIMSE1 value and prompt you for a new value. Once you enter the new value, the previous extension-line or dimension-line prompt will be repeated.

If DIMSE1 is off, the first extension line will be drawn for the current dimension line. If DIMSE1 is on, the extension line will not be drawn. You can turn DIMSE1 on or off in response to any dimensioning prompt, except when you are prompted for the dimension text. You can also change the DIMSE1 value using the SETVAR command or AutoLISP.

Examples

When AutoCAD draws dimensions in a drawing, extension lines are often drawn over previous extension lines. This is particularly true when you use the CONTINUE subcommand. To avoid having duplicate extension lines in this case, use the DIMSE1 variable to suppress the first extension line for subsequent dimensions.

```
Dim: CONTINUE
Second extension line origin: DIMSE1
Current value <Off> New value: ON
Second extension line origin: (select next dimension point)
Dimension text <4.5>:
```

Tips   The DIMSE1 variable is one that you should definitely include on your screen menu whenever you are using the dimensioning subcommands. You will find it helpful to be able to turn this variable on and off quickly.

The initial DIMSE1 value is determined from the prototype drawing (default value is 0: off, extension lines are drawn). The current value is saved with the drawing file.

DIMSE2

Overview   The DIMSE2 (Dimension Suppress Extension-line 2) dimensioning system variable controls whether AutoCAD draws the second extension line when adding dimensions to a drawing. DIMSE2 may be on or off.

Procedure   As you use the dimensioning subcommands to add dimensions to a drawing, AutoCAD prompts you for the first and second extension-line origins and then the dimension-line location. You can enter DIMSE2 in response to any of these prompts. AutoCAD will display the current DIMSE2 value and prompt you for a new value. Once you enter the new value, the previous extension-line or dimension-line prompt will be repeated.

If DIMSE2 is off, the second extension line will be drawn for the current dimension line. If DIMSE2 is on, the extension line will not be drawn. You can turn DIMSE2 on or off in response to any dimensioning prompt, except when you are prompted for the dimension text. You can also change the DIMSE2 value using the SETVAR command or AutoLISP.

Examples   When you are dimensioning to an existing object, such as the interior of a room, it is not necessary to have AutoCAD draw extension lines. In this example, DIMSE2 is turned on (suppressing the second extension line) while the command is active.

```
Dim: HORIZONTAL
First extension line origin or RETURN to select: ENDPOINT
of: (select point A)
Second extension line origin: DIMSE2
```

```
Current value <Off> New value: ON
Second extension line origin: NEAREST
to: (select point B)
Dimension line location: (select point C)
Dimension text <10'-0">: ( RETURN )
```

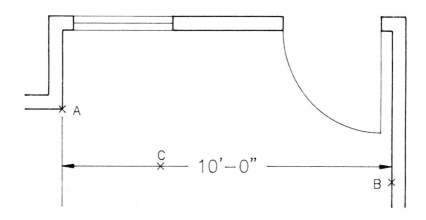

Tips

See DIMSE1 for further information.

The initial DIMSE2 value is determined from the prototype drawing (default value is 0: off, extension lines are drawn). The current value is saved with the drawing file.

DIMSHO

[ADE-1] [v2.6, r9]

Overview

The DIMSHO (Dimension Show) dimensioning system variable controls whether associative dimensions are constantly updated as they are being dragged. DIMSHO may be on or off.

Procedure

The DIMSHO variable has meaning only if associative dimensioning is enabled (that is, DIMASO is on). If an object and its associative dimension are dragged as they are edited, the displayed associative dimension is updated constantly while it is being dragged if DIMSHO is on. (For more information, see Associative Dimensions.)

You can change the DIMSHO variable at any time directly from the dim prompt or by using the SETVAR command or AutoLISP.

Examples Turn DIMSHO on only when you need to see the continuously updated dimension of an object as you change it.

```
Dim: DIMSHO
Current value <Off> New value: ON
```

Warnings When DIMSHO is on, the computer's response may slow down significantly when objects are being dragged.
DIMSHO is not available in AutoCAD version 2.5.

Tips The initial value of DIMSHO is determined from the prototype drawing (default value is 0, off). The current value is saved with the drawing file.

DIMTAD

[ADE-1] [all versions]

Overview The DIMTAD (Dimension Text above Dimension-line) dimensioning system variable controls whether AutoCAD places linear dimension text above the dimension line or breaks the dimension line in two and places it inside the dimension line. DIMTAD can be on or off.

Procedure If DIMTAD is off, AutoCAD divides linear dimension lines in two and places the dimension text inside the gap. If DIMTAD is on, the dimension line is not broken. Instead, AutoCAD places the dimension text above the dimension line.
You can change the DIMTAD variable directly from the dim prompt or by using the SETVAR command or AutoLISP.

Examples The following drawing illustrates the difference in how dimension text is placed, depending on the setting of DIMTAD.

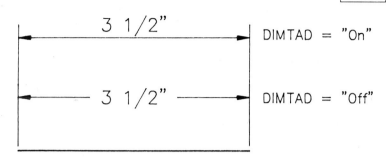

Tips   The initial value of DIMTAD is determined from the prototype drawing (default value is 0, off, dimension text placed within a broken dimension line). The current value is saved with the drawing file. Edit the prototype drawing to change the DIMTAD setting to produce dimension text that meets the standard way in which dimensions are commonly drawn for your particular application.

DIMTIH

[ADE-1] [all versions]

Overview   The DIMTIH (Dimension Text Inside Horizontal) dimensioning system variable controls the orientation of linear dimensioning text that is placed between the extension lines. DIMTIH can be on or off.

Procedure   If DIMTIH is on, all dimension text that is placed using a linear dimensioning subcommand and that AutoCAD determines can be placed between the extension lines is drawn horizontally (it will read from left to right). If DIMTIH is off, the dimension text is drawn in alignment with the dimension line (the text rotation angle equals the angle of the dimension line).

You can change DIMTIH directly from the dim prompt or by using the SETVAR command or AutoLISP.

Examples   The following drawing illustrates the difference in how dimension text is placed, depending on the setting of DIMTIH.

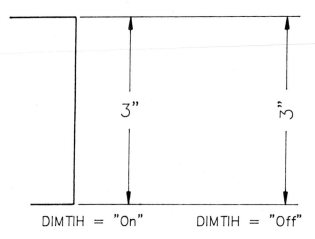

DIMTIH = "On"   DIMTIH = "Off"

Tips   DIMTIH affects only text drawn between the extension lines. For text drawn outside the extension lines, see DIMTOH.

The initial value of DIMTIH is determined from the prototype drawing (default value is 1, on, text is always horizontal). The current value is saved with the drawing. The American National Standards Institute (ANSI) prefers horizontal text, but many professions do not use this orientation. Edit the prototype drawing to establish dimension text orientation appropriate to the conventional method for your application.

DIMTM   [ADE-1] [all versions]

Overview   The DIMTM (Dimension Tolerance Minus) dimensioning system variable establishes a minus tolerance value used when either DIMLIM (limits dimensioning) or DIMTOL (tolerance dimensioning) is on.

Procedure   The DIMTM value is the number of units subtracted from the measured linear dimension and displayed as the lower limit when DIMLIM is on. If DIMTOL is on, the DIMTM value is displayed as the minus tolerance value. DIMTM has no effect on the drawing unless DIMLIM or DIMTOL is on.

You can change DIMTM directly from the dim prompt or by using the SETVAR command or AutoLISP.

208

Examples   In the following drawing, DIMTM is set to 0.02. The left drawing shows how this minus value is displayed when limits dimensioning is on. The lower dimension is the result of subtracting 0.02 from the actual measured dimension of 4.75. The right drawing shows the effect of this same DIMTM value when tolerance dimensioning is used. The 0.02 value is shown as the minus tolerance value. (For additional information, see DIMLIM, DIMTOL, and DIMTP.)

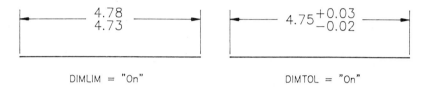

DIMLIM = "On"   DIMTOL = "On"

Tips   The initial value of DIMTM is determined from the prototype drawing (default value is 0.0000 unit). The current value is saved with the drawing file.

DIMTOH

Overview   The DIMTOH (Dimension Text Outside Horizontal) dimensioning system variable controls the orientation of linear dimensioning text that is placed outside the extension lines. DIMTOH can be either on or off.

Procedure   If DIMTOH is on, all dimension text that is placed using a linear dimensioning subcommand and that AutoCAD determines cannot be placed between the extension lines is drawn horizontally (it will read from left to right). If DIMTOH is off, the dimension text is drawn in alignment with the dimension line (the text rotation angle equals the angle of the dimension line).

You can change DIMTOH directly from the dim prompt or by using the SETVAR command or AutoLISP.

Examples   The following drawing illustrates the difference in how dimension text is placed, depending on the setting of DIMTOH.

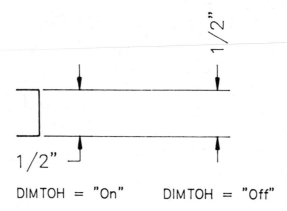

DIMTOH = "On"   DIMTOH = "Off"

Tips

DIMTOH affects only text drawn outside the extension lines. For text drawn between the extension lines, see DIMTIH.

The initial value of DIMTOH is determined from the prototype drawing (default value is 1, on, text is always horizontal). The current value is saved with the drawing. The American National Standards Institute (ANSI) prefers horizontal text, but many professions do not use this orientation. Edit the prototype drawing to establish dimension text orientation appropriate to the conventional method for your application.

DIMTOL

[ADE-1] [all versions]

Overview

The DIMTOL (Dimension Tolerance) dimensioning system variable determines whether AutoCAD will append an upper and a lower tolerance value to the default dimension text. DIMTOL can be either on or off.

Procedure

If DIMTOL is on, a plus and a minus tolerance value are added to the default dimension text for linear dimensions. The upper tolerance value is determined by the DIMTP variable. The lower tolerance value is the DIMTM value. The measured dimension is also subject to adjustment by the DIMLFAC global scale factor.

You can change the DIMTOL variable directly from the dim prompt or by using the SETVAR command or AutoLISP.

210

Examples   In the following drawing, DIMTOL has been set to on. The DIMTM value is set to 0.02. The DIMTP value is set to 0.03. The actual measured dimension is 4.75 inches.

Dimension text <4.75>: (RETURN)

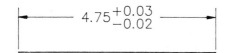

Warnings   If you set the DIMTOL variable on, the DIMLIM variable (which controls the display of limit dimensions) is set to off. DIMTOL and DIMLIM cannot be on at the same time.

Tips   The number of decimal places displayed in the dimension is determined by the number of decimal places specified in the UNITS command and saved in the LUPREC system variable.

The displaying of tolerance dimensions is independent of the angle bracket (< >) method. Any prefix or suffix added using the < > feature will be added before or after the limit dimensions. (For more information on the < > feature, see Dimension Text.)

If alternate dimensions are enabled (DIMALT is on), the alternate dimensions will also be displayed along with their tolerance values. If a suffix is specified in the DIMPOST (or DIMAPOST) variable, the suffix will be appended to the plus and minus tolerance values.

The initial setting of DIMTOL is determined from the prototype drawing (default value is 0, off). The current value is saved with the drawing.

DIMTP   [ADE-1] [all versions]

Overview   The DIMTP (Dimension Tolerance Plus) dimensioning system variable establishes a plus tolerance value that is used when either DIMLIM (limits dimensioning) or DIMTOL (tolerance dimensioning) is on.

Procedure   The DIMTP value is the number of units added to the measured linear
                dimension and displayed as the upper limit when DIMLIM is on. If
                DIMTOL is on, the DIMTP value is displayed as the plus tolerance value.
                DIMTP has no effect on the drawing unless DIMLIM or DIMTOL is on.
                    You can change DIMTP directly from the dim prompt or by using the
                SETVAR command or AutoLISP.

Examples   In the following example, DIMTP is set to 0.03. The left drawing shows how
                this added value is displayed when limits dimensioning is on. The upper
                dimension is the result of adding 0.03 to the actual measured dimension of
                4.75. The right drawing shows the effect of this same DIMTM value when
                tolerance dimensioning is used. The 0.03 value is shown as the plus
                tolerance value. (For additional information, see DIMLIM, DIMTM, and
                DIMTOL.)

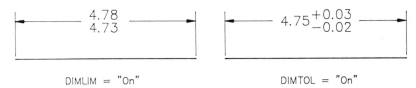

         DIMLIM = "On"   DIMTOL = "On"

Tips   The initial value of DIMTP is determined from the prototype drawing (de-
                fault value is 0.0000 unit). The current value is saved with the drawing file.

DIMTSZ   [ADE-1] [all versions]

Overview   The DIMTSZ (Dimension Tick Size) dimensioning system variable controls
                the size of the tick mark that is drawn instead of AutoCAD's standard arrow
                (or any user-defined arrow; see DIMBLK) if the DIMTSZ value is not 0.

Procedure   If the value of DIMTSZ is 0, tick marks are not drawn. AutoCAD draws
                arrows based on the variable DIMBLK. If DIMTSZ has a value other than 0,
                that value determines the size of the tick mark that is drawn instead of
                arrows. The size is defined in drawing units and defines the number of
                units that the tick mark extends above the dimension line when measured
                perpendicular to it.

You can change the DIMTSZ value at any time directly from the dim prompt or by using the SETVAR command or AutoLISP. You can enter values as a distance in any format acceptable by the current UNITS. You can also enter distances by using the screen cursor to select two points in the drawing. The DIMTSZ value will be the distance between the two points.

Examples   The following drawing illustrates how AutoCAD determines the size of the tick marks. DIMTSZ is measured in drawing units. In this drawing, DIMTSZ equals 1 unit.

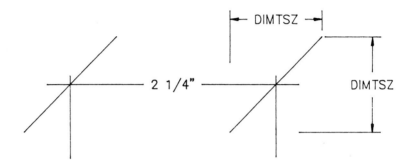

Tips   The initial value of DIMTSZ is determined from the prototype drawing (default value is 0.0000, tick marks not drawn). The current value is saved with the drawing file. If you commonly draw tick marks rather than arrows, edit the prototype drawing so that tick marks are the default whenever you begin a new drawing.

DIMTXT   [ADE-1] [all versions]

Overview   The DIMTXT (Dimension Text) dimensioning system variable determines the height of all subsequently entered dimensioning text.

Procedure   The value contained in the DIMTXT variable represents the height of dimensioning text, in drawing units, subject to multiplication by the DIMSCALE factor (see DIMSCALE).

213

You can change the DIMTXT value at any time directly from the dim prompt or by using the SETVAR command or AutoLISP. You can enter values as a distance in any format acceptable by the current UNITS. You can also enter distances by using the screen cursor to select two points in the drawing. The DIMTXT value will be the distance between the two points.

Examples   The following drawing illustrates the method AutoCAD uses to size dimension text. In this example, DIMTXT equals 0.18 unit and DIMSCALE is 1.

Tips   The initial value of DIMTXT is determined from the prototype drawing (default value is 0.1800 unit). The current value is saved with the drawing file. You can edit the prototype drawing to adjust the dimension text height to suit the scale at which your drawings are usually plotted. Or you can simply adjust the DIMSCALE factor, which will adjust a host of dimensioning variables. For more information, see DIMSCALE.

DIMZIN   [ADE-1] [v2.5, v2.6, r9]

Overview   The DIMZIN (Dimension Zero Inches) dimensioning system variable determines the way in which AutoCAD displays the inches portion of a dimension.

Procedure   When the dimension can be displayed in feet and inches, you can make AutoCAD display default dimension values in several different ways. If DIMZIN equals 0, the feet portion is omitted if the dimension is less than 12 inches. Inches are not displayed if the inches portion of the dimension text string is zero.

D

If DIMZIN equals 1, the feet and inches portions of the dimension string are included even if they are zero. If DIMZIN equals 2, the feet portion of the dimension is always included, but inches are not displayed if they equal exactly zero. If DIMZIN equals 3, inches are always displayed but feet are included only if they are not zero.

You can change the DIMZIN variable at any time directly from the dim prompt or by using the SETVAR command or AutoLISP.

Examples   The following table illustrates how dimensions appear, depending on the setting of the DIMZIN variable.

DIMZIN VALUE	EXAMPLES			
0	1/4″	8″	2′	2′–8 1/4″
1	0′–0 1/4″	0′–8″	2′–0″	2′–8 1/4″
2	0′–0 1/4″	0′–8″	2′	2′–8 1/4″
3	1/4″	8″	2′–0″	2′–8 1/4″

Warnings   In AutoCAD version 2.5, the DIMZIN variable allows only an on or off setting. When off, zero inches are dropped (as in 3′); if DIMZIN is on, zero inches are included (3′–0″).

Tips   The initial value of DIMZIN is determined from the prototype drawing (default value is 0). The current value is saved with the drawing file. Edit the prototype drawing to change the DIMZIN value so that dimensions are always displayed in the form appropriate for your use. By changing the value in the prototype drawing, DIMZIN will be set correctly whenever you begin a new drawing.

DIST   [all versions]

Overview   The DIST (Distance) command displays the distance and the angle between two points you select.

Procedure First enter the DIST command, then respond to AutoCAD's prompts to select the two points between which you want to calculate the distance. The distance between the two points, the angle from the first point to the second point, and the change in X and Y that the distance represents are displayed. The way in which the distances and the angle are displayed is determined by the current UNITS.

Examples Besides listing the distance and the angle between two selected points, the DIST command also displays the distance between them as separate horizontal (Delta X) and vertical (Delta Y) dimensions.

```
Command:  DIST
First point:  (select point A)
Second point:  (select point B)
Distance = 0'-3 5/8"    Angle = 34
Delta X = 0'-3"   Delta Y = 0'-2",    Delta Z = 0'-0"
```

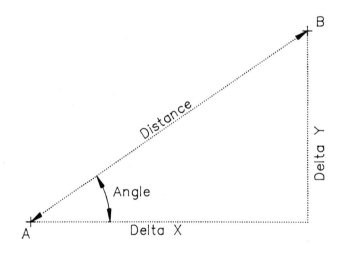

Tips Generally, you will use object snap to select the first point and the second point when using the DIST command. The DIST command is a necessary command for creating precise drawings.

The distance computed by the most recent use of the DIST command is stored in the system variable DISTANCE and can be read with the SETVAR command or with AutoLISP.

Sometimes you need to measure the length and the height of an object in a drawing. Because the DIST command returns the X-distance and the Y-distance as well as the actual distance and angle, it is possible to cut your measuring down to one step. Just measure the distance between two opposite corners and then use the resultant X- and Y-distances.

You can measure the distance in the Z-direction (and the Delta Z reported) when you are measuring in three-dimensional space (only in AutoCAD version 2.6 and release 9 when ADE-3 is included).

DISTANCE
[v2.5, v2.6, r9]

Overview   The system variable DISTANCE stores the last distance returned by the DIST command.

Procedure   The DISTANCE value is a real number that you can read only by using AutoCAD's SETVAR command or the (getvar) AutoLISP function. DISTANCE is a read-only variable.

Examples   The following example illustrates the use of SETVAR to return the DISTANCE value.

```
Command: SETVAR
Variable name or ?: DISTANCE
DISTANCE = 24.0000 (read only)
```

Tips   The DISTANCE variable is not saved. It is of use only within AutoLISP routines that must deal with the last measured distance.

DIVIDE
[ADE-3] [v2.5, v2.6, r9]

Overview   The DIVIDE command is used to visually divide an arc, circle, line, or polyline into a specified number of equal parts. Markers are placed along the selected object at the division points. The markers may be AutoCAD points or any predefined block. The selected object is not altered in any way. The markers are simply added to the drawing.

Procedure   Enter the DIVIDE command and select the object by pointing. You cannot use the Window, Last, or Crossing method to select the object to divide. AutoCAD then prompts for the number of equal parts that you want to divide the object into or for the name of a block to use as a marker. You can enter any number from 2 to 32767.

```
Command: DIVIDE
Select object to divide: (select object)
<Number of segments>/Block: (enter number)
```

AutoCAD then places point entities at the division points. Later you can use the object snap NODE to select these points.

If you enter B for block, AutoCAD prompts for the name of the block you want to use as a marker. The block you name must have already been defined in the drawing. Next, AutoCAD asks whether you want each insertion of the block to be aligned with the object. If you answer yes, each insertion of the block is aligned with its X-axis tangent to the object being divided. If you respond no, each insertion of the block is aligned as originally defined. Finally, AutoCAD asks for the number of segments to divide the selected object.

```
Command: DIVIDE
Select object to divide: (select object)
<Number of segments>/Block: B
Block name to insert: (enter name of existing block)
Align block with object? <Y>: (enter Y or N or press Return)
Number of segments: (enter number)
```

AutoCAD places the block at the division points. Each occurrence of the block is treated as an individual insertion; later you can use the INSERT object snap to select the division points.

Examples   A particularly useful application of the DIVIDE command is to divide the plan view of a large window into equal sections of glass, placing a mullion (defined as a block in advance) at each division point. Or you can draw a large room and define the walls that subdivide it into several smaller rooms as a block. Then, using the DIVIDE command and selecting the line representing the long wall of the room, you can divide the large room into a number of equally sized smaller rooms.

The keystroke sequence in the following example divides a window using a mullion block (named MULL) that was previously saved.

```
Command: DIVIDE
Select object to divide: (select point A)
<Number of segments>/Block: B
Block name to insert: MULL
Align block with object? <Y> ( RETURN )
Number of segments: 4
```

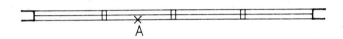

Warnings   If the object you select to divide is not an arc, circle, line, or polyline, AutoCAD will respond

```
Cannot divide that entity
```

and the DIVIDE command will terminate.

If you select to insert a block at the division points, the block must already exist in the drawing. The DIVIDE command will not load a block from a disk file.

If you use simple points as the markers for the DIVIDE command, you may not be able to see these points in the drawing. However, AutoCAD has placed the points into the drawing.

The DIVIDE command, when used on spline curved polylines, sees only the curve itself, not the frame (see PEDIT; release 9 only).

Tips   When the points used for the DIVIDE command are not readily visible in the drawing, you can change the way points are displayed in the drawing by using the SETVAR command to alter the style of the displayed points (PDMODE) and/or the size at which they are displayed (PDSIZE). Remember, however, that when you plot, AutoCAD plots points exactly the way they are displayed in the drawing. So unless you want points plotted as other than simple dots, change the PDMODE and PDSIZE back to 0.

It is often useful to define a special block that is nothing more than two lines in an X. When you use the DIVIDE command, you can specify this special block, possibly called *X*, as the marker to denote the divisions.

Any points or blocks added to the drawing by the DIVIDE command can be removed from the drawing with ERASE previous until you select a new object in response to an AutoCAD command.

DONUT *or* DOUGHNUT

[ADE-3] [v2.5, v2.6, r9]

Overview

The DONUT command can also be spelled DOUGHNUT. Most users prefer DONUT because it has fewer letters. This command causes AutoCAD to draw a solid filled circle or ring. The entity drawn is actually a closed polyline and can be edited with the PEDIT command. The solid filling of the circle or ring is controlled by the status of the Fill mode.

Procedure

From the AutoCAD Command prompt, enter either *DONUT* or *DOUGH-NUT*. AutoCAD prompts for the inside and outside diameters of the doughnut. You are then prompted for the center of the doughnut. Auto-CAD continues to prompt for more doughnuts until you press Return to cancel the command. To create a solid filled circle, enter a value of 0 for the inside diameter of the doughnut.

The most recent settings for the inside and outside diameters are the default values. As with most other AutoCAD commands, you can enter a value at each prompt, or you can point to show AutoCAD the dimensions.

```
Command: DONUT or DOUGHNUT
Inside diameter <current>: (enter value)
Outside diameter <current>: (enter value)
Center of doughnut: (select point)
Center of doughnut: ( RETURN )
```

Examples

The first example and drawing illustrate the command used to draw a solid filled circle (inside diameter is 0).

```
Command: DONUT
Inside diameter <0'-0 1/2">: 0
```

```
Outside diameter <0'-1">: ( RETURN )
Center of doughnut: (select center point)
Center of doughnut: ( RETURN )        (to end the command)
```

The second example and drawing illustrate the command used to draw a solid filled ring (inside diameter is not 0).

```
Command: DONUT
Inside diameter <0'-0">: 1/2
Outside diameter <0'-1">: ( RETURN )
Center of doughnut: (select center point)
Center of doughnut: ( RETURN )        (to end the command)
```

Warnings The value for the outside diameter must be larger than that for the inside diameter.

Tips Turn the Fill mode toggle off to save time when you are regenerating a drawing that has many solid filled objects.

Solid filled circles and rings are useful in charts and are easier to manipulate and add to a drawing than circles that have been filled using the HATCH command. They also take up less room in memory and always appear and plot as solid filled objects no matter how much the drawing is enlarged.

DRAGMODE

[ADE-2] [all versions]

Options

On   Honor drag requests when applicable
Off Ignore drag requests
A   Set Auto mode; drag whenever possible

Overview   AutoCAD's Drag feature allows you to see the results of many commands before completing the command. For example, you can see the size and the placement of a block while you are inserting it. This visual Drag feature is controlled by the DRAGMODE command. The current DRAGMODE setting is stored in the DRAGMODE system variable.

Procedure   AutoCAD permits three settings for DRAGMODE: on, off, and auto. When DRAGMODE is off, the Drag feature is completely disabled. Any attempt to drag an object is ignored. When DRAGMODE is on, commands that allow dragging may be used with the Drag feature, but you must type in the subcommand DRAG at the proper time within the active command. When DRAGMODE is set to Auto, AutoCAD causes the Drag feature to be activated during all commands at the proper times. You can change the current DRAGMODE by using the SETVAR command or with the (setvar) AutoLISP function.

Examples   To change the setting for DRAGMODE, simply enter the command at the AutoCAD command prompt, then enter the desired setting. The current setting for DRAGMODE is always shown as the default.

```
Command: DRAGMODE
ON/OFF/Auto <current>: (enter the desired mode)
```

Warnings   There are times when you will not want DRAGMODE set to auto. When you are going to insert a very large block or manipulate a large number of objects at one time using one of AutoCAD's editing commands, you should first change the DRAGMODE setting to something other than auto. AutoCAD must generate a partial display of an object or objects when a command uses the Drag feature. With large or complex objects, this may slow down the reaction of the AutoCAD program.

Tips   The initial DRAGMODE setting is determined from the prototype drawing (default value is 2, DRAGMODE auto). The current value is saved with the drawing. A value of 0 turns DRAGMODE off, a value of 1 turns it on.

Normally, setting DRAGMODE to auto will save keyboard steps, since you won't have to enter the subcommand DRAG in the middle of other commands to activate the dragging feature. In the case of a situation such as the one in the previous warning, you can change the DRAGMODE setting while in the middle of another command by using the transparent operation of the SETVAR command (see SETVAR).

There are two other system variables, DRAGP1 and DRAGP2, that control aspects of the operation of the Dragging feature. These variables control how much of the object or objects AutoCAD displays when visually dragging them and how often AutoCAD checks the location of the cursor while dragging. A smaller value causes AutoCAD to display less of the objects and to update the image more often.

DRAGP1

[ADE-2] [v2.5, v2.6, r9]

Overview   The AutoCAD manual defines the DRAGP1 system variable as the "regendrag input sampling rate." It is an integer value that is saved within AutoCAD's current configuration file. DRAGP1 controls how often an object's image is redrawn as it is being dragged across the screen.

Procedure   Normally, AutoCAD's default setting for this variable works well and you will not need to adjust it. If, however, your cursor's movements are not steady or you just want to experiment, use the SETVAR command or the (getvar) and (setvar) AutoLISP functions to change the DRAGP1 variable. A larger value causes more of an object's image to be redrawn between cursor movements. (See also DRAGP2.)

Examples   The following example shows how to use the SETVAR command to change the value of DRAGP1.

```
Command: SETVAR
Variable name or ? : DRAGP1
New value for DRAGP1 <10>: (enter your value)
```

Tips   You do not need to change the DRAGP1 value unless you often drag very large objects within your drawing. In that case, you may want to reduce DRAGP1.

The DRAGP1 value is saved in AutoCAD's configuration file ACAD.CFG. The initial value of DRAGP1 is 10.

DRAGP2

[ADE-2] [v2.5, v2.6, r9]

Overview   The DRAGP2 system variable is a companion to DRAGP1. The AutoCAD manual defines it as the "fast-drag input sampling rate." DRAGP2 controls how often AutoCAD checks the cursor position when an object is being dragged.

Procedure   A larger DRAGP2 value causes AutoCAD to check the cursor position less often. A smaller value causes the dragged image to be redrawn more often. Normally the default value is sufficient and you will not need to change this variable. If you want to experiment, however, you can read the value using the SETVAR command or AutoLISP's (getvar) function and change the value using either SETVAR or the (setvar) function.

Examples   The following example shows how to use the SETVAR command to change the value of DRAGP2.

```
Command: SETVAR
Variable name or ? : DRAGP2
New value for DRAGP2 <25>: (enter your value)
```

Tips   You do not need to change this value unless you often drag very large objects within your drawing. In that case, you may want to increase DRAGP2.

   The DRAGP2 value is saved in AutoCAD's configuration file ACAD.CFG. The initial value of DRAGP2 is 25.

Drawing

[all versions]

Overview   A drawing is a drawing file. Drawing files always have the file extension .DWG. When you save a drawing file, AutoCAD generally also saves a backup file (see Backup). To draw in AutoCAD means to use AutoCAD's entity creation and editing commands to manipulate entities in a drawing file while within AutoCAD's drawing editor.

Procedure To create or edit a drawing, you must first load the AutoCAD program (see ACAD) and then enter the drawing editor. Enter the drawing editor by selecting either main menu task 1 (Begin a NEW drawing) to create a brand new drawing file or task 2 (Edit an EXISTING drawing) to alter a drawing that was created previously.

Examples If you select task 1, you will be asked for the name you want to give the new drawing. AutoCAD will then load its drawing editor. Generally the screen you see will have the status line across the top, a command line across the bottom, a screen menu along the right side of the screen, and a blank drawing area (the drawing area may have objects already drawn, depending on your prototype drawing; see Prototype Drawing).

If you select task 2, you will be asked to provide the name of an existing drawing. AutoCAD will load the drawing editor and then present your drawing in the form it was in when you last saved it.

Tips Keep your drawing files as small as possible. Make good use of AutoCAD's block features. Don't draw multiple lines on top of each other. When drawing a sheet that is a composite of several individual drawings, create the individual drawings as separate drawing files. Use the INSERT command to create the final composite sheet.

DTEXT [ADE-3] [v2.5, v2.6, r9]

Overview The DTEXT (Dynamic Text) command allows you to enter text into a drawing. It operates the same as the TEXT command, except that the text appears on the screen as you enter it. The DTEXT command lets you enter multiple lines of text in one step.

Procedure When the command is started, a cursor box corresponding to the height and width factors of the text is displayed on the screen, and each letter appears on the screen as it is typed. When you press Return, the cursor moves down to the next logical line on the screen. To complete the DTEXT command, press Return a second time (without entering any text on a given line). As you enter text on the screen, the text is also echoed on the command line following the text prompt.

```
Command: DTEXT
Start point or Align/Center/Fit/Middle/Right/Style: (enter point or select)
Height <current height>: (enter value or press Return)
Rotation angle <current value>: (enter value or press Return)
Text: (enter text)
Text: (enter text)
Text: ( RETURN )  (to end command)
```

Examples

At the command prompt, enter *DTEXT*. AutoCAD issues the same prompts as for the TEXT command (see TEXT for a complete description). You can select the text justification and text style. You are then prompted for the starting point, the text height, and the text rotation angle. At the text prompt, you begin entering text, which appears both on the screen and in the prompt area. When you press Return to end a line, the screen cursor moves down and you are presented with a new text prompt. To end the DTEXT command, you must enter a Return on a new line, without having entered any text after the text prompt. AutoCAD then erases the text it had temporarily placed on the screen and reenters it as actual text. Figure 46 shows how the text appears on the screen.

```
Command: DTEXT
Start point or Align/Center/Fit/Middle/Right/Style: (select point A)
Height <0'-0 1/8">: 3/16
Rotation angle <0>: ( RETURN )
Text: This is the first line
Text: This is the second.
Text: ( RETURN )  (to end the command)
```

Warnings

Special text-formatting characters, such as %%u (underline) and %%p (plus-or-minus symbol), appear on the screen as the actual formatting characters. They are translated into the appropriate symbols, however, when the DTEXT command is completed.

Regardless of the text justification setting, the DTEXT command initially draws the text on the screen as left-justified. When the DTEXT command is completed, however, the text displays properly. Aligned text is not particularly useful with the DTEXT command.

Pressing Ctrl-C cancels the DTEXT command, and any text you have entered is lost. Text is not actually written to the drawing until you press Return the final time to complete the command.

```
Layer 0 Ortho                    0'-0 11/16",0'-2 3/16"      AUTOCAD
                                                            * * * *
                                                            SETUP

                                                            BLOCKS
                                                            DIM:
         This is the first line.                           DISPLAY
       A x                                                  DRAW
         This is the sec□                                   EDIT
                                                            INQUIRY
                                                            LAYER:
                                                            SETTINGS
                                                            PLOT
                                                            UTILITY

                                                            3D

                                                            SAVE:



Rotation angle <0>:
Text: This is the first line.
Text: This is the sec
```

Figure 46

The menu bar and the pull-down menu features are not accessible from within the DTEXT command once the text rotation angle has been set. This applies only to AutoCAD release 9 (see Advanced User Interface).

Tips   The DTEXT command enables you to see errors as you are entering them. You can use Backspace to go back and correct an error, regardless of the number of lines you have entered. When you backspace over a particular line of text, the screen cursor box goes back to the previous line and *deleted* displays on the command line. Because any letters you backspace over have to be reentered, this feature is useful only if the error occurred on the current or previous line. Otherwise, it is usually easier to leave the error in until you have finished entering the text and then use the CHANGE command to correct the error.

You can cancel the current line of text without cancelling the DTEXT command by pressing Ctrl-X.

You can use the cursor to point to a new location on the screen at the start of each line of text. This makes it easy to enter several notes in one short step. You cannot, however, use the object snap feature to locate any line of text except the very first while you are using the DTEXT command.

227

DWGNAME

The DTEXT command is particularly useful in conjunction with the fit-text justification, since you can visually determine when the text being entered nears the right margin. In this way, you can approximate fully justified text (justified left and right margins) and avoid having some lines of text looking too compressed.

To skip a line while you are entering text with the DTEXT command, enter a space with the Spacebar before pressing Return. Because the space is considered a character, the Return will not end the command. Instead, the cursor box will skip down a line.

If you are using AutoCAD release 9 (with the standard menu ACAD.MNU), you can select the DTEXT command from the pull-down menu under the draw menu bar. The menu repetition in release 9 causes the command to repeat automatically until you cancel it by pressing Ctrl-C or select another command from the menu (only with AutoCAD release 9; see Advanced User Interface and Repeating Commands).

DWGNAME

[v2.5, v2.6, r9]

Overview   The DWGNAME system variable contains the name of the drawing file currently loaded into the drawing editor. It is a read-only variable, which means you can only observe the name, not alter it.

Procedure   To return the DWGNAME system variable, use either the SETVAR command or the (getvar) AutoLISP function.

Examples   The following method returns the DWGNAME variable using the SETVAR command:

```
Command: SETVAR
Variable name or ?: DWGNAME (AutoCAD returns the drawing name)
```

The following AutoLISP routine saves the drawing name to the variable s. Typing !s at AutoCAD's command prompt then causes the drawing name to be displayed:

```
(setq s (getvar "DWGNAME"))
```

228

Tips   The DWGNAME system variable is not saved. It is for use within AutoLISP routines while the drawing is loaded. (See also DWGPREFIX.)

DWGPREFIX

Overview   The drive designation and the directory path used to load the current drawing are stored in the DWGPREFIX system variable. Like the DWGNAME variable, this is a read-only variable.

Procedure   Use the SETVAR command or (getvar) function to return the current drawing's drive and path.

Examples   The following method returns the DWGPREFIX variable using the SETVAR command:

```
Command: SETVAR
Variable name or ?: DWGPREFIX (AutoCAD returns the drawing name)
```

The following AutoLISP routine saves the drawing prefix to the variable *s*. Typing *!s* at AutoCAD's command prompt then causes the drawing prefix to be displayed:

```
(setq s (getvar "DWGPREFIX"))
```

Tips   The DWGPREFIX system variable is not saved. It is for use within AutoLISP routines while the drawing is loaded. (See also DWGNAME.)

DXBIN

Overview   The DXBIN command is a special command. It is used to load a binary DXB file, produced by another program such as Autodesk's CAD/camera, into an AutoCAD drawing.

Procedure   From the AutoCAD command prompt, enter *DXBIN*. At the prompt, enter the name of the DXB file you want to import. DXB files have the extension .DXB. Do not include the extension when entering the file name. You can include a drive designation and/or a path to a particular subdirectory.

Examples   The following command sequence loads the CAD/camera drawing saved as DETAIL-1.DXB in the \SCANNER subdirectory on drive D.

```
Command: DXBIN
DXB file: D:\SCANNER\DETAIL-1
```

Warnings   AutoCAD assumes that the extension is .DXB and will reject the command if you include the extension.

Tips   There is one particularly powerful use for the DXBIN command. It can be used to insert an AutoCAD drawing that had previously been saved to disk using the special ADI plotter-driver. This feature is extremely useful when you are using AutoCAD's three-dimensional capabilities. It allows you to combine a three-dimensional image with a two-dimensional drawing. Making use of this feature requires several steps.

First, configure AutoCAD to use the ADI plotter-driver and select the DXB File Output option. (See Configuring AutoCAD.) Create the three-dimensional drawing and use the VPOINT command (see VPOINT) so that the drawing appears on the screen exactly as you want. Use the PLOT command. When using the ADI plotter-driver with the DXB File Output option, AutoCAD automatically sends the plot output to a file rather than a plotter. This is similar to plotting to a file. (See PLOT.) You must provide a plot file name or accept the default value. The file created will have the file extension .DXB.

Next, end the current drawing and load a new drawing (or the drawing that you wish to insert the three-dimensional drawing into). When Auto-CAD's command prompt appears, enter DXBIN. Then enter the file name of the plot file previously created. The three-dimensional image from your previous drawing will appear in the current drawing. The new entities (from the three-dimensional drawing), however, are not three-dimensional entities in the current drawing. Instead, they are simple line entities, just a two-dimensional representation of the original three-dimensional objects.

This technique takes some practice, but not much. You should try it a few times with a simple drawing before using it on a large one. Other than using the DXBIN command for this ability of merging two- and three-dimensional images in one drawing file, the only other use for the DXBIN command is in conjunction with CAD/camera or another program that uses the DXB files. When using these programs, you should consult the program's manual.

DXF

Overview   The DXF (Drawing Interchange File) feature is a standardized method of representing an AutoCAD drawing file in a readable form. The DXF format has been adopted by most other CAD packages as a industry standard of drawing-information exchange.

Procedure   The DXF format consists of an ASCII file containing coded text information about a drawing. The DXFOUT command generates a DXF file of an existing AutoCAD drawing. The DXFIN command reads any valid DXF file into AutoCAD, creating a new AutoCAD drawing.

   This book does not attempt to give you a thorough understanding of the DXF format. However, some explanation of this feature may prompt you to further investigate the capabilities of the DXF format. There are many external programs that read DXF format files to generate bills of materials or to perform mathematical analysis of drawn objects or that create AutoCAD drawings from data provided by other programs.

Examples   The DXF format has a standard general structure. A DXF file can be broken down into four general sections:

1. HEADER, which contains variable names and values

2. TABLES, which contain definitions of named objects such as layers, linetypes, styles, and views

3. BLOCKS, which contain block definitions

4. ENTITIES, which contain all drawing entities, including block references

Every DXF file ends with an end-of-file (EOF) marker.

The DXF file is composed of pairs of entries, with each pair occupying two lines. The first line is a group code, which identifies what type of entry follows on the next line. Each of the four sections starts with the group code 0 and the entry SECTION. Each section ends with the group code 0 and the entry ENDSEC. Other pairs of entries identify the features in the individual sections.

Group codes identify the format of the entry on the next line. The following formats are used:

GROUP CODE	FORMAT OF NEXT LINE
0–9	String value
10–59	Floating point value
60–79	Integer value
999	Remarks (a comment within the file)

The actual group code number further represents specific features within the file. For example, a code of 0 always marks the beginning of a section or an entry. The line that follows always identifies the type of entry. A code of 10 always indicates that the next line contains an X-coordinate value. A code of 6 indicates that a linetype name follows.

The following program lines are from an actual DXF file and describe the drawing that follows the list.

PROGRAM	EXPLANATION
0	Start of an entry
SECTION	Entry is the beginning of a section of the file
2	Says that what follows is the name of the section
ENTITIES	Identifies section as the entities section
0	Start of an entry
LINE	Entry is line entity type
8	Next entry is the layer name
0	Line is drawn on layer 0
10	Next entry is from point X-coordinate
1.0	X-coordinate is 1.0
20	Next entry is from point Y-coordinate
1.0	Y-coordinate is 1.0
11	Next entry is to point X-coordinate
7.0	X-coordinate is 7.0

PROGRAM	EXPLANATION
21	Next entry is to point Y-coordinate
1.0	Y-coordinate is 1.0
0	Start of an entry
CIRCLE	Entry is circle entity type
8	Next entry is the layer name
0	Circle is drawn on layer 0
10	Next entry is centerpoint X-coordinate
4.0	X-coordinate is 4.0
20	Next entry is centerpoint Y-coordinate
1.0	Y-coordinate is 1.0
40	Next entry is radius of circle
3.0	Radius is 3.0
0	Start of an entry
ENDSEC	Entry is the end-of-section marker
0	Start of an entry
EOF	Entry is the end-of-file marker

This DXF file describes the following drawing:

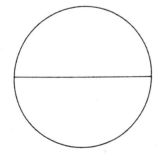

Tips   Appendix C of the AutoCAD manual thoroughly describes all the features of the DXF file format. As an additional aid, you could generate a DXF file of a relatively simple drawing. Print out a hard copy of the file and refer to it as you read about DXF files in the AutoCAD manual.

DXFIN

[all versions]

Overview   The DXF (Drawing Interchange File) format is an integral part of Auto-CAD. A DXF file of a drawing is an ASCII text file that contains all the information contained within the drawing. Many other CAD packages include the ability to convert their drawings into AutoCAD-compatible drawings or AutoCAD drawings into their format through the use of the DXF format. The DXFIN command allows DXF files to be put into an Auto-CAD drawing.

A DXF file may appear complex to the novice, but the format is relatively easy to understand. It is particularly easy to write programs that can read, manipulate, or even create DXF files.

Procedure To load an entire DXF file into an AutoCAD drawing, you must start with an empty drawing file. From AutoCAD's main menu, select task 1, which creates a new drawing. Then issue the DXFIN command as the first command. AutoCAD prompts for the name of the DXF file. DXF files have the file extension .DXF. Do not include the file extension. You may, however, include a drive designation and/or a path to a particular subdirectory.

```
Command: DXFIN
File name: (enter name of DXF file)
```

If any errors are encountered in the DXF file, an error message appears on the command line, telling you the type of error and the line on which it occurred. If no errors are encountered, AutoCAD performs an automatic ZOOM ALL, after which you are free to edit the drawing normally.

It is also possible to import DXF file information into an existing drawing. However, only the drawing entities within the DXF file are entered into the AutoCAD drawing. AutoCAD displays the message

```
Not a new drawing -- only ENTITIES section will be input.
```

If errors are encountered during this type of DXFIN, the entire DXFIN is cancelled and the drawing restored to its state prior to the DXFIN command.

Examples The following example loads the DXF file named SCHEDULE from the subdirectory \INTERCHG on drive D into the current drawing.

```
Command: DXFIN
File name: D:\INTERCHG\SCHEDULE
```

Warnings   Do not include the file extension .DXF when you enter the file name. AutoCAD assumes that the file name has this extension and will cancel the command if you include a file extension name.

Tips   If you do get an error when importing a DXF file, you can use a word processor to go into the DXF file, locate the line that contains the error, and correct the error. (This book does not attempt to teach you the complete DXF file structure, but for more information see DXF.)

Understanding the use of the DXFIN command is the first step in being able to exchange your drawing files with those of associates using other CAD packages.

A particularly interesting use of the DXFIN command is its ability to fix drawing files that have been inadvertantly damaged in some way, perhaps by an internal error or a crashed disk. You must first create a DXF file using the DXFOUT command and then edit the resulting file using a word processor or text editor. Once any errors in the file have been corrected or removed, you can use the DXFIN command to reload the file into Auto-CAD's drawing editor. (For more information, see DXFOUT.)

DXFOUT

[all versions]

Option

E   Output-selected entities only

Overview   The DXF (Drawing Interchange File) format is a particularly powerful aspect of the AutoCAD program. The feature enables many other programs to convert their drawing files to and from AutoCAD. The DXFOUT command allows the creation of a DXF file from an existing drawing. Once the DXF file is created, it can be used by a third-party program or as input to another CAD program that can read AutoCAD's DXF files.

Procedure   At the command prompt, enter *DXFOUT* and specify a valid file name as the DXF output file. AutoCAD automatically appends the file extension .DXF. Do not include an extension name, although you may include a device designation and/or a valid path to a specific subdirectory. The DXFOUT command then asks for the numerical precision you wish to use and whether you want to output only specific entities. If you enter an *E* to indicate that you want to include only specific entities, you must then select the entities to be included. Only the objects you select will be included. Once you have selected the entities, press Return to indicate that you have finished selecting objects. Then respond to the numerical precision prompt.

Examples   The following example creates a DXF file, named SCHEDULE, of the entire drawing. The file will be sent to the \INTERCHG subdirectory on drive D. The default value of six-decimal-place accuracy is selected.

```
Command: DXFOUT
File name: D:\INTERCHG\SCHEDULE
Enter decimal places of accuracy (0 to 16)(or Entities)<6>: ( RETURN )
```

Warnings   If a file with the name you specify in response to the prompt for a file name already exists, the old file will be deleted and the new file created by the DXFOUT command will replace it.

Tips   The DXFOUT command is the most popular method for creating a file that can be used in most conversion programs, to convert a drawing either to another CAD package or to an earlier version of AutoCAD.

Versions of AutoCAD prior to version 2.0 used a different DXF format. To output a DXF file that can be used by these earlier versions, respond to the prompt for a file name with *name,OLD*.

The DXFOUT command has a particularly interesting use in recovering drawings corrupted by an internal error or a crashed disk. If you encounter an error while trying to load a drawing file, try loading it from the main menu again. As soon as you are in the drawing editor, press Ctrl-C. Immediately use the DXFOUT command. You may then be able to use a word processor or a text editor to look at the resulting DXF file to locate and correct an internal error. This method can be quite tedious, however, and requires a thorough understanding of the DXF file format. For more information, consult the AutoCAD manual.

ELEV

Overview The ELEV (Elevation) command allows you to set the current elevation and extrusion thickness. Once set, this value controls the elevation and thickness of subsequent entities. Elevation and thickness have meaning only within AutoCAD's three-dimensional visualization. Viewed in two-dimensional plan view, objects appear the same whether or not they have an elevation and thickness. In fact, taken a step further, all objects within AutoCAD that have not had an elevation and thickness assigned have an elevation of 0 and a thickness of 0.

The elevation is understood as being the base coordinate of an object in the Z-plane. The thickness is then the distance that the object is extruded above or below its base, along the Z-axis. Thus, a circle with an elevation of 1 and a thickness of 3, would be seen as a circle when you are looking at a two-dimensional plan view. But with AutoCAD's three-dimensional visualization, that circle would be seen for what it is, namely, a cylinder with a base at $Z = 1$ and a length of 3 units in the Z-direction.

Procedure The current elevation and thickness are assumed any time AutoCAD accepts a three-dimensional coordinate in response to a command. You can change the elevation and the thickness at any time by entering the ELEV command at the command prompt.

```
Command: ELEV
New current elevation <current value>: (enter value)
New current thickness <current value>: (enter value)
```

Any value is valid. A negative elevation places the base below the Z-base plane. A negative thickness extrudes the object downward from the base elevation. A positive thickness causes the object to be extruded up.

Release 9 Only If you are using AutoCAD release 9, can you change the functions of the ELEV command from the Entity Creation Modes dialog box displayed by the DDEMODES command (available only in AutoCAD release 9; see Advanced User Interface and DDEMODES).

Examples

Different entities appear differently when three dimensions are applied, as illustrated in the following drawing. A point with an elevation and a thickness becomes a line parallel to the Z-axis. A circle becomes a cylinder. A line becomes a Z-plane. A square becomes a cube.

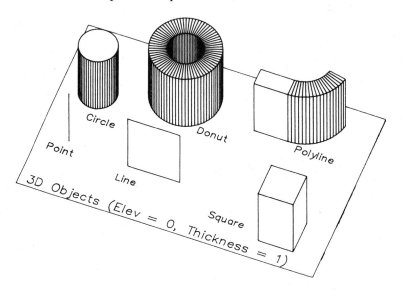

Warnings

Several things will happen if an elevation other than 0 is set:

- The BASE command will use the current elevation as the Z-base coordinate for subsequent insertions of the drawing.
- Entities grouped using the BLOCK command will use the current elevation as the base coordinate for future insertions of that block.
- The INSERT command will insert blocks at the current elevation.
- Objects must be on the current elevation to be selected with the OSNAP command.
- Objects grouped and written to a disk file using the WBLOCK command will use the current elevation as the base coordinate.

If an elevation and/or thickness other than 0 is set, any object subsequently drawn will have that elevation and thickness.

Tips   Setting an elevation and thickness has meaning only if you are going to view the object in three dimensions. Don't use the ELEV command unless you will also need to use three-dimensional views. The extra information enlarges your drawing file.

You can alter the elevation and thickness established for any existing object with the CHANGE command (see CHANGE).

The current elevation set by the ELEV command is stored in the ELEVATION system variable. The current thickness is stored in the THICKNESS system variable. You can read and alter both with the SETVAR command or AutoLISP. These two system variables are saved with the individual drawing file.

The initial settings for the elevation and the thickness are determined by the prototype drawing.

ELEVATION   [ADE-3] [all versions]

Overview   The ELEVATION system variable stores the current three-dimensional elevation value.

Procedure   The ELEV command is normally used to set the current three-dimensional elevation value (see ELEV). The most recent value is stored in the ELEVATION system variable as a real number. You can use the SETVAR command or the AutoLISP functions (getvar) and (setvar) to read and change the ELEVATION value. Changing the value affects only subsequent objects.

You can also change the current ELEVATION setting from the Entity Creation Modes dialog box displayed by the DDEMODES command (available only in AutoCAD release 9; see DDEMODES).

Examples   The following command sequence can be used to change the ELEVATION value. In this example, the current value, 0'-0", is changed to 3'-0".

```
Command: SETVAR
Variable name or ? <>: ELEVATION
New value for ELEVATION <0'-0">: 3'0"
```

Tips   The initial ELEVATION value is determined from the prototype drawing (default value is 0.0000, no elevation applied). The current value is saved with the drawing.

ELLIPSE   [ADE-3] [v2.5, v2.6, r9]

Options

C   Specifies center point rather than first axis end point
R   Specifies eccentricity via rotation rather than second axis
I   Draws isometric circle in current ISOPLANE

Overview   The ELLIPSE command provides several methods with which to draw ellipses. The ellipses are not a separate entity.

Procedure   In AutoCAD, rather than create a separate entity for drawing ellipses, the program draws a polyline that closely approximates an ellipse. There are three methods available for drawing ellipses:

1. By specifying one of the axes of the ellipse and the ellipse's eccentricity
2. By specifying the center of the ellipse and the two axes
3. By causing AutoCAD to draw the ellipse, which actually represents a circle drawn within an isometric plane

In the case of the first two methods, you are drawing on a standard two-dimensional drawing. In the third case, you are actually using AutoCAD's ISOPLANE command to draw in one of the three isometric planes.

Examples   **Axis and Eccentricity**   This method of constructing ellipses allows you to specify either the major (longer) or minor axis of an ellipse. AutoCAD then accepts either the distance from the midpoint of the first axis described to the end point of the second axis or the angle at which the ellipse, if viewed as a circle, would be rotated into the third dimension (the eccentricity). The second method is similar to the way in which ellipses are described on ellipse templates.

In the first method, you are prompted to select two points that represent the end points of one axis of the ellipse. Next you enter a distance. The distance is a measurement from the midpoint of the axis just described to the end point of the second axis, perpendicular to the first axis. You can enter a distance at the keyboard. You can also point on the screen, but the distance described by pointing will be used by AutoCAD as the distance measured perpendicular from the midpoint of the first axis. AutoCAD will determine whether the axis you are describing is the major or the minor axis by examining the distance between the first pair of end points to the distance specified at the third prompt. If the first distance is greater, you have defined the major axis. The angle from the first point selected to the second point is the angle at which the ellipse will be drawn. This is illustrated in the following drawing. The distance from point A to point B defines the major axis.

```
Command: ELLIPSE
<Axis endpoint 1>/Center: (select point A)
Axis endpoint 2: (select point B)
<Other axis distance>/Rotation: (select point C or enter distance)
```

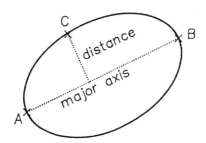

In the second method, after specifying the first and second points and defining the major axis and its angle, you are asked to enter a rotation angle. The rotation angle must be between 0 and 89.4 degrees. AutoCAD treats the distance between the first two points as the diameter of a circle. This circle is then rotated into the third dimension by the specified rotation angle. A rotation angle of 0 would cause AutoCAD to draw a circle. Imagine the computer screen as a flat plane with a circle drawn on it. Now rotate that circle about the axis formed by a line through the two points that you have just entered. You have just visualized what AutoCAD is doing. If you want, rather than entering the number of degrees of rotation, you can show AutoCAD how far to rotate the circle into the third dimension by pointing with the cursor. AutoCAD will use a rotation angle that is measured from the midpoint of the major axis relative to the angle of that axis. The following drawing illustrates this method. The distance from point A to point B defines the major axis. The plane on which the circle is drawn is then rotated 60 degrees, yielding the desired ellipse.

```
Command: ELLIPSE
<Axis endpoint 1>/Center: (select point A)
Axis endpoint 2: (select point B)
<Other axis distance>/Rotation: R
Rotation around major axis: (select rotation angle) (in this case, 60)
```

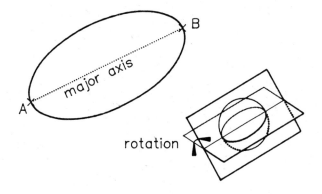

Center and Two Axes This method of constructing an ellipse requires the center of the ellipse, the end point of one axis, and the length of the other axis. The center of the ellipse is the intersection of the major and minor axes. AutoCAD first requests the center point. In response, you can either enter a coordinate or simply point on the screen. Next, you must provide the end point of an axis. The angle of the ellipse is determined by the angle from the center point to this end point. At this point, the prompts are identical to the axis-and-eccentricity method. You can select to specify either the distance measured from the center of the ellipse to the end point of the second axis, measured perpendicular to the first axis, or you can specify a rotation angle into the third dimension. Once again, if you elect to provide a distance, AutoCAD will decide if the first axis defined is the major or minor axis. If you provide a rotation, the first axis automatically becomes the major axis. The prompts for the distance method result in the first drawing and are as follows:

242

```
Command: ELLIPSE
<Axis endpoint 1>/Center: C
Center of ellipse: (select center point A)
Axis endpoint: (select point B)
<Other axis distance>/Rotation: (select point C or enter distance)
```

Using the center-and-two-axes method with a rotation angle for the last prompt creates the second drawing. The prompts are as follows:

```
Command: ELLIPSE
<Axis endpoint 1>/Center: C
Center of ellipse: (select center point D)
Axis endpoint: (select point E)
<Other axis distance>/Rotation: R
Rotation around major axis: (select rotation angle) (in this case, 60)
```

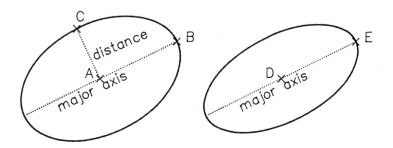

Isometric Circles If you have the Snap mode set to isometric, the first prompt in response to the ELLIPSE command will include an additional selection, that of isocircle. The isocircle ellipse method permits the user to draw circles in the three planes of AutoCAD's isometric projection. The plane in which the ellipse will be drawn is governed by the current isoplane. That is, the ELLIPSE command will draw in only one of the three isometric construction planes at a time, and the plane used at any given time will be the currently selected plane. You can toggle between isoplanes using the ISOPLANE command, the isometric plane toggle key, or the SETVAR command. (For a further discussion of the isometric planes, see ISOPLANE.)

The ellipse drawn is a representation of a circle drawn on one of the isometric planes. Because it makes more sense to visualize the ellipse as a circle drawn on one of these planes, AutoCAD prompts for the center of the circle. You can then provide either the radius or the diameter of the circle. Both of these methods are illustrated in the following drawing. To use the radius method, simply enter a radius or use the cursor to point on the screen.

```
Command: ELLIPSE
<Axis endpoint 1>/Center/Isocircle: I
Center of circle: (select point A)
<Circle radius>/Diameter: (enter radius or select a point)
```

To specify a diameter, enter D or diameter at the last prompt.

```
Command: ELLIPSE
<Axis endpoint 1>/Center/Isocircle: I
Center of circle: (select point B)
<Circle radius>/Diameter: D
Circle diamter: 1
```

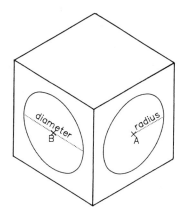

The ELLIPSE command was used with AutoCAD's Snap mode set to isometric to create the drawing of a machine tool lathe in Appendix D (see Figure D-7).

Warnings

Remember that if you are specifying a distance to describe the second axis of an ellipse, the distance will always be measured from the midpoint of the first axis to the end point of the second axis in a direction perpendicular to the first axis. If you are pointing with the cursor, this will be done regardless of the direction of the point used to select the distance.

When drawing isocircles, you must be in the isoplane in which you want to draw before you select the center point. Otherwise, the ellipse will be drawn in the wrong plane.

Tips   Because the ellipse drawn is actually a closed polyline, you can use the PEDIT command to edit the ellipse. You can also use any other commands that edit polylines, such as BREAK, EXTEND, OFFSET, and TRIM. For example, the OFFSET command can be used to draw concentric ellipses. The BREAK and TRIM commands can be used to create elliptical arcs.

You can use the isoplane toggle key or the transparent mode of the SETVAR command to change isoplanes while in the middle of the ELLIPSE command.

If DRAGMODE is set, you can visually drag an ellipse during its creation. During their construction, ellipses are particularly difficult to visualize. Dragging them is helpful.

END

[all versions]

Overview   The END command saves the current drawing and then returns you to AutoCAD's main menu.

Procedure   Typing the END command at AutoCAD's command prompt causes the drawing to be saved using the same name you specified when you started the editing session. Drawing files all have the file extension .DWG. The previous copy of the drawing file will have its file extension changed from .DWG to .BAK. When AutoCAD has finished saving and renaming these two files, you are returned to AutoCAD's main menu.

Examples   Once you have typed the command in response to the command prompt and pressed Return, AutoCAD will end your drawing session.

Command: **END**

Warnings   As you are editing a drawing, AutoCAD saves your drawing as a temporary file. When you finish the drawing with the END command, AutoCAD renames the temporary file to the file name you specified when you started your editing session, after first changing the file extension of the previous copy of the drawing file to .BAK. Although it does not happen often, things can go wrong when you end your drawing. AutoCAD must have enough disk space to save both the drawing file and the backup file.

If there is not enough room, AutoCAD may issue an error message that the disk is full and indicate that the backup file will not be saved. If AutoCAD cannot give your drawing file the name you specified, it will try naming it *dwg.$$$*. If that doesn't work, it will then try the name ef.$ac. Whatever the case, AutoCAD will display a message indicating what is happening and your drawing *will be saved*.

Tips

Do not abbreviate the ENDPOINT object snap override as END. You could make the mistake of selecting this override in response to the command prompt, in which case AutoCAD would interpret it as the END command and end your drawing session. Always use ENDP as an abbreviation for the object snap and reserve END only for the END command.

If you are using AutoCAD release 9 (with the standard menu ACAD.MNU), you can select the END command from the File pull-down menu (only with AutoCAD release 9; see Advanced User Interface).

Entities

[all versions]

Overview

Every object within an AutoCAD drawing is composed of one or more drawing entities. Entities are the basic parts from which all complex objects within a drawing are created. Each AutoCAD drawing command creates a specific entity type. The editing commands simply manipulate the entities that you have previously placed in the drawing, scaling them, copying them, or combining them to create a finished drawing.

Procedure

The AutoCAD program works with twelve different entity types. All objects in a drawing consist of combinations of one or more of the following entities:

- Arcs
- Blocks
- Circles
- Lines
- Points
- Polylines [ADE-3]
- Shapes

- Solids
- Text
- Traces
- 3DLines [ADE-3] [v2.6, r9]
- 3DFaces [ADE-3] [v2.6, r9]

Examples   Each entity type has an associated command for its creation. For example, the LINE command creates line entities, the CIRCLE command creates circle entities, and so on.

Tips   See the individual commands for more information.

Entity Selection

[all versions]

Overview   Entity selection refers to the methods used to respond to the prompt to select objects, which is normally presented by every editing command.

Procedure   Each AutoCAD editing command (ERASE, BREAK, ARRAY, MOVE, and so on) requires that you first select an object or objects before the command actually executes the desired action. This necessary preliminary step is referred to as entity selection. There are five basic methods available for selecting entities: crossing, last, pointing, previous, and window.

    Most commands do not actually execute until you press Return, which indicates that you have completed selecting all of the entities that you want the command to act on. This is called building a selection set. In such cases, before you press Return you have the option of adding other objects to or removing objects from the selection set you are building. (There are additional entity selection methods available if you are using AutoCAD release 9, as described under the next section.)

Examples   **Selecting Objects**   Whenever AutoCAD requires you to select an object, the normal cross-hair is replaced by a pick box. The pick box is an indication that AutoCAD is waiting for you to select objects. The simplest method of entity selection is to move the pick box until it is over the desired object and then to press the pick button. You could also enter either an absolute or a relative coordinate from the keyboard. AutoCAD immediately scans the drawing file until it finds the object you have pointed to. When the object is found, it is added to the current selection set and is highlighted on the screen.

When AutoCAD prompts you to select an object, you can also respond with the letter corresponding to one of the other available methods of entity selection. Entering a *C* instructs AutoCAD to use the crossing method (see Crossing); *W* indicates the window method (see Window). You can immediately select the last object added to the drawing that is currently displayed on the screen by entering *L* for the last object method.

Sometimes you will want to reselect the same objects you have just selected for the previous editing command (say you want to rotate the objects that you have just moved). Entering *P* in response to the prompt to select objects will cause the previous selection set to be highlighted again.

Whenever you add items to the selection set, AutoCAD highlights the objects added and displays a message telling you how many objects you have selected, how many were found, and how many of those objects were already part of the selection set.

Each time you select an object or objects, AutoCAD must scan the drawing file until it locates the objects. In a large drawing, this can be time consuming. To speed things up, AutoCAD provides a multiple-selection option. By entering *M* in response to the prompt to select objects, you can use the pointing method to select any number of objects, one at a time. When you have finished selecting objects in this fashion, press Return. Only after you press Return will AutoCAD scan the drawing file for the selected objects. It will then report how many were found and highlight them. You can then continue to select objects using any method or press Return to complete entity selection.

Adding/Removing Objects from the Selection Set   As you build a selection set, you may inadvertently pick an object that you do not want as part of the selection set. When this happens, simply enter *R* in response to the prompt to select objects. The prompt will immediately change to

```
Remove objects:
```

You can now select any of the highlighted objects that you do not want to be part of the current selection set, again using any of the methods described previously. If you want to continue adding entities to the selection set, press *A* in response to the prompt to remove objects. The prompt to select objects will reappear. Any entities subsequently selected will again be highlighted, indicating that they have been added to the set. You can toggle back and forth between the Add and Remove modes as often as you wish.

You can also use the Undo option to back up through the selection set in the reverse order that the entities were added. Each time you enter *U* in response to the prompt to select objects, you remove the most recently added object from the selection set.

When you have completed entity selection, press Return in response to the prompt to either select or remove objects. The editing command will then complete its work.

Release 9 Only　　If you are using AutoCAD release 9, there are three additional entity selection options that you can use. If you enter *Box* in reply to the prompt to select objects, you are immediately prompted to select the first corner and second corners. If after selecting the first corner, however, you move the cursor to the left, AutoCAD immediately assumes you will make a crossing selection. Moving the cursor toward the right indicates a window selection method. If your graphics card supports the Advanced User Interface (see Advanced User Interface), AutoCAD further indicates you are making the selection by crossing method by displaying the rubberband box with a dashed linetype. The windowing method of entity selection is indicated by a solid rubberband box line.

Entering *Auto* or *AU* in response to the prompt to select objects causes AutoCAD to prompt again to select objects. If you select a blank area of the screen, AutoCAD immediately goes into the box selection mode.

Normally, AutoCAD continues to prompt you to select objects until you press Return a final time. In this way, you can manipulate the selection set, adding or subtracting objects as required. If you respond to the prompt to select objects by entering *Single* or *SI*, no selection-set manipulation is allowed. AutoCAD will prompt again for you to select objects. You can then use any valid selection method, such as crossing, box, or auto, but as soon as an object is selected, entity selection is complete.

Tips　　Most editing commands allow you to use one or more methods of entity selection, even within the same command. For example, you can select some entities by pointing to them and then select others by windowing before actually erasing them. There are some commands, however, that require a particular method of entity selection to be used. Refer to the individual command descriptions for further information.

When in doubt, enter *?* in response to a prompt to select or remove objects. A list of acceptable responses will be displayed.

Pressing Ctrl-C causes the current command to cancel, removing all selected entities from the selection set and turning off the highlighting.

Environment Variables

[all versions]

Overview   Environment variables, a DOS-level feature, are special computer system variables. They must be configured before you actually load AutoCAD. These variables all exist and are given default values whenever you load AutoCAD by entering the ACAD command. You can assign environment variables values other than their default values by using the DOS command SET to establish their values explicitly.

AutoCAD's environment variables consist of the following:

VARIABLE	USAGE
ACAD	Its value is the name of the directory that contains AutoCAD's support files.
ACADCFG	Its value is the name of the directory that contains the desired AutoCAD configuration file.
ACADFREERAM	Reserves a portion of memory as a working storage area.
ACADLIMEM	Controls what portion of Lotus/Intel/Microsoft Expanded Memory Specification (EMS) memory AutoCAD will use for its extended I/O page space.
ACADXMEM	Controls what portion of an AT's extended memory AutoCAD will use for its extended I/O page space.
LISPHEAP	Determines the size of the memory area where AutoLISP functions and variables are stored.
LISPSTACK	Determines the size of the memory area where AutoLISP function arguments and partial results are stored and controls the number of levels that LISP functions can be nested.

Procedure The setting of any of these environment variables must be done before you load AutoCAD. Once AutoCAD is running, they cannot be changed. You must exit from AutoCAD before you can change their values.

The environment variable values are established through the use of the DOS command SET. Normally you will set some or all of these variables at once by using a BATCH command. The procedure for setting each variable is described in the next section.

Examples **ACAD** When you insert a file, change menus, or select a text font, Auto-CAD searches for a drawing, menu, shape, or font file first in the current directory and then in the directory in which the AutoCAD program files are located (the system directory). Setting the ACAD environment variable causes AutoCAD to always look in the set subdirectory *before* it searches the AutoCAD system directory. To make AutoCAD search the subdirectory \DRAWINGS\DETAILS before it searches the subdirectory containing the AutoCAD program, you would set this value to the ACAD variable (all letters can be uppercase or lowercase):

```
C>SET ACAD=c:\drawings\details
```

ACADCFG Normally when you load AutoCAD, it looks in the AutoCAD system directory for its configuration files. If you need to maintain more than one configuration for AutoCAD (for example, if you have two different plotters or use two different menu systems), rather than reconfigure AutoCAD each time you change configurations you can keep the two different sets of configuration files in separate subdirectories. You can determine which configuration will be used during a particular editing session by setting the ACADCFG environment variable. Whatever directory name is set to this variable will determine where AutoCAD looks for its configuration files when it loads. To set ACADCFG to a configuration saved in the subdirectory \PLOTTER1, you would set the ACADCFG variable as follows:

```
C>SET ACADCFG=c:\plotter1
```

ACADFREERAM The ACADFREERAM variable determines the amount of RAM reserved for AutoCAD to use as a work area. If you don't set this variable, the default value is 14K bytes. Any value between 5 and 24 is valid. Too low a value may result in AutoCAD displaying a message that it is out of RAM and returning to the DOS prompt. Too high a value results in wasted memory. According to Autodesk, you should experiment, but a value of 20 is usually acceptable. To reserve 20K bytes for working storage area, use the following DOS SET command:

C>**SET ACADFREERAM=20**

ACADLIMEM [Release 9 Only] If you are using memory that conforms to the Lotus/Intel/Microsoft Expanded Memory Specification (EMS), the ACADLIMEM variable allows you to determine what portion of the available expanded memory will be used for extended I/O page space. If you do not set this variable, AutoCAD will use all the expanded memory that is available except for any memory that has been previously assigned to a RAM disk or another program. If you use the SHELL command from within AutoCAD to run another program that uses EMS, you may want to set this variable to reserve some expanded memory for use by that program.

There are four settings for the ACADLIMEM variable:

1. If you do not set the ACADLIMEM variable or it is set to *all*, AutoCAD uses all available expanded memory (except memory previously assigned to a RAM disk or other program).

C>**SET ACADLIMEM=ALL**

2. If ACADLIMEM is set to 0 or NONE, no expanded memory will be used by AutoCAD.

C>**SET ACADLIMEM=0**

3. If ACADLIMEM is set to a positive number, AutoCAD will use up to that many 16K pages of EMS memory if they are available. If the number provided is greater than 512, AutoCAD assumes the number to represent bytes and allocates enough 16K pages to acquire the specified number of bytes of expanded memory. If the number is preceded by *0x*, AutoCAD assumes the number to be in hexadecimal. If the number has a suffix of *K*, AutoCAD recognizes the value to be in kilobytes (1024 bytes).

If a given system has 1 megabyte of EMS memory installed (64 pages), all of the following four methods would allocate 16 pages (256K bytes) of expanded memory to AutoCAD, making 48 pages available:

METHOD	EXPLANATION
C> **SET ACADLIMEM=16**	16 pages 16K = $16 \times 16 \times 1024 = 262144$ bytes
C> **SET ACADLIMEM=262144**	262144 bytes
C> **SET ACADLIMEM=256K**	$245K \times 1024 = 262144$ bytes
C> **SET ACADLIMEM=0x40000**	40000 hex = 262144 bytes

If the same system is used and ACADLIMEM is set to 72, AutoCAD will use all 64 available pages and ignore the fact that 8 pages are missing.

4. If ACADLIMEM is set to a negative number, up to that many 16K pages of EMS memory will be left unused by AutoCAD. If AutoCAD cannot find that number of 16K pages to leave unused, however, no expanded memory will be used. Again, if the number is greater than 512, AutoCAD assumes that the number represents bytes. If the number has an *0x* prefix, the number is assumed to be in hexadecimal; if it has a *K* suffix, it is assumed to be in kilobytes.

In the following example, the same 64 pages of expanded memory have been installed. All four methods would leave 48 pages available.

```
C>SET ACADLIMEM=-48
C>SET ACADLIMEM=-786432
C>SET ACADLIMEM=-768K
C>SET ACADLIMEM=-0xC0000
```

If the same system is used and ACADLIMEM is set to -72, AutoCAD will not use any of the available expanded memory, because the number of pages to be left unused exceeds the number of pages available. In this situation, AutoCAD will display the message

```
Expanded memory disabled.
```

ACADXMEM [Version 2.6, Release 9]   If you are using a PC/AT-type computer, the ACADXMEM variable allows you to determine what portion of the available extended memory AutoCAD will use for extended I/O page space. If you have other programs that use extended memory, the proper setting of this variable will allow AutoCAD to coexist with them. If no value for this variable is provided, AutoCAD will use all of the extended memory that is available, except for any memory used by the standard VDISK RAM disk driver.

There are four formats for setting this variable:

1. C>**SET ACADXMEM**=start
2. C>**SET ACADXMEM**=start,size
3. C>**SET ACADXMEM**=,size
4. C>**SET ACADXMEM**=none

Start is a memory address, and *size* is the amount of memory to be assigned for extended I/O page space. Both numbers can be represented in either hexadecimal or decimal formats. If expressed in decimal format, the value must be expressed as the number of kilobytes, with the number value followed by the letter *K* (for example, 1024K). The value must be between 0100000 and 01000000 (hexadecimal) or between 1024K and 16384K (decimal).

The first format specifies the absolute lowest memory address that AutoCAD can use for extended I/O page space. If this address is below the ending address of the configured VDISK, AutoCAD's area will automatically start at the end of the VDISK buffer. The following would set the starting address to 1644K:

C>**SET ACADXMEM**=0x1a0000

or

C>**SET ACADXMEM**=1644K

The second format establishes the starting address and the size of Auto-CAD's area. To set aside an area of extended memory starting at address 1644K and extending to 1900K (256K bytes in size), you could use either of the following:

C>**SET ACADXMEM**=0x1a0000,0x40000

or

C>**SET ACADXMEM**=1644K,256K

The third method establishes the size of the extended memory area to be used by AutoCAD but allows AutoCAD to determine the starting address. Thus, you could limit AutoCAD's extended I/O page space to 256K bytes by setting the variable as follows:

C>**SET ACADXMEM**=,0x40000

or

C>**SET ACADXMEM**=,256K

The fourth method disables AutoCAD's use of extended memory entirely by setting the ACADXMEM variable to a value that does not exist. The message

```
Extended memory disabled
```

will be displayed in this case, unless there is no extended memory in the system or it has all been used by VDISK.

LISPHEAP and LISPSTACK   Most AutoLISP functions end up using a great deal of the LISPHEAP area because users do not make judicious use of variable names. In addition, the more AutoLISP functions that are loaded, the more LISPHEAP area that is used. The LISPHEAP variable has a default value of 5,000 bytes. Generally, this number will be too low. A number approaching 40,000 is usually more appropriate. If you are using a third-party package that utilizes AutoLISP, the accompanying manual will usually recommend a setting. The total of LISPHEAP and LISPSTACK cannot be greater than 45,000 bytes. To set these variables (they are usually set in tandem), use the SET command:

```
C>SET LISPHEAP=40000
C>SET LISPSTACK=5000
```

Warnings   The ACADXMEM environment variable is not available in AutoCAD version 2.5.

The ACADLIMEM environment variable is available only in AutoCAD release 9.

Tips   Use batch files to set all of the appropriate environment variables and to load AutoCAD rather than type each SET command at the DOS prompt.

AutoCAD can use up to 4 megabytes of extended or expanded memory. If the system contains both types of memory, AutoCAD will utilize the expanded memory first.

ERASE

[all versions]

Overview   The most basic AutoCAD editing command is the ERASE command. This command allows you to remove an object or objects from the drawing. Generally, the object is permanently removed from the drawing, although there are ways to restore an object that is accidentally removed.

255

Procedure

To remove an unwanted object, simply enter the ERASE command and select the object. You can pick any number of objects for erasure before completing the command. You can use any combination of AutoCAD's entity selection methods to select objects. You can point to them with the cursor, window them using the window or crossing mode, select the previous selection set, and/or select the last object entered into the drawing. (For more information, see Entity Selection.)

You can add or remove objects from the selection set prior to completing the ERASE command. Once you press Return the final time, all of the entities selected will be removed from the drawing.

Examples

```
Command: ERASE
Select objects: (select desired objects)
Select objects: (RETURN) (to complete command)
```

Warnings

Any objects that you erase are removed from the drawing.

Tips

During the current editing session, the OOPS command will restore all of the objects removed from the drawing by the last ERASE command, regardless of the number of commands that came after it. However, OOPS will restore only the objects from the most recent ERASE command. The PLOT and PRPLOT commands mark a practical end to the current editing session, because objects erased prior to plotting a drawing cannot be restored after the plotting.

You can save drawing display time while working in a drawing by erasing a particularly large block, as long as you remember exactly where the erased block was inserted and you don't purge the block definition from the drawing. For example, if you draw a portion of a drawing and then save it as a block with an insertion point of 0,0, you can then temporarily erase that block. As long as you don't use the PURGE command to remove the block's definition from the drawing file (see PURGE), you can reinsert that block when you are ready to plot the drawing, using the 0,0 insertion point. The block's description is still saved with the drawing file even though it isn't part of the drawing.

If you are using AutoCAD release 9 (with the standard menu ACAD.MNU), you can select the ERASE command from the Edit pull-down menu. This situation uses the Single and Auto selection modes. The menu item repetition in release 9 causes the command to repeat automatically until you cancel it by pressing Ctrl-C or select another command from the menu (only with AutoCAD release 9; see Advanced User Interface and Entity Selection).

EXPERT

[v2.5, v2.6, r9]

Overview

The EXPERT system variable allows you to disable the "are you sure?" type of prompts. Its value can be 0, 1, 2, or 3.

Many of AutoCAD's commands will ask you if you are sure when you attempt to do certain things. For example, if you attempt to turn off the current layer, AutoCAD will prompt:

```
Really want layer xxx (the CURRENT layer) off? <N>:
```

Generally, this type of warning is a good thing. If you were to turn the current layer off and then continued drawing objects, you would not be able to see what you had drawn until you turned the layer back on. To avoid this type of confusing situation, AutoCAD prompts you appropriately.

There may be times, however, when you don't want AutoCAD to display these types of prompts. The EXPERT system variable permits you to turn these "are you sure?" prompts off.

Procedure

You can change the value of EXPERT by using the SETVAR command or the AutoLISP (setvar) function. A value of 1 suppresses the "are you sure?" prompts normally presented when you turn the current layer off or when AutoCAD asks "About to regen, proceed?" A value of 2 suppresses the warning message normally issued when you are about to reuse a block name already defined or overwrite an existing drawing using the SAVE or WBLOCK command, as well as those warnings suppressed by an EXPERT value of 1. A value of 3, in addition to suppressing the warnings already listed, suppresses the warning issued if you attempt to load a linetype that has already been loaded or create a new linetype in a file that already contains that linetype's definition. A value of 0 is the default condition and causes all warning messages to be displayed normally.

Examples

This example attempts to turn off the current layer (in this case, layer 1). AutoCAD then issues the appropriate prompt. The SETVAR command is used to change the value of the EXPERT system variable.

```
Command: LAYER
?/Make/Set/New/ON/OFF/Color/Ltype/Freeze/Thaw: OFF
Layer name(s) to turn Off: 1
Really want layer 1 (the CURRENT layer) off? <N>:  ( RETURN )
?/Make/Set/New/ON/OFF/Color/Ltype/Freeze/Thaw: ( RETURN )

Command: SETVAR
Variable name or ?: EXPERT
New value for EXPERT <0>: 1

Command: LAYER
?/Make/Set/New/ON/OFF/Color/Ltype/Freeze/Thaw: OFF
Layer name(s) to turn Off: 1
?/Make/Set/New/ON/OFF/Color/Ltype/Freeze/Thaw: ( RETURN )

Command:
```

Warnings   In AutoCAD version 2.5, EXPERT can be set only to values of 0 or 1. When set to a value of 1, only the prompts

```
About to regen, proceed?
```

and

```
Really want layer xxx (the CURRENT layer) off <N>:
```

are suppressed. No other prompt suppression is provided. When EXPERT is set to a value of 0, all prompts are issued.

Tips   Whenever you enter any drawing, the initial value of the EXPERT variable will be whatever value has been set during the current editing session. When you first load AutoCAD, the initial value is always 0. AutoCAD retains the current EXPERT value until you end the program (using task 0 from the main menu).

EXPLODE

[ADE-3] [v2.5, v2.6, r9]

Overview   The EXPLODE command reduces a complex object such as a block, a polyline, or an associative dimension back into the individual entities that actually made up the complex object. A block, once exploded, appears exactly the same on the screen. The difference is that the individual parts that make up the whole can now be edited individually. An associative dimension appears the same except that all of the individual entities are placed on the 0 layer. When a wide polyline is exploded, however, the widths of all the segments return to zero. It is also possible to explode hatch patterns.

Procedure   You can use the EXPLODE command on only one object at a time. To begin, enter EXPLODE at the command prompt. When AutoCAD prompts, simply select the object that you want to explode. Select the object by pointing or using the last, window, or crossing method of entity selection.

Examples   Regardless of the type of entity you want to explode, the command prompts are the same.

```
Command: EXPLODE
Select block reference, polyline, or dimension: (select object)
```

Warnings   You can use the EXPLODE command only on blocks, polylines, and associative dimension entities.

You cannot select an object to explode using the previous method because a selection set may contain more than one object. You can select only one object to explode at a time.

If a block was inserted with different values for the X-, Y-, and Z-scale factors, AutoCAD will not allow the block to be exploded. The message

```
X, Y, and Z scale factors must be equal.
```

will be displayed.

If a block containing attribute information is exploded, the attribute values are lost but the attribute definitions are retained and will be displayed.

You can explode a hatch pattern, but the entities that make up the hatch pattern will be placed on the 0 layer with color and linetype specified as BYBLOCK.

You can explode an associative dimension, but the components that make up the dimension (lines, arrows, text) will be placed on the 0 layer with color and linetype specified as BYBLOCK.

When you explode a block containing a 3DLine or 3DFace, the Z values will be scaled by the Z-scale factor of the block reference and adjusted by the Z-height of the block reference less the height of the block definition.

You cannot explode blocks that have been inserted with the MINSERT command.

If you explode a polyline that has width or tangent information, the tangent information is lost and the individual segments will be drawn as arcs or lines (zero width).

The EXPLODE command affects only one nesting level at a time. If the block that you explode contains other nested blocks, you will have to explode each nested block separately.

Tips Use the EXPLODE command only when you need to edit the individual entities that made up a block or polyline. Otherwise, the individual entities will take up more room in your drawing file than the blocks or polylines of which they were part.

Only the UNDO command can restore an exploded object to its original condition. You can restore an object you have just exploded by entering U at the next command prompt. This will restore lost attributes, polyline width information, and associative dimensions.

EXTEND

[ADE-3] [v2.5, v2.6, r9]

Overview The EXTEND command lengthens an object so that it ends precisely at a predefined boundary. Only arcs, lines, and polylines can be extended. The EXTEND command will be valid only if the object to be lengthened can physically touch the boundary object when it is lengthened. Only arcs, circles, lines, and polylines can be used as boundary objects.

Procedure When you start the EXTEND command, you must first select one or more boundary edges. As just mentioned, only certain objects can be extended, and only certain objects can be used as boundary objects. You can select the boundary object or objects by using any valid entity selection method —

pointing, windowing, last, previous, and so on. Select as many boundary edges as you want. Each valid boundary object, once selected, remains highlighted until the EXTEND command is completed. When you have finished selecting boundary edges, press Return in response to the prompt

```
Select boundary edge(s)...
```

AutoCAD then prompts you to select the objects you want to extend. You must select each object to be lengthened individually by pointing. You cannot select them by windowing. The prompt will be displayed after each object is lengthened, which allows you to continue to select additional objects to extend to the selected boundaries. When you have finished extending objects, press Return a final time to conclude the command.

Examples   In the following drawing, the EXTEND command is used to lengthen the two horizontal lines in both directions so that they touch the two vertical lines.

```
Command: EXTEND
Select boundary edge(s)...
Select objects: (select point A)
Select objects: (select point B)
Select objects: (RETURN)
Select object to extend: (select point C)
Select object to extend: (select point D)
Select object to extend: (select point E)
Select object to extend: (select point F)
Select object to extend: (RETURN)
```

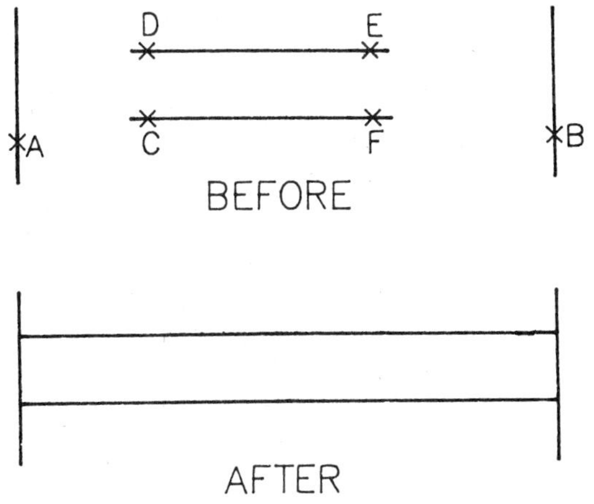

BEFORE

AFTER

Warnings   If the object selected is not a valid boundary object, AutoCAD will reject the boundary and it will not be highlighted. If none of the objects selected is valid as a boundary, when you press Return, AutoCAD responds

```
"No edges selected."
```

and you will be returned to the command prompt.

   If an object selected to be lengthened is not a valid object or the object will not extend to touch a boundary edge, AutoCAD responds

```
Cannot EXTEND this entity.
```

You cannot extend a closed polyline. If you attempt to do so, AutoCAD displays an appropriate message and returns you to the command prompt.

Tips   You can only point to objects to extend. If the pick box is too large or too small to select an object satisfactorily, you can change its size while within the EXTEND command by using the Transparent mode of the SETVAR command (see SETVAR).

   If you select multiple boundary edges, an object will lengthen only until it intersects the nearest boundary edge. You can continue to extend the object until it hits the next boundary edge by selecting that object again.

   If you select multiple boundary edges or you are lengthening an arc, it is possible that the object could be extended in more than one direction. AutoCAD determines which direction to extend the object based on which end of the object to be extended is closest to the point you used to select the object. The following drawing illustrates this procedure.

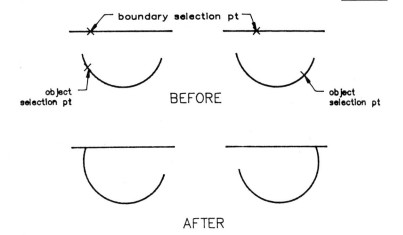

boundary selection pt

object
selection pt

object
selection pt

BEFORE

AFTER

When you are extending wide polylines, the centerline of the polyline determines the point to which the polyline will be extended. If the polyline is not perpendicular to the boundary, some of the width of the polyline may extend beyond the boundary.

When you are extending a wide polyline with a tapering width, the taper will continue. If this would cause the endwidth to be negative, the degree of taper will be varied so the polyline ends in a point at the boundary edge. If the polyline gets wider in the direction toward the boundary edge, the width simply increases at the same rate, with the resulting polyline being that much wider where it touches the boundary edge. Both of these conditions are illustrated in the following drawing.

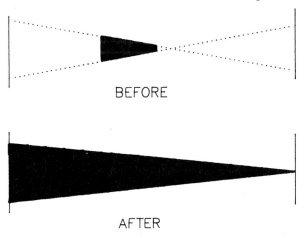

BEFORE

AFTER

If you are using AutoCAD release 9 (with the standard menu ACAD.MNU), you can select the EXTEND command from the Edit pull-down menu. This situation uses the Auto selection mode to select the first boundary edge(s). The menu item repetition of release 9 causes the command to repeat automatically until you cancel it by pressing Ctrl-C or selecting another command from the menu. (See Advanced User Interface and Entity Selection.)

Extents

[all versions]

Overview

The actual portion of AutoCAD's graphics screen occupied by a drawing represents the drawings extents. The extents may be less than a drawing's limits. If limit checking is off, the drawing extents may exceed the drawing's limits. The extents of a drawing can be represented as the smallest rectangle that will contain all the elements of the drawing.

The actual portion of the drawing currently shown on the computer screen represents the display extents or view extents.

Procedure

The drawing extents are recalculated whenever the drawing is regenerated. If you execute a ZOOM — Extents, the display extents equal the drawing extents.

Examples

You can see the extents of your drawing by executing a ZOOM command and selecting the Extents option. If the drawing uses less than its allotted limits, the image displayed is a smaller area. If you execute a ZOOM — All command, AutoCAD displays the drawing out to its limits, as shown in Figure 47.

Tips

The lower left corner and upper right corner of the drawing extents are stored in the EXTMIN and EXTMAX system variables, respectively, and are displayed by the STATUS command in the drawing uses value (see EXTMAX, EXTMIN, and STATUS).

The lower left corner and upper right corner of the display extents are stored in the VSMIN and VSMAX system variables, respectively (see VSMAX and VSMIN).

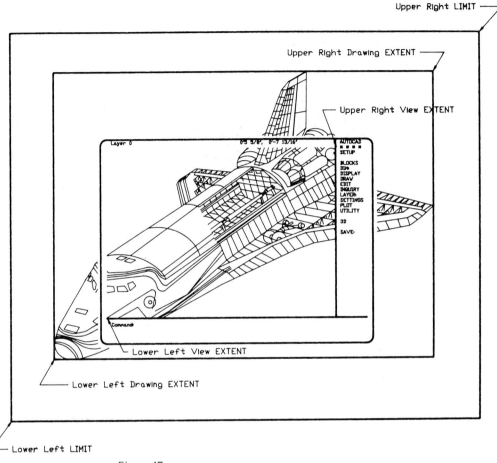

Upper Right LIMIT

Upper Right Drawing EXTENT

Upper Right View EXTENT

Lower Left View EXTENT

Lower Left Drawing EXTENT

Lower Left LIMIT

Figure 47

EXTMAX

[all versions]

Overview   The coordinate of the upper right corner of a drawing's extents is stored in the system variable EXTMAX (Extents Maximum). AutoCAD determines this point based on changes made to the current drawing. When the STATUS command reports the coordinates for the area that the drawing uses, EXTMAX is the upper right drawing uses value reported.

265

Procedure   EXTMAX is a read-only variable. You can observe it by using the SETVAR command or (getvar) AutoLISP function. You cannot change EXTMAX. Changing the area that the drawing uses, however, will result in a change in the EXTMAX value. The value is saved along with the drawing file.

Examples   The EXTMAX value for the prototype drawing is as follows:

```
Command: SETVAR
Variable name or ?: EXTMAX
EXTMAX = -1.0000E+20,-1.0000E+20 (read only)
```

Tips   The initial value of EXTMAX is determined from the prototype drawing (default value is -1.0000E + 20,-1.0000E + 20). The current value is used to set the drawing extents when the drawing is first loaded into AutoCAD's drawing editor.

EXTMIN

[all versions]

Overview   The EXTMIN (Extents Minimum) variable is the corresponding lower left coordinate of the drawing uses extent. It is the other coordinate that is reported by the STATUS command when that command lists the area that the drawing uses.

Procedure   Like EXTMAX, EXTMIN is a read-only variable that is saved along with the drawing file. You can observe it by using the SETVAR command or (getvar) AutoLISP function. You cannot change EXTMIN. Changing the area that the drawing uses, however, will result in a change in the EXTMIN value. The value is saved along with the drawing file.

Examples   The EXTMIN value for the prototype drawing is as follows:

```
Command: SETVAR
Variable name or ?: EXTMIN
EXTMIN = 1.0000E+20,1.0000E+20 (read only)
```

Tips The initial value of EXTMIN is determined from the prototype drawing (default value is $1.0000E + 20, 1.0000E + 20$). The current value is used to set the drawing extents when the drawing is first loaded into AutoCAD's drawing editor.

Fast Zoom Mode [v2.5, v2.6, r9]

Overview The Fast Zoom mode feature allows AutoCAD to execute many PAN and ZOOM commands at redraw speed rather than having to regenerate the entire drawing.

To increase the speed with which AutoCAD can redraw the screen after a display control command (such as ZOOM or PAN), AutoCAD maintains both the screen coordinates of the drawing and the actual drawing coordinates. The screen coordinates are saved as integers; the actual drawing coordinates are calculated using a floating-point database. The floating-point database enables AutoCAD drawings to be almost infinitely large and to permit a zoom ratio of 1,000,000 to 1 but would require the drawing to be regenerated with every display control command. The screen coordinates, on the other hand, are saved as a virtual screen.

Whenever a drawing is regenerated, the floating-point database is converted into screen-pixel integer coordinates and saved as a virtual screen. When Fast Zoom mode is on, this virtual screen is several times larger than the actual display. In this way, when you zoom or pan the drawing, the drawing will be redrawn on the screen from the virtual screen rather than requiring a regeneration, as long as you have not moved outside the virtual screen or magnified the drawing to such a degree that curves will not be drawn using enough vectors. If Fast Zoom mode is off, the virtual screen is always the same size as the actual screen, and most display commands will cause a regeneration.

Procedure The Fast Zoom mode is actually controlled by the VIEWRES command (see VIEWRES). When you execute the VIEWRES command, the first question that you are prompted to answer is

```
Do you want fast zoom? <Y>:
```

Answering yes turns Fast Zoom mode on, and answering no turns it off.

Examples   In the following two examples, the ZOOM command is used to enlarge the same area of the drawing. (See Figure 48.) In the first example, the Fast Zoom mode is disabled, causing AutoCAD to regenerate the drawing. The VIEWRES command is then used to turn the Fast Zoom mode on. In the second example, no regeneration is required.

```
Command: ZOOM
All/Center/Dynamic/Extents/Left/Previous/Window/<Scale(X)>: W
First corner: (select point A)
Other corner: (select point B)
About to regen -- proceed? <Y>

Command: VIEWRES
Do you want fast zooms? <N>: Y
Enter circle zoom percent (1-20000) <100>: ( RETURN )
Regenerating drawing.

Command: ZOOM
All/Center/Dynamic/Extents/Left/Previous/Window/<Scale(X)>: W
First corner: (select point A)
Other corner: (select point B)
```

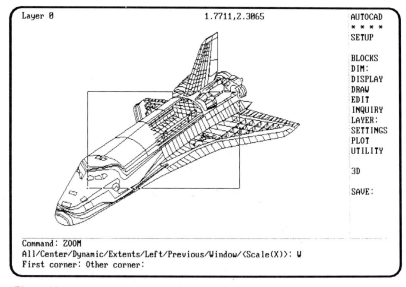

Figure 48

Warnings   Fast Zoom mode must be on to perform transparent PAN, ZOOM, and VIEW commands.

Tips   In most cases, having the Fast Zoom mode on allows AutoCAD to perform PAN and ZOOM commands at redraw rather than regen speed. Sometimes, however, the virtual screen will contain too many objects and object selection will be severely affected (slowed down). To alleviate this problem, you can do one of two things. You can zoom in to the desired view and regenerate the drawing so that both the actual screen and the virtual screen are now smaller, or you can turn Fast Zoom mode off. When Fast Zoom mode is off, the virtual screen contains the same number of objects as the actual screen. Object selection therefore speeds up at the expense of frequent drawing regeneration.

The initial setting of the Fast Zoom mode feature is determined from the prototype drawing. The current setting is saved with the drawing file.

FILES

[all versions]

Overview   The FILES command temporarily suspends drawing editing and loads the File Utilities menu. You can load this same menu from AutoCAD's main menu by selecting task 6. The File Utilities menu allows you to list, delete, rename, and copy computer files. Since this is an important aspect of working with a computer, AutoCAD provides access to the same File Utilities menu from within the drawing editor (while you are working on a drawing) using the FILES command.

Procedure   You can access the File Utilities menu (Figure 49) at any time from the command prompt by entering the command FILES.

Command: **FILES**

AutoCAD will then display the following File Utility menu.

To execute a task, enter any menu selection and press Return. Task 0, exiting from the file utility menu (returning to the drawing editor), is the default selection.

```
              A U T O C A D
   Copyright (C) 1982,83,84,85,86,87 Autodesk, Inc.
   Version X.0.5B (7/31/87) IBM PC
   Advanced Drafting Extensions 3
   Serial Number:   xx-xxxxxx

   File Utility Menu

     0.  Exit File Utility Menu
     1.  List Drawing files
     2.  List user specified files
     3.  Delete files
     4.  Rename files
     5.  Copy file

   Enter selection (0 to 5) <0>:
```

Figure 49

Examples   **Task 1, Listing Drawing Files**   Task 1 displays a list of all files with the file extension .DWG. AutoCAD prompts

Enter drive or directory:

Press Return to list drawings in the current directory or specify a drive designation and/or a subdirectory name to display drawings saved in a different drive or directory. If there are too many drawings to display at one time, AutoCAD will pause at each full screen and allow you to display additional screens when you are ready.

Task 2, Listing Other Files   Task 2 prompts you to provide a file name for which to search:

Enter file search specifications:

You can use the wildcard characters ? and * to search for all files matching certain designations. A ? wildcard matches any character that occurs in that position. The * wildcard matches any character that comes after it until the period separating the file name from the file extension. Again, you can include a drive designation and/or a subdirectory. Otherwise, AutoCAD searches in the current directory. Thus, to list all the slide files in the subdirectory named ACAD, you would respond to the prompt

```
Enter file search specifications:
```

with

```
/ACAD/*.SLD
```

Task 3, Deleting Files   Task 3 lets you delete specified files. AutoCAD prompts

```
Enter file deletion specification:
```

You can enter a file name explicitly (you must include the extension). You can also use wildcards. If you specify only one file or the wildcard matches only one file name, that file is deleted and AutoCAD displays a message that it has deleted a file. If the wildcard matches more than one file, Auto-CAD prompts you for each file name matched before deleting it. You must respond before a file is deleted. The default is no. For example, if you type *.BAK* in response to the prompt to specify a file to delete, you could delete all the backup files. But you can retain whichever ones you want by pressing Return (to indicate no) to the AutoCAD prompt to delete specific files.

```
Delete DRAWING1.BAK? <N>: Y
Delete DRAWING2.BAK? <N>: ( RETURN )
Delete DRAWING3.BAK? <N>: Y
```

In this example, the files DRAWING1.BAK and DRAWING3.BAK would be deleted, but DRAWING2.BAK would be retained.

Task 4, Renaming Files   Task 4 allows you to change the name of any file. Renaming a file changes only the name. You cannot use task 4 to move a file from one drive to another, but it does permit you to move a file to a different subdirectory. If you do not specify a directory, the current directory is assumed. AutoCAD prompts for the current file name and the name you want to change it to. You must enter valid file names. The current name must already exist and the new name must not yet exist. You can rename only one file at a time. Wildcards are not valid.

```
Enter current filename: (enter an existing filename)
Enter new filename: (enter a valid file name)
```

Task 5, Copying Files   Copying a file creates a duplicate of an existing file. You can copy a file to a different name within the current directory or to another directory or drive. If you do not specify a drive or directory, the current directory is assumed. You can copy only one file at a time. Wildcards are not valid when you are copying files. AutoCAD's dialog looks like this:

Source filename: **(enter an existing filename)**
Destination filename: **(enter a valid filename)**

Warnings
Once a file is deleted, it is gone. The UNDO command will not restore it. Some file recovery programs can be used successfully on files that have been deleted as long as no other file has overwritten the original file's location on the disk. However, because of the way AutoCAD makes use of temporary files as it is running, the recovered file may not be usable.

Do not delete one of AutoCAD's temporary files (.$ac and .$a) or Auto-CAD may not be able to successfully save your drawing.

If you supply an existing file name in response to the prompt for the destination filename when copying a file, the existing file is deleted and the copied file takes its place.

Tips
When your hard disk is full, you may want to delete some of the backup drawing files that AutoCAD creates whenever a drawing is saved.

AutoCAD uses special file extensions to indicate particular types of files. The extensions and their corresponding file types are listed here as a convenience.

EXTENSION	FILE TYPE
BAK	Drawing file backup
DWG	Drawing file
DXB	Binary drawing interchange file
DXF	Drawing interchange file
DXX	Attribute extract file (DXF format)
FLM	Filmroll file for use with AutoSHADE
IGS	IGES interchange file
LIN	Linetype library file
LSP	AutoLISP program file
LST	Printer plot output file
MNU	Menu source code file
MNX	Compiled menu file
OLD	Original version of converted drawing file
PAT	Hatch-pattern library file
PLT	Plot output file

SCR	Command script file
SHP	Shape/font-definition source file
SHX	Shape/font-definition compiled file
SLB	Slide library file [r9]
SLD	Slide file
TXT	Attribute extract or template file (CDF/SDF format)
$ac	AutoCAD temporary file
$a	AutoCAD temporary file

If you are using AutoCAD release 9, all files produced by AutoCAD are totally compatible and transportable between different operating systems and machines (for example, between UNIX- and DOS-based machines). Release 9 can read most files produced by earlier versions of AutoCAD. However, compiled menu files (.MNX files) from earlier versions of Auto-CAD are not compatible with release 9. They must be recompiled from the original menu file (.MNU). In addition, files created or edited with Auto-CAD release 9 are not compatible with earlier versions of AutoCAD.

FILL

[all versions]

Options

On	Solids, traces, and wide polylines filled
Off	Solids, traces, and wide polylines outlined

Overview   The FILL command allows you to determine whether solids, traces, and wide polylines are displayed and plotted as filled objects. The command acts as a toggle, turning the Fill mode on or off.

Procedure   Objects that are filled often require more time to display on the screen or to plot than objects that are not filled. You may be able to greatly speed up redrawing, regenerating, and plotting by turning interior object filling off when it is not required. The FILL command allows you to do so at any time. When FILL is off, only the outlines of solids, traces and polylines are displayed or plotted.

When you enter the FILL command at AutoCAD's command prompt, the current setting is displayed and you are prompted to enter a new setting. Press Return to keep the current setting.

Examples

From AutoCAD's command prompt, you can change the current setting of FILL by entering the FILL command. AutoCAD displays the current setting. In this example, the Fill mode is currently on; the FILL command is used to turn it off. The drawing illustrates the difference in the way objects are displayed.

```
Command: FILL
ON/OFF <ON>: OFF
```

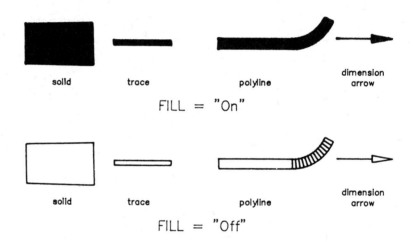

Warnings

Changing the current setting for FILL has no effect until the drawing is regenerated (see REGEN).

When you are viewing an object in AutoCAD's three-dimensional visualization mode, objects are filled only when you are looking at the plan view and hidden lines have not been removed.

Tips

The initial setting of FILL is governed by the prototype drawing (default value is on). The setting is controlled by the FILLMODE system variable. The current setting of FILLMODE is saved with the drawing. (For more information, see FILLMODE.)

The filling of arrowheads drawn by AutoCAD's dimensioning commands is also controlled by the FILL command.

FILLET

Options

P   Fillets an entire polyline [ADE-3]
R   Sets fillet radius

Overview   The FILLET command connects two lines, arcs, or circles with an arc of a specified radius (a fillet). If your copy of AutoCAD includes ADE-3, you can also fillet polylines. If the two objects intersect or extend beyond their intersection, they are trimmed back from their intersection and an arc is drawn to connect them. If the two objects do not intersect, they are first extended until they intersect and AutoCAD then trims them back and draws the connecting fillet arc. If the two objects are parallel or concentric, AutoCAD rejects the command. In the case of two circles, the circles are connected with a fillet arc of the specified radius. The circles are not broken. When both objects to be filleted are on the same layer, the fillet arc is drawn on that layer. Otherwise, the arc is drawn on the current layer. The same applies for color and linetype.

You can fillet two lines; two arcs; two circles; an arc and a line; a circle and a line; an arc and a circle; a polyline and an arc, line, or circle; two segments of a polyline; or an entire polyline.

Procedure   Begin by entering the FILLET command. AutoCAD assumes that you will select two objects. You do not need to press Return if you select objects.

```
Command: FILLET
Polyline/Radius/<Select two objects>: (point to two objects)
```

To set the fillet radius, enter *R* or *Radius* in response to AutoCAD's prompt.

```
Command: FILLET
Polyline/Radius/<Select two objects>: R
Enter fillet radius <current value>: (enter value)
```

You may enter a value or show AutoCAD the radius by using the cursor to select two points on the screen. The radius will then be the distance between those two points. The fillet radius thus set will affect all future fillets until the radius is again changed. It has no effect on objects filleted previously. Once you enter a fillet radius, you need to reactivate the FILLET command by pressing Return again or entering FILLET at the command prompt.

If the objects selected cannot be filleted, because either they are parallel or the fillet radius is too large, AutoCAD displays an appropriate error message.

You can fillet an entire polyline at once by responding to AutoCAD's prompt with a *P* or *Polyline.* If the fillet radius is not 0, a fillet arc of the specified radius is added at each vertex where two line segments meet. If two line segments are separated by an arc segment and the two line segments do not diverge as they approach the arc segment, the arc segment is removed and replaced with a fillet arc. If the fillet radius is 0, no fillet arcs are inserted. Any fillet arcs entered by a previous fillet command, however, are removed, as well as arc segments separating two line segments that don't diverge as they approach the arc.

```
Command: FILLET
Polyline/Radius/<Select two objects>: P
Select polyline: (point to polyline)
```

When the polyline fillet is complete, AutoCAD displays a report indicating the number of line segments that were filleted and any special conditions that prevented a fillet arc from being added.

Examples

For the following examples, the fillet radius is first set to 0.25 using the FILLET command with the R option. Then, in the first example, the two lines are filleted by simply pointing to them. In the Before drawing, the solid lines indicate the original lines. The dotted lines show how the fillet is calculated by first extending the lines and then drawing the radius. The drawing labeled After shows the result.

```
Command: FILLET
Polyline/Radius/<Select two objects>: R
Enter fillet radius <0>: .25

Command: FILLET
Polyline/Radius/<Select two objects>: (select point A and point B)
```

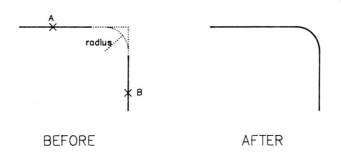

BEFORE   AFTER

When you are filleting objects to an arc or a circle, there often will be more than one fillet arc possible. AutoCAD selects which fillet to draw by drawing the one closest to the two points used to select the objects to be filleted. This is illustrated in the following example and drawings.

```
Command: FILLET
Polyline/Radius/<Select two objects>: (select the points indicated)
```

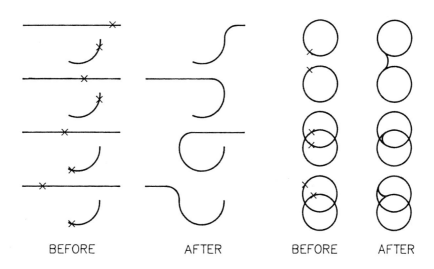

BEFORE   AFTER   BEFORE   AFTER

In the next example and drawing, the same 0.25 fillet radius is still in effect. To fillet the entire polyline at once, the P option is selected and one point on the polyline is chosen by pointing.

277

```
Command: FILLET
Polyline/Radius/<Select two objects>: P
Select polyline: (select point E)
4 lines were filleted
```

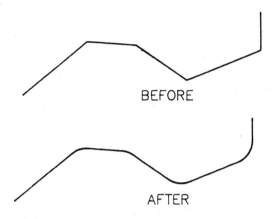

BEFORE

AFTER

Warnings   You can select the objects to be filleted using a window or crossing, but AutoCAD will select which objects within the window to fillet by selecting the last two objects. This may not always be the two objects you wanted.

Arcs and circles can be filleted only by pointing.

If the two objects to be filleted extend beyond their intersection, the shorter segment extending beyond is removed by the FILLET command.

If the two segments of a polyline, separated by one segment, are selected for filleting, the segment separating the two selected is deleted and replaced with the fillet arc.

If the intersection point of the two chosen objects is outside the drawing limits and LIMITS checking is on, AutoCAD will not fillet the objects (see LIMITS).

If you fillet a spline-curved polyline, only the curve is filleted, not the frame (see PEDIT; release 9 only).

Tips   If you make a mistake, you can restore the last object filleted to its previous condition by entering *U* at the next command prompt (see UNDO).

If you select a fillet radius of 0, the objects will be extended or trimmed to meet precisely at their intersection.

The fillet radius is stored in the system variable FILLETRAD. This variable is saved with the drawing and can be changed with the SETVAR command or AutoLISP. The initial value is determined from the prototype drawing (default value is 0).

FILLETRAD

[ADE-1] [all versions]

Overview   The FILLETRAD system variable stores the current fillet radius.

Procedure   You can observe or change the FILLETRAD value using the SETVAR command or the (getvar) and (setvar) AutoLISP functions, respectively. Changing the fillet radius with the FILLET command also changes the FILLETRAD value (see FILLET).

Examples   The FILLETRAD variable is provided mainly as a means for AutoLISP routines to adjust the fillet radius as they are operating. To retrieve the current FILLETRAD value in AutoLISP, the code might read

```
(setq f (getvar "FILLETRAD"))
```

You can use the SETVAR command to observe or change the FILLETRAD value directly from the command prompt. In this example, the value is changed from its initial value of 0 to 0.25.

```
Command: SETVAR
Variable name or ?: FILLETRAD
New value for FILLETRAD <0.0000>: .25
```

Tips   The initial value of FILLETRAD is determined from the prototype drawing (default value is 0.0000). The current value is saved with the drawing file.

FILLMODE

[all versions]

Overview   The FILLMODE system variable saves the current object fill setting that determines whether Fill mode is on or off. When FILLMODE is on, polylines, solids, and traces are solid filled.

Procedure   Normally the FILL command is used to control whether Fill mode is on or off. The FILLMODE variable allows you to control the Fill mode setting from within AutoLISP. A FILLMODE value of 1 turns Fill mode on; a value of 0 turns it off. The FILLMODE variable can be observed or changed with the SETVAR command or the (getvar) and (setvar) AutoLISP functions.

Examples   The FILLMODE variable is provided mainly as a means for AutoLISP routines to determine the current Fill mode setting. To retrieve the current FILLMODE value in AutoLISP, the code might read

```
(setq f (getvar "FILLMODE"))
```

You can use the SETVAR command to observe or change the FILLMODE value directly from the command prompt. In this example, the value is changed from its initial value of on to off.

```
Command: SETVAR
Variable name or ?: FILLMODE
New value for FILLMODE <ON>: OFF
```

Warnings   Changing the current value of FILLMODE has no effect until the drawing is regenerated.

Tips   The initial FILLMODE setting is determined from the prototype drawing (default value is 1, FILLMODE on). The current value is saved along with the drawing file.

FILMROLL

[ADE-3] [v2.6, r9]

Overview   The FILMROLL command produces files for use by the AutoShade rendering package.

Procedure   To produce a file for use by the AutoShade program, enter the FILMROLL command at AutoCAD's command prompt. The command then prompts you for a file name, giving the current drawing file as the default. You can include a drive designation and a directory path if you want. Do not include a file extension. The extension .FLM is assumed.

Examples   Command: **FILMROLL**
Enter the filmroll file name <*default*>:

Tips   Further information on the use of this command is available in the AutoSHADE manual. The FILMROLL command is not relevant except when used in conjunction with the AutoShade rendering program.

　　If you are using AutoCAD release 9 (with the standard menu ACAD.MNU), you can activate the FILMROLL command from the displayed icon menu by selecting Ashade from the Options pull-down menu (only with AutoCAD release 9; see Advanced User Interface and Entity Selection).

Fonts

[all versions]

Overview   A font is a description of the pattern used by AutoCAD to draw text characters in a drawing. The AutoCAD package has five different fonts: TXT, SIMPLEX, COMPLEX, ITALIC, and MONOTXT. Many other fonts are available from third-party developers. It is also possible to create your own text fonts, although the effort involved in doing so is not inconsequential.

Release 9 Only　AutoCAD release 9 has twenty-five text fonts. The original SIMPLEX, COMPLEX, and ITALIC fonts are still supplied but are similar to the new ROMANS, ROMANC, and ITALICC fonts, respectively. The TXT and MONOTXT fonts are also included. The new fonts are derived from characters designed by Dr. Allen V. Hershey of the U.S. Naval Weapons Laboratory. The names and brief descriptions of the twenty new fonts follow:

FONT	DESCRIPTION
ROMANS	Roman simplex (sans serif, single stroke)
SCRIPTS	Script simplex
GREEKS	Greek simplex
ROMAND	Roman duplex (sans serif, double stroke)
ROMANC	Roman complex (serif, double stroke)
ITALICC	Italic complex

SCRIPTC	Script complex
GREEKC	Greek complex
CYRILLIC	Cyrillic, alphabetical
CYRILTLC	Cyrillic, transliteration
ROMANT	Roman triplex (serif, triple stroke)
ITALICT	Italic triplex
GOTHICE	Gothic English
GOTHICG	Gothic German
GOTHICI	Gothic Italian
SYASTRO	Astronomical symbols
SYMAP	Mapping symbols
SYMATH	Mathematical symbols
SYMETEO	Meteorological symbols
SYMUSIC	Music symbols

Procedure   Text fonts are a special type of AutoCAD shape file, created through the use of a line editor or word processor capable of creating a file with only ASCII characters. The actual creation of a font file is beyond the scope of this book. The font file, once created, must be compiled from .SHP to .SHX file type using the AutoCAD main menu task 7, which compiles a shape/font description file.

The STYLE command determines which text font is used with each user-defined text style (see STYLE). A text style combines the specific text font file with other user-definable factors such as height and width factors, obliquing angles, and whether the text is placed in the drawing backward, upside-down, vertically, or right-reading.

Release 9 Only   The Options pull-down menu in AutoCAD release 9 (with the standard menu ACAD.MNU), has a selection called Fonts. Selecting this pull-down item displays an icon menu of all the standard Auto-CAD release 9 fonts as a series of three screens. (See Figure 50.) Selecting one of the fonts from these icon menus executes the STYLE command, selecting that font and using the font name as the text style name. You are then prompted for the other style factors (width, obliquing angle, and so on). These icon menus are available only if your graphics display supports the Advanced User Interface (see Advanced User Interace).

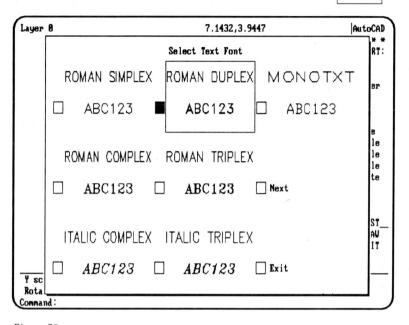

Figure 50

Examples   The following two drawings show examples of AutoCAD text fonts. Figure 51 shows the fonts supplied with the AutoCAD program (versions 2.5 and 2.6) and a selection of fonts designed by third-party developers that are available for sale. Figure 52 illustrates the additional text fonts supplied with AutoCAD release 9.

Tips   Make good use of the available text fonts to suit your particular application. Contact your AutoCAD dealer for third-party fonts. You can also find a listing of some additional fonts in the *AutoCAD Applications Guide,* available from Autodesk.

AutoCAD Version 2.5 & 2.6 Fonts

FONT = TXT
AaBbCcDdEeFfGgHhIiJjKkLlMmNnOoPpQqRrSsTtUuVvWwXxYyZz0123456789
FONT = SIMPLEX
AaBbCcDdEeFfGgHhIiJjKkLlMmNnOoPpQqRrSsTtUuVvWwXxYyZz0123456789
FONT = COMPLEX
AaBbCcDdEeFfGgHhIiJjKkLlMmNnOoPpQqRrSsTtUuVvWwXxYyZz0123456789
FONT = ITALIC
AaBbCcDdEeFfGgHhIiJjKkLlMmNnOoPpQqRrSsTtUuVvWwXxYyZz0123456789
FONT = MONOTXT
AaBbCcDdEeFfGgHhIiJjKkLlMmNnOoPpQqRrSsTtUuVvWwXxYyZz0123456789

Third—Party Fonts

(1) text fonts by GEOCAD, Pound Ridge, NY
(2) text fonts by SYMBOL GRAPHICS, INC., Corona, CA
(3) text fonts by GRAPHIC COMPUTER SERVICES, Western Australia

[1] ABCDEFGHIJKLMNOPQRSTUVWXYZ0123456789

ABCDEFGHIJKLMNOPQRSTUVWXYZ0123456789

ABCDEFGHIJKLMNOPQRSTUVWXYZ0123456789

AaBbCcDdEeFfGgHhIiJjKkLlMmNnOoPpQqRrSsTtUuVvWwXxYyZz0123456789

[2] **AaBbCcDdEeFfGgHhIiJjKkLlMmNnOoPpQqRrSsTtUuVvWwXxYyZz0123456789**

AaBbCcDdEeFfGgHhIiJjKkLlMmNnOoPpQqRrSsTtUuVvWwXxYyZz0123456789

AaBbCcDdEeFfGgHhIiJjKkLlMmNnOoPpQqRrSsTtUuVvWwXxYyZz0123456789

AaBbCcDdEeFfGgHhIiJjKkLlMmNnOoPpQqRrSsTtUuVvWwXxYyZz0123456789

AaBbCcDdEeFfGgHhIiJjKkLlMmNnOoPpQqRrSsTtUuVvWwXxYyZz0123456789

AaBbCcDdEeFfGgHhIiJjKkLlMmNnOoPpQqRrSsTtUuVvWwXxYyZz0123456789

[3] AaBbCcDdEeFfGgHhIiJjKkLlMmNnOoPpQqRrSsTtUuVvWwXxYyZz0123456789

AaBbCcDdEeFfGgHhIiJjKkLlMmNnOoPpQqRrSsTtUuVvWwXxYyZz0123456789

AaBbCcDdEeFfGgHhIiJjKkLlMmNnOoPpQqRrSsTtUuVvWwXxYyZz0123456789

AaBbCcDdEeFfGgHhIiJjKkLlMmNnOoPpQqRrSsTtUuVvWwXxYyZz0123456789

AaBbCcDdEeFfGgHhIiJjKkLlMmNnOoPpQqRrSsTtUuVvWwXxYyZz0123456789

AaBbCcDdEeFfGgHhIiJjKkLlMmNnOoPpQqRrSsTtUuVvWwXxYyZz0123456789

Figure 51

AutoCAD Release 9 Fonts

FONT = ROMANS
AaBbCcDdEeFfGgHhIiJjKkLlMmNnOoPpQqRrSsTtUuVvWwXxYyZz0123456789

FONT = SCRIPTS
AaBbCcDdEeFfGgHhIiJjKkLlMmNnOoPpQqRrSsTtUuVvWwXxYyZz0123456789

FONT = GREEKS
ΑαΒβΧχΔδΕεΦφΓγΗηΙιϑϑΚκΛλΜμΝνΟοΠπΘϑΡρΣσΤτΥυ∈ΩωΞξΨψΖζ0123456789

FONT = ROMAND
AaBbCcDdEeFfGgHhIiJjKkLlMmNnOoPpQqRrSsTtUuVvWwXxYyZz0123456789

FONT = ROMANC
AaBbCcDdEeFfGgHhIiJjKkLlMmNnOoPpQqRrSsTtUuVvWwXxYyZz0123456789

FONT = ITALICC
AaBbCcDdEeFfGgHhIiJjKkLlMmNnOoPpQqRrSsTtUuVvWwXxYyZz0123456789

FONT = SCRIPTC
AaBbCcDdEeFfGgHhIiJjKkLlMmNnOoPpQqRrSsTtUuVvWwXxYyZz0123456789

FONT = GREEKC
ΑαΒβΧχΔδΕεΦφΓγΗηΙιϑϑΚκΛλΜμΝνΟοΠπΘϑΡρΣσΤτΥυ∈ΩωΞξΨψΖζ0123456789

FONT = CYRILLIC
АаБбВвГгДдЕеЖжЗзИиЙйКкЛлМмНнОоПпРрСсТтУуФфХхЦцЧчШшЩщ0123456789

FONT = CYRILTLC
АаБбЧчДдЕеФфГгХхИиЩщКкЛлМмНнОоПпЦцРрСсТтУуВвШшЖжЙйЗз0123456789

FONT = ROMANT
AaBbCcDdEeFfGgHhIiJjKkLlMmNnOoPpQqRrSsTtUuVvWwXxYyZz0123456789

FONT = ITALICT
AaBbCcDdEeFfGgHhIiJjKkLlMmNnOoPpQqRrSsTtUuVvWwXxYyZz0123456789

FONT = GOTHICE
AaBbCcDdEeFfGgHhIiJjKkLlMmNnOoPpQqRrSsTtUuVvWwXxYyZz0123456789

FONT = GOTHICG
AaBbCcDdEeFfGgHhIiJjKkLlMmNnOoPpQqRrSsTtUuVvWwXxYyZz0123456789

FONT = GOTHICI
AaBbCcDdEeFfGgHhIiJjKkLlMmNnOoPpQqRrSsTtUuVvWwXxYyZz0123456789

FONT = SYASTRO
⊙✷☿'♀'♂⊕☌♂♅♃♄♆⋂♉∈Ψ→Β↑☾←↉↓✳∂Ω∇℧∽~Υ'Ϫ'Ϫ'Ϫ☊☋§♏†☌‡♍∃✶⚹⚛☌≈☉0123456789

FONT = SYMAP
⬯☗▲⌖☆☝⌁✢✦+☆☓⬭◯◯◯◯◯⬭⬮⬯☖☐⊥↔⌂⌐☐☓☰❄☘♣♠♡◇◇☗☘♠⚘Δ♣0123456789

FONT = SYMATH
ℵ←'↓‖∂‖∇±√∓∫×∮·∞÷§=†≠‡≡∃<∏>∑≤(≥)∝[~]√{⊂}∪(⊃)∩√∈∫→≈↑≌0123456789

FONT = SYMETEO
〰〜〜⌒〜〜〜⌒〜〜⌒〜⌒〜〜⌒〜⟋〜⟋◖⟍⟋⟋◗〜⟋◗〜⟋⟍〜⟋⊘0123456789

FONT = SYMUSIC
··ʾʾ♩♩♩♪♪♪♪oooo●●●##♮♮bbb┅╌╌╌✗↓⌐✝✝⌐𝄞𝄞☺☹:♩)𝄐‖𝄚𝄢☉✿··⌐⌐⊕⌣♂⌣⌢↝↶♮♩♩Ψℙ0123456789

Figure 52

F

Function Keys

[all versions]

Overview

The function keys enable you to toggle certain modes on or off transparently (that is, while other commands are active).

Procedure

AutoCAD assigns special control functions to most of the function keys on the computer keyboard. The following functions are assigned for MS-DOS systems:

KEY	FUNCTION	COMMENTS
F1	Flip screen graphics text	No effect in dual-screen configuration
F6	Toggle coordinate display on/off	Can also use Ctrl-D
F7	Toggle grid on/off	Can also use Ctrl-G
F8	Toggle Ortho mode on/off	Can also use Ctrl-O
F9	Toggle snap on/off	Can also use Ctrl-B
F10	Toggle tablet on/off	Can also use Ctrl-T

To toggle any of these modes on or off, or to flip from the graphics screen to the text screen, simply press the appropriate function key. Pressing the key one time toggles to the opposite mode setting. Pressing the key again returns to the original mode setting.

Examples

To toggle ORTHO mode on and off, simply press the F8 function key. If ORTHO mode is currently off, pressing the key the first time will toggle it on. Pressing it again will toggle it back off.

Tips

You will know when any of the modes affected by the toggle keys have been changed. They are reflected in either the appearance of the computer screen or are so noted on AutoCAD's status line (at the top of the graphics screen). A report of this toggling also appears on the command line.

'GRAPHSCR

[all versions]

Overview

The GRAPHSCR command causes AutoCAD to flip from the text screen to the graphics screen when you are running AutoCAD on a single-screen system. If you are using a dual-screen system, the GRAPHSCR command has no effect. This command is included in AutoCAD's command set for the explicit purpose of incorporating it into menus and script files.

Procedure   Although the GRAPHSCR command is usually used in menus or script files, it is possible to use the command to flip the display to the graphics screen directly from AutoCAD's command prompt.

Command: **GRAPHSCR**

If the text screen was previously displayed, AutoCAD would now be flipped back to the graphics screen.

Examples   There are various reasons to use the GRAPHSCR command in a menu or script file. Usually you want to force AutoCAD back into the Graphics mode after a command that automatically flips to the text screen, such as the LIST command.

Tips   Just as the Flip Screen key (usually the F1 key) can flip between the graphics screen and the text screen without affecting a command while it is active, you can use the GRAPHSCR command transparently while another AutoCAD command is active, by preceding the command with an apostrophe ('GRAPHSCR).

GRID

[all versions]

Options

On	Turns grid on
Off	Turns grid off
S	Locks grid spacing to snap resolution
A	Sets aspect [ADE-2]
number	Sets grid spacing
numberX	Sets grid spacing to multiple of snap

Overview   The GRID command lets you display a reference grid of dots and change the spacing and orientation of these dots. The grid is used only for reference. It is never plotted.

Procedure

```
Command: GRID
Grid spacing(X) or ON/OFF/Snap/Aspect <current value>:
```

Selecting ON turns the grid dots on using the current spacing. Selecting OFF turns them off. Entering a value sets the grid spacing to a multiple of drawing units. (For example, entering 5 would place the grid dots at every fifth drawing unit—every 5 inches if architectural units have been selected.) You can also lock the grid dots to the current snap spacing by entering S (for *snap*), by entering a grid spacing value of 0, or by setting the grid to a multiple of the snap resolution by entering a value followed by X. For example, 5X would place grid dots at every fifth snap point. Later, if the snap spacing is changed, the grid spacing will readjust accordingly. (See SNAP.)

The Aspect option allows you to set different grid spacing for the X- and Y-axes. Here again, a specific value would set the grid spacing to exactly that number of drawing units. A value followed by X would set the grid spacing to a spacing that is a multiple of the snap spacing. When you use the Aspect option, AutoCAD prompts for the horizontal and the vertical spacing.

Release 9 Only In AutoCAD release 9, the functions of the GRID command are duplicated by selections within the dialog box displayed by the DDRMODES command. While this dialog box is displayed, you can turn the displaying of the grid on and off, change the X- and Y-grid spacing, and even change the snap base and the snap rotation angle. This dialog box is available only if your graphics display supports the Advanced User Interface (see Advanced User Interface and the DDRMODES).

Examples

In the following example, the current snap spacing is 1 unit. The GRID command is used to select a horizontal grid spacing of ten times the SNAP spacing and a vertical grid spacing of five times the SNAP spacing. Figure 53 illustrates how this grid would appear on the screen.

```
Command: GRID
Grid spacing(X) or ON/OFF/Snap/Aspect <10.0000>: A
Horizontal spacing(X) <10.0000>: 10X
Vertical spacing(X) <10.0000>: 5X
```

Warnings

If the grid spacing is so small (or you zoom out) that the grid dots would be too densely packed on the screen, AutoCAD warns:

```
Grid too dense to display.
```

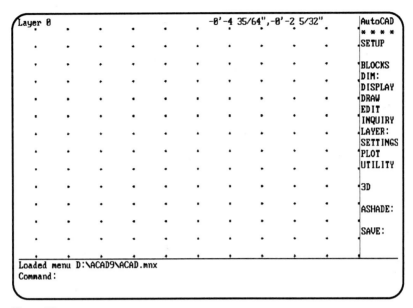

Figure 53

If the current snap style is isometric, the Aspect option will not be usable.

Tips

The initial grid spacing and the on/off setting of the grid are determined by the prototype drawing. They are controlled by the system variables GRID-MODE and GRIDUNIT (see GRIDMODE and GRIDUNIT).

The grid is displayed only within the drawing's limits even if you are zoomed out beyond the limits (see LIMITS). This is a handy way to see the limits of the drawing as illustrated in Figure 54.

You can turn the grid on and off using the GRIDMODE system variable. A GRIDMODE value of 0 turns the grid display off, and a value of 1 turns it on. The spacing of the grid dots can be changed with the GRIDUNIT system variable. Both the X- and Y-spacing can be specified, separated by a comma. Both of these system variables can be changed with the SETVAR command or AutoLISP.

The rotation angle, base point, and variable X- and Y-spacing of the grid display are controlled by the SNAP command. These settings are always multiples of the corresponding snap settings (see SNAP). Figure 55 shows a rotated grid.

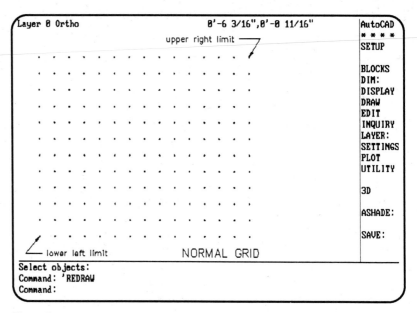

Figure 54

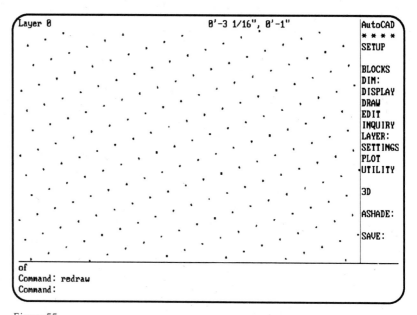

Figure 55

You can also use a grid toggle key to turn the grid on and off without affecting an active command. On most computers, the function key F7 is the grid toggle key. You can also toggle the grid by using Ctrl-G (on all systems). These transparent toggles can also be included in menus.

GRIDMODE

[all versions]

Overview   The GRIDMODE system variable determines whether a reference grid is displayed on the screen.

Procedure   The GRIDMODE variable is an integer, either 0 or 1. If GRIDMODE equals 0, the reference grid is off; if it equals 1, it is on. The GRID command normally controls the GRIDMODE setting, which can also be changed with the SET-VAR command or from within AutoLISP with the (getvar) and (setvar) functions.

Examples   The following AutoLISP routine retrieves the current GRIDMODE setting, storing it in the variable *gm*:

```
(setq gm (getvar "GRIDMODE"))
```

While another command is active, you can use the transparent 'SETVAR command to change the current GRIDMODE setting. The grid is not visible until you also execute a transparent 'REDRAW command.

```
To point: 'SETVAR
>>Variable name or ?: GRIDMODE
>>New value for GRIDMODE <0>: 1
Resuming LINE command.
To point: 'REDRAW
Resuming LINE command.
To point:
```

Tips   The initial GRIDMODE setting is determined by the prototype drawing (default value is 0, grid is off). The current value is saved with the drawing file.

You can also change the current GRIDMODE setting from in the dialog box displayed by the DRRMODES command (only with AutoCAD release 9; see DDRMODES and Advanced User Interface).

GRIDUNIT

[all versions]

Overview   The GRIDUNIT system variable controls the X- and Y-spacing of the dots within the reference grid.

Procedure   Normally you would set the grid spacing using the GRID command. The GRIDUNIT variable allows you to control the grid spacing while within another command by using the 'SETVAR command or from within AutoLISP by using the (getvar) and (setvar) functions.

Examples   The following AutoLISP routine retrieves the current GRIDUNIT setting, storing it in the variable *gu*:

```
(setq gu (getvar "GRIDUNIT"))
```

While another command is active, you can use the transparent 'SETVAR command to change the current GRIDUNIT setting. The change is not visible until you also execute the transparent 'REDRAW command.

```
To point: 'SETVAR
>>Variable name or ? <>: GRIDUNIT
>>New value for GRIDUNIT <0.0000,0.0000>: 2,2
Resuming LINE command.
To point: 'REDRAW
Resuming LINE command.
To point:
```

Tips   The initial GRIDUNIT setting is determined from the prototype drawing (default value is 0.0000,0.0000, grid spacing matches snap spacing). The current value is stored as a point value and saved with the drawing file.

You can also change the current GRIDUNIT setting from the dialog box displayed by the DRRMODES command (only with AutoCAD release 9; see DDRMODES and Advanced User Interface).

HATCH

[ADE-1] [all versions]

Options

I   Ignores internal structure
N   Normal style, turns hatch lines on and off as internal structure is encountered
O   Hatch outermost portion only
U   Specifies a hatch pattern "on the fly"

Overview   The HATCH command allows AutoCAD to quickly fill an area of a drawing with a predetermined pattern. A library of hatch patterns is provided with the AutoCAD program. You can also design your own hatch patterns and save them in a .PAT file. Using the HATCH command to cross-hatch or add patterned fill to a drawing is considerably faster than doing the same repetitive task by hand. You can specify the pattern to use, the scale for the hatch pattern, and the angle at which the pattern will be drawn. It is also possible to define simple patterns at the spur of the moment.

Procedure   You must first enclose the object or objects to be hatched in a boundary. Several options permit you to control the way in which AutoCAD determines how to hatch the selected objects. Normally the program begins hatching starting at the outer boundary. If there are no objects selected within this boundary, everything within the boundary is hatched. If, however, you have selected other objects within the outer boundary that form another closed boundary inside the first, AutoCAD hatches only between the first and second boundaries. Then it skips between the second and third, hatches between the third and fourth, and so on.

    You can force AutoCAD to hatch everything within the outer boundary, regardless of any inner boundaries, by telling the HATCH command to ignore any internal structure. Or you can force the HATCH command to hatch only within the outer boundary until it encounters the first internal boundary and then not to hatch any further. Samples of these hatching options follow.

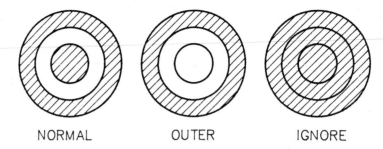

NORMAL   OUTER   IGNORE

The HATCH command first prompts for the name of the hatch pattern you want to use. AutoCAD then prompts for the scale of the hatch pattern and the angle at which it is to be inserted. Last, AutoCAD prompts you to select the object or objects you want to hatch.

```
Command: HATCH
Pattern (? or name/U,style) <default>:
Scale for pattern <default>:
Angle for pattern <default>:
Select objects:
```

You can select objects using any method—windowing, pointing, and so on. You can respond to the scale and angle prompts by entering values, or you can use the screen cursor to select points on the screen to define the scale and the angle.

When you enter the name of a hatch pattern, you can instruct AutoCAD to use one of the hatching options (ignore internal structure, hatch outermost boundary only, or normal) by following the hatch pattern name with a comma and the appropriate option. For example, the middle drawing in the previous illustration was created by specifying the pattern *LINE,O*, telling AutoCAD to use the line pattern and to hatch only the outermost boundary.

You can also define simple hatch patterns as you proceed by entering *U* in response to the pattern prompt. AutoCAD will create a hatch pattern composed of lines at whatever angle and spacing you specify. You can also tell AutoCAD to double-hatch the area selected with lines at 90 degrees to the first set.

```
Command: HATCH
Pattern (? or name/U,style) <default>: U
Angle for crosshatch lines <default>:
```

Spacing between lines <*default*>:
Double hatch area? <*default*>: (*enter Y or N*)

Examples   Several HATCH command examples follow. In the first example, Auto-CAD's brick hatch pattern is used to fill the wall elevation at a scale factor of 10.6667 and a 0-pattern angle. (Note: The windows were drawn on a separate layer and turned off during the HATCH command so that they would not be considered.)

```
Command: HATCH
Pattern (? or name/U,style) <>: BRICK
Scale for pattern <1.0>: 10.6667
Angle for pattern <0>: ( RETURN )
Select objects: W
First corner: (select point A)
Other corner: (select point B)
Select objects: ( RETURN )
```

In the second example, a hatch pattern is designed "on the fly." An angle of 45 degrees is chosen with a spacing of 3 units between the lines. The pattern is double-hatched.

```
Command: HATCH
Pattern (? or name/U,style): U
Angle for crosshatch lines <0>: 45
Spacing between lines <1.0000>: 3
Double hatch area? <N> Y
Select objects: (select point C on circle)
Select objects: ( RETURN )
```

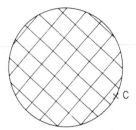

Figure 56 shows all of AutoCAD's standard hatch patterns. The shading and shadows used to create the elevation of the private residence in Appendix D (see Figure D-6) would have been very impractical to draw by hand. With the help of AutoCAD's HATCH command, they were created very quickly. By placing the hatch pattern on its own layer, the shadows can be turned off and the same elevation drawing used to create the architect's working drawings when it comes time to actually build the house.

Warnings If you do not provide a closed boundary, the hatch pattern may not fill the area properly. Lines may extend outside the boundary or not fill the bounded area completely.

Hatching takes time and increases the size of the drawing file. Add hatch patterns toward the end of your drawing process, so you do not slow down the loading, redrawing, and regenerating of drawings. As an alternative, when it is not possible to add hatching later, put the hatch patterns on a layer that can be frozen while you proceed with other drawing and editing.

Hatch patterns are normally drawn and saved within the drawing as blocks. It is possible, however, to explode a hatch pattern into its individual entities or to define the hatch pattern initially as individual entities by preceding the pattern name with an asterisk, for example, *ESCHER. If defined with an asterisk, the individual entities will be drawn on the current layer. But if the hatch pattern is exploded, the entities will be placed on the 0 layer. Either of these two instances may make the drawing more difficult to edit. You should take appropriate steps to avoid unnecessary problems. This may entail creating a special layer for hatch patterns or not drawing anything else on the 0 layer. Since hatch patterns, when

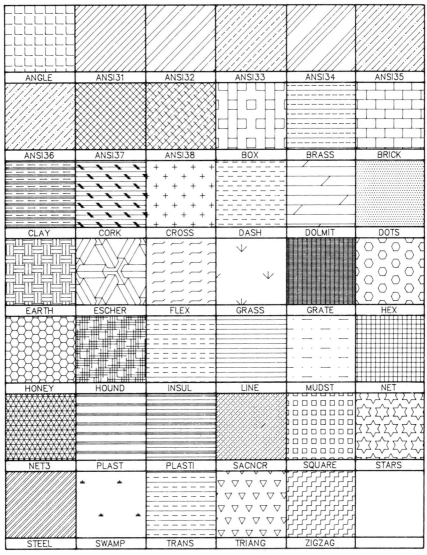

Figure 56

entered or changed to individual entities, can place hundreds of entities into a drawing with one command, these steps will ensure that you can always edit them easily by isolating them on their own layer.

Since you ultimately will plot the drawing at a particular scale, you should always think in terms of what scale a hatch pattern should be drawn at to yield a readable pattern once the drawing is plotted. Don't use a smaller scale than you can reasonably expect to see in your final plot.

It is not possible to use a hatch pattern to fill a polyline, trace, or solid. These can be only filled solid or not filled, as determined by the status of the Fill mode (see FILL).

If you use the HATCH command on a spline curved polyline, the command will "see" only the curve, not the frame (see PEDIT; release 9 only).

Tips   AutoCAD's standard hatch patterns are stored in the file ACAD.PAT. If you are using a hatch pattern other than one supplied with the program, AutoCAD first looks for the pattern in the ACAD.PAT file and then in a file with the pattern name and the file extension .PAT. If the pattern has been saved with a different file extension name, you can provide the extension name explicitly. You can also specify a drive and subdirectory designation.

If you cannot remember the name of the hatch pattern, entering a question mark (?) will bring up a listing of the hatch patterns provided with the program, along with a description of each pattern.

If text, trace, shape, or solid entities are selected as part of the selection set when AutoCAD prompts you to select objects, the hatch pattern will not pass through them, as long as the option to ignore the internal structure was not selected. Attributes and text have an invisible "box" around them that prevents hatch patterns from being drawn over them, as long as they are selected when you are using the HATCH command.

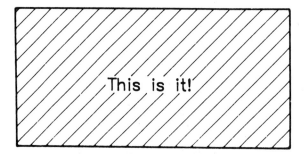

You can cancel the HATCH command at any time by pressing Ctrl-C. Any hatching already entered will remain part of the drawing.

If the hatch pattern was entered as a block, you can use ERASE — Last to remove the hatch pattern just entered.

If the hatch pattern was entered as individual entities (preceding the pattern name with an asterisk), you can remove the hatch pattern just entered with the UNDO command (see UNDO).

Even if you need to have the hatch pattern drawn as individual entities, to edit them later, you would be wise to enter the pattern first as a block and then, after you are sure the pattern will be drawn correctly, erase the block pattern and enter it again with an asterisk before the pattern name.

You can select an entire block to be hatched simply by pointing to any part of the block in response to the prompt to select objects. All of the hatching options will be applied to the block's internal structure.

Drawing a closed polyline is a good method to define the boundary of an area that will later be hatched.

AutoCAD determines the starting point for a hatch pattern from the snap base point. This point is normally the 0,0 point, but you can change it using the SNAP command's Rotate option, the SETVAR command, or AutoLISP to change the SNAPBASE system variable.

If you repeat the HATCH command immediately by pressing Return at the next command prompt, AutoCAD skips the pattern, scale, and angle prompts, assuming that you mean to use the same hatch pattern again. You are simply prompted to select objects. If you need to specify a new hatch pattern, enter the HATCH command explicitly at the command prompt.

If you are using AutoCAD release 9 (with the standard menu ACAD.MNU), you can activate the HATCH command by selecting the Hatch selection from the Draw pull-down menu. (See Figure 57.) Making this selection causes a series of icon menus to be displayed. You can select the desired hatch pattern visually from these icon menus. Selecting the hatch pattern from the icon menu activates the HATCH command using the selected pattern. To reselect the previous hatch pattern or to define a pattern on the fly, select the Previous/User item from the first icon menu. You can page through the available hatch patterns by selecting the next item. (This feature is available only with AutoCAD release 9; see Advanced User Interface and Icon Menus).

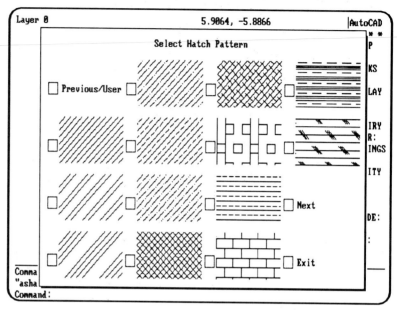

Figure 57

'HELP *or* '?

[all versions]

Overview AutoCAD's HELP command allows you to obtain a summary of information about any command. The HELP command is transparent, that is, you can obtain help even while another command is active. The HELP command can display a listing of every valid command, or you can use it to view individual commands.

Procedure From the command prompt, enter *HELP* or a question mark (?). Pressing Return at this point brings up a list of every AutoCAD command (in case you have forgotten a command name). Entering the name of a command at this point will bring up a help screen about that particular command.

```
Command: HELP (or ?)
Command name (RETURN for list):
```

You can also get help for a particular command while it is active by preceding your call for help with an apostrophe. Say, for example, that you are currently in the middle of drawing lines with the LINE command. Typing 'HELP or '? in response to any prompt will bring up a help screen about the LINE command.

If there is more than one screen of help text for a given command, AutoCAD will tell you to

```
Press RETURN for further help.
```

Examples The following example illustrates the method of obtaining help while within the active command LINE. The screen in Figure 58 is then displayed.

```
Command: LINE
From point: '? (or 'HELP)
```

Tips On a single-screen system, calling for help causes AutoCAD to automatically flip to the text screen. You will have to flip back to the graphics screen to continue viewing your drawing.

You can stop the HELP command at any point by pressing Ctrl-C.

```
The  LINE  command allows you to draw straight lines.

Format:     LINE  From point:  (point)
            To point:  (point)
            To point:  (point)
            To point:  ...RETURN to end line sequence

To erase the latest line segment without exiting the LINE command,
enter "U" when prompted for a "To" point.

You can continue the previous line or arc by responding to the
"From point:" prompt with a space or RETURN.  If you are drawing
a sequence of lines that will become a closed polygon, you can
reply to the "To point" prompt with "C" to draw the last segment
(close the polygon).

Lines may be constrained to horizontal or vertical by the ORTHO command.

See also:   Section 4.1 of the Reference Manual.

Press RETURN to resume LINE command.
```

Figure 58

One particularly handy feature of the HELP command is that the help screens always list the section in the AutoCAD manual where the command can be found. Thus, the help screen serves as a fast index to the AutoCAD manual.

The text displayed by AutoCAD's HELP command is stored in a file called ACAD.HLP. You can edit this file if you want. The format is very simple. Each help screen is preceded by a line of text containing a backslash (\) and the command name. Any lines of text that come after that line are displayed as part of the help screen for that command until the next \COMMAND or the end of the file is encountered. A file named ACAD.HDX is used as an index, enabling you to locate the proper help screen quickly. If you change the ACAD.HLP file, delete the ACAD.HDX file. AutoCAD will create a new ACAD.HDX file the next time you use AutoCAD.

HIDE

[ADE-3] [all versions]

Overview   AutoCAD normally displays three-dimensional views as a "wire-frame" display. That is, the object is seen with all of its lines in view. Surfaces are transparent. The HIDE command causes AutoCAD to eliminate all of the lines that would be hidden behind other objects or surfaces when seen from the current viewpoint. Because the drawing must be mathematically reconstructed, AutoCAD thus regenerates the drawing whenever the HIDE command is activated.

Procedure   There are no prompts for the HIDE command. You simply execute the command by entering it at the command prompt.

Command: **HIDE**

The screen remains blank while AutoCAD counts the hidden lines being removed and then displays

Removing hidden line: *number*

Examples   The following drawings illustrate the difference in appearance of three-dimensional objects before and after a HIDE command.

302

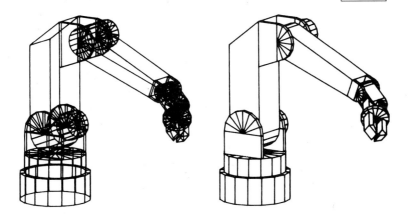

Warnings

Lines will remain hidden only until the drawing is regenerated.

The HIDE command takes a long time. Use it judiciously. Freeze layers containing objects that you really don't need to be concerned with. Zoom in to eliminate unneccesary objects.

The HIDE command considers objects on layers that are turned off. Thus, an object that is invisible because its layer is off may still obscure other objects in the drawing. If you really don't need to consider objects on a particular layer, freeze that layer.

Don't draw more detail than can reasonably be visible at the scale at which the drawing will be plotted. The more detail, the longer it takes to hide.

The HIDE command is not a substitute for removing hidden lines when plotting. If you want AutoCAD to hide lines in your plotted drawings, you must explicitly instruct AutoCAD to do so.

Tips

Circles, solids, traces, and wide polylines are treated as solid objects. Their surfaces hide other objects that are behind them.

Hidden lines can be displayed with a different color rather than hidden completely. If you create a layer with the identical name as an existing layer, except that the name is preceded with *HIDDEN* (for example, WALLS and HIDDENWALLS), any entities on that layer that are hidden by the HIDE command will be displayed on the "hidden" layer. Hidden layers can be any color and can be turned on and off. These hidden layers affect only the display of three-dimensional views. Objects do not get moved from one layer to another, and this procedure has no effect on plotting with hidden lines removed.

If you are using AutoCAD release 9 (with the standard menu ACAD.MNU), you can select the HIDE command from the icon menu displayed by selecting 3DView from the Display pull-down menu (only with AutoCAD release 9; see Advanced User Interface and Entity Selection).

HIGHLIGHT

[ADE-3] [v2.5, v2.6, r9]

Overview   Under normal AutoCAD operation, when you select an object in response to an AutoCAD prompt, the object is highlighted to indicate its selection. The HIGHLIGHT system variable allows you to turn this highlighting on and off.

Procedure   AutoCAD's normal object highlighting occurs when the value of the variable HIGHLIGHT is 1 and is suppressed when the value is 0. No other values are allowed. You can change the variable using the SETVAR command or the AutoLISP functions (getvar) to return the value and (setvar) to change the value.

Examples   The following example illustrates the use of the SETVAR command to turn highlighting off.

```
Command: SETVAR
Variable name or ?: HIGHLIGHT
New value for HIGHLIGHT <1>: 0
```

Tips   The HIGHLIGHT value always is 1 initially (highlighting is on). If you change the value, the new value will be retained only while you are in the current drawing. The current value reverts to 1 when you return to the main menu. The next time you load any drawing, the HIGHLIGHT value will once again be 1.

Icon Menu

[ADE-3] [r9]

Overview   A feature of the Advanced User Interface, icon menus are user-customizable screens that pop up over AutoCAD's graphics screen when called from a menu macro. Icon menus are made up of AutoCAD slide files. Icon menu screens have four, nine, or sixteen slide files. Icon menus allow you to select AutoCAD commands and macros from a graphics list. The icon menu function is an extension of AutoCAD's standard menu facilities.

Procedure   When an icon menu is activated, the normal cross-hair screen cursor is replaced by an arrow cursor. These icons are presented on the menu, each with an associated pick-button box. Moving the screen cursor until the arrow is within one of the button boxes causes the button box to be highlighted and that icon selection to appear with a box around it for confirmation. To select that icon, press the pick button. The only way to exit from an icon menu is to pick something from it.

An icon menu is implemented by providing a menu section with the header ***ICON. Each item in the icon section of the menu is defined in a fashion similar to a screen menu. Each icon submenu within the ***ICON section begins with a submenu name, such as **HATCH1. The first line of the icon menu section is its title. This title is the one displayed on the icon menu screen and it is enclosed in square brackets. Each succeeding line is composed of a slide file or slide library file enclosed in square brackets. The slide file name must be written exactly as it would be in response to the VSLIDE command (see VSLIDE). The remainder of each line consists of a standard AutoCAD menu macro. If the text within the square brackets begins with a blank space, however, the actual text appears as a label in the icon menu rather than a slide file. This is how simple text such as "Next" and "Exit" are provided in icon menus.

Icon menus can be called by other menu commands using the $I= construction, which is similar to the $S= used for calling screen menus. The $I= cannot be entered from the keyboard. Icon menus can contain up to sixteen individual selections (sixteen lines, plus the title line and the menu name). Additional selections are ignored. A blank line indicates the end of an icon menu. In this way, when a "next" option calls a second icon menu, if the second menu has fewer lines, the additional lines from the first menu can remain on the screen.

Icon menus with four or fewer items display four icon boxes. Icon menus with between five and nine items are displayed on a nine-icon screen. Icon menus with ten to sixteen selections are displayed on a sixteen-item screen.

Examples   The standard menu file ACAD.MNU provided with AutoCAD release 9 provides several examples of icon menus. For example, selecting Hatch from the Draw pull-down menu displays a three-page icon menu of all of AutoCAD's standard hatch patterns. Selecting 3D View from the Display pull-down menu pops up an icon menu for selecting a preconfigured viewpoint. Selecting Ashade, 3D Objects, or Fonts from the Options pull-down menu also causes icon menus to be displayed. The Ashade icon menu is shown in Figure 59.

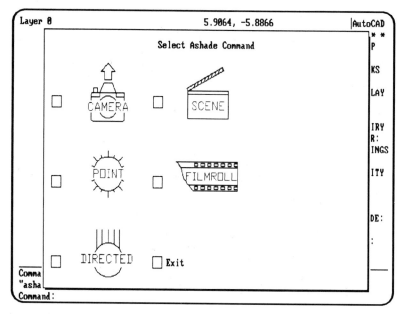

Figure 59

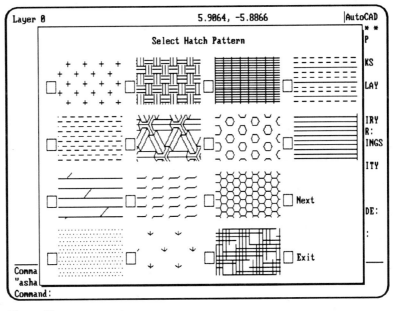

Figure 60

The icon menu in Figure 60 is displayed when you select Hatch from the Draw pull-down menu. The icon menu portion of the actual menu file follows.

PROGRAM	EXPLANATION
***icon	Icon menu section of file
**hatch2	Icon menu name
[Select Hatch Pattern]	Icon menu title
[acad(cross)]^c^chatch cross	First icon
[acad(dash)]^c^chatch dash	Second icon
[acad(dolmit)]^c^chatch dolmit	Third icon
[acad(dots)]^c^chatch dots	Fourth icon
[acad(earth)]^c^chatch earth	Fifth icon
[acad(escher)]^c^chatch escher	Sixth icon
[acad(flex)]^c^chatch flex	Seventh icon
[acad(grass)]^c^chatch grass	Eighth icon
[acad(grate)]^c^chatch grate	Ninth icon
[acad(hex)]^c^chatch hex	Tenth icon
[acad(honey)]^c^chatch honey	Eleventh icon
[acad(hound)]^c^chatch hound	Twelfth icon

```
[acad(insul)]^c^chatch insul        Thirteenth icon
[acad(line)]^c^chatch line          Fourteenth icon
[ Next]$i=hatch3 $i=*               Fifteenth icon; calls next menu screen
[ Exit]^c^c                          Cancels the command
```

Warnings   Icon menus are available only with AutoCAD release 9 when ADE-3 is
               included. In addition, the icon menus can be used only if the graphics card
               you are using supports the Advanced User Interface features. Otherwise,
               any menu calls to the icon menu section of the menu file will be ignored.

               You can exit from icon menus only by selecting something from them.
               All keyboard input, other than using the keyboard arrow keys for moving
               the arrow cursor, are ignored. For this reason, always provide an exit icon
               for canceling an icon menu when you change your mind.

Tips   You can move the arrow cursor within icon menus by using the keyboard
               arrow keys or the pointing device.

               When you are preparing slides for use in icon menus, keep them simple.
               Slides display at redraw speed. Complex slides take longer to display.

ID   [all versions]

Overview   The ID command displays the coordinates of any selected point. By issu-
               ing the ID command, you basically are asking AutoCAD, "What is the
               precise coordinate of this point?" AutoCAD responds by displaying the X-,
               Y-, and Z-coordinates of the selected point.

Procedure   At the command prompt, enter the ID command. AutoCAD prompts you
               to select a point. You can simply use the screen cursor to select the desired
               point. Generally you will use AutoCAD's Object Snap mode to select a
               specific point, since you are looking for a specific coordinate. The coordi-
               nate displayed will be the absolute coordinate, referenced to the draw-
               ing's 0,0,0 coordinate.

               As an alternative, you could tell AutoCAD the absolute or relative coor-
               dinate of a point. In this case, if BLIPMODE is on, the program places a blip
               marker at the specified point. When you redraw the drawing, the blip

disappears. This is a handy way to get your bearings in a drawing, to see exactly where a particular coordinate is.

Examples

```
Command: ID
Point: (select a point)
X = 4.7072      Y = 4.9148      Z = 0.0000
```

The coordinates display in the current Units format.

Tips

If you are pointing to a point in two-dimensional space, the Z-coordinate will display as the current elevation. But if you use object snap to select a three-dimensional object, AutoCAD will display the actual Z-coordinate of the selected point, regardless of the current elevation.

IGESIN

[ADE-3] [v2.5, v2.6, r9]

Overview

The IGES (Initial Graphics Exchange) file format is a standard file format used to transfer graphic computer files between otherwise incompatible systems. The IGES format was developed by a joint effort of the American National Standards Institute (ANSI), the National Bureau of Standards (NBS), and the U.S. Air Force. Version 1.0 of the standard was adopted in January 1980. Originally designed to communicate two- and three-dimensional data, primarily for the manufacturing fields, IGES has undergone several changes over the years. Different computer graphics systems implement different versions and even different features of the IGES standard. Therefore, not all of the data contained in the file on one system may make it through the IGES translation process into another system. Still, IGES is one of the few industry standards.

The IGESIN command converts an IGES file created by another program into an AutoCAD drawing.

Procedure

To use the IGESIN command, select task 1 to create a new drawing from AutoCAD's main menu. Then, before adding any entities to the drawing, execute the IGESIN command.

```
Command: IGESIN
File name: (enter the IGES file name)
```

You can specify any valid IGES file name, including the file extension. You can also include a drive designation and a subdirectory name.

Examples The following example reads in an IGES file named DRAWING.IGS from the \EXCHANGE subdirectory on the D drive.

```
Command: IGESIN
File name: D:\EXCHANGE\DRAWING.IGS
```

Tips If AutoCAD detects something in the IGES file that it cannot deal with (remember, not all features in the IGES file may be compatible with Auto-CAD), an error message is displayed. In any event, the partial drawing already translated into AutoCAD remains.

Even though AutoCAD release 9 now provides for the creation of spline curves, spline curves read into AutoCAD using the IGESIN command still result only in the generation of a regular polyline (as in previous versions; [release 9]).

IGESOUT

[ADE-3] [v2.5, v2.6, r9]

Overview The IGESOUT command creates an IGES (Initial Graphics Exchange) file from the current AutoCAD drawing. AutoCAD can create an IGES-format file from any AutoCAD drawing. The IGES file, once created, can be converted into any other computer graphics program, if an appropriate translation program is available. Since IGES is a nationally adopted format, many programs currently available have some IGES translation capabilities. In addition, many independent software developers have written IGES translators.

Because different programs implement different versions of the IGES standard and because a particular program may not have a feature compatible with every feature in AutoCAD, not every object in your drawing may be converted through the IGES format when brought into the other program.

Procedure The IGESOUT command allows you to generate an IGES file from any AutoCAD drawing at any point in the drawing's creation. To create the IGES file, enter the IGESOUT command at AutoCAD's command prompt.

```
Command: IGESOUT
File name <default>: (enter the file name) (or press Return to accept the default)
```

AutoCAD provides the current file name as the default file name when it creates an IGES file. It adds the file extension .IGS. If you enter a different file name, you can specify a drive designation and a subdirectory name. Do not include a file extension, however, since .IGS is always assumed. When the IGES file has been generated, AutoCAD returns the command prompt.

Examples The following example creates an IGES file of the current drawing (named SAMPLE), saving it to the file DRAWING.IGS in the subdirectory \EXCHANGE on the D drive.

```
Command: IGESOUT
File name <SAMPLE>: D:\EXCHANGE\DRAWING
```

Warnings If a file with the same name as the IGES file name already exists, it will be deleted and replaced by the new file.

Tips If you are using AutoCAD release 9, the IGESOUT command outputs spline curves created using the Spline curve fit option of the PEDIT command. Only the curve itself is output, however, not the frame. (Release 9 only; see PEDIT.)

INSBASE

[all versions]

Overview When you create a drawing, AutoCAD establishes a reference base point for subsequent insertions of that drawing. The 0,0 coordinate point is initially established as the base point. You can use the BASE command to change this reference point coordinate. The value established by the BASE command is stored as the system variable INSBASE (Insertion Base) and is saved with the drawing file.

Procedure You can read the INSBASE system variable using the SETVAR command or the (getvar) AutoLISP function to check the coordinate being used as the base point. You can also change this reference point by changing the coordinate using the SETVAR command or (setvar) AutoLISP function (in addition to using the BASE command).

Examples In the following example, the SETVAR command is used to read the current INSBASE value.

Command: **SETVAR**
Variable name or ?: **INSBASE**
New value for INSBASE <0.0000,0.0000>:

Tips   The initial INSBASE value is determined from the prototype drawing (default value is 0.0000,0.0000). The current value is saved with the drawing file.

INSERT

[all versions]

Options

name	Loads file *name* as a block
name = f	Creates block *name* from file *f*
*name	Retains individual part entities
?	Lists names of defined blocks
C	As reply to X-scale prompt, specifies scale via two points (corner specification of scale)
XYZ	As reply to X-scale prompt, readies INSERT for X-, Y-, and Z-scales [ADE-3]

Overview   The INSERT command inserts a previously defined block or another entire drawing into the current drawing. A copy of the block is inserted with its base point placed at a user-specified insertion point. You can also specify X-, Y-, and even Z-scaling factors (if inserting a three-dimensional object) and an angle at which the block will be rotated from its original orientation. You can enter the scale factors and rotation angle from the keyboard or use the screen cursor to show AutoCAD the scaling and rotation.

Procedure   To insert an object into the current drawing, use the INSERT command.

Command: **INSERT**
Block name (or ?): **(enter name of block)**
Insertion point: **(select insertion point)**
X scale factor <1>/Corner/XYZ: **(enter number or select point)**
Y scale factor (default=X): **(enter number)**
Rotation angle <0>: **(enter number or select point)**

The default X-scale factor is 1. Inserting a block at that scale factor inserts the block at the same size at which it was originally created. The Y-scale factor, by default, is the same as the X-scale factor, either 1 or whatever scale you entered in response to the X-scale prompt. The rotation angle uses the insertion point as the reference point about which the block will be rotated.

You can create a mirror image of the original block by inserting the block with a negative X- and/or Y-scale factor. A negative X-scale factor and a positive Y-scale factor will mirror the object about a vertical line. A positive X-scale factor and a negative Y-scale factor will mirror the object about a horizontal line. A negative X-scale factor and a negative Y-scale factor is the same as rotating the object 180 degrees.

You can specify the X- and Y-scale factors graphically by using the screen cursor to define two corners of a box. The length of the box specifies the X-scale factor, and the height of the box specifies the Y-scale factor. You can use this corner method by entering a point in response to the prompt for the X-scale factor. AutoCAD determines the size of the box by measuring from the insertion point to the point specified. You can also enter a C in response to the prompt. In that case, AutoCAD prompts you for the two corner points making up the box. In either case, if the second point is toward the upper right from the first point or insertion point, AutoCAD inserts the block with a positive scale factor. If the second point is below or to the left, AutoCAD uses a negative scale factor.

The scale factors used for the corner method of indicating the insertion scale are determined by the number of drawing units making up each side of the box. For example, if the block being inserted is a square, drawn initially as 1 unit × 1 unit, and the box specified by the corner method is 3 units × 3 units, the block will be inserted at three times its original size.

Normally, the block inserted is a copy of the original block. It is treated as a single entity, that is, you cannot edit the individual entities that make up the block. Of course, you can always use the EXPLODE command to reduce the block to its individual parts. You can also insert the block as individual entities by preceding the block name with an asterisk (*). In this case, AutoCAD will prompt only for the insertion point, a scale factor, and an insertion angle. You cannot specify individual X- and Y-scaling factors.

```
Command: INSERT
Block name (or ?): *name
Insertion point: (select point)
Scale factor <1>: (enter number)
Rotation angle <0>: (enter rotation angle)
```

If you specify a block name that is not defined within the current drawing, AutoCAD searches the disk in the current directory and any other directories specified in a DOS PATH command. If the program finds a drawing with the same name as the block name specified, it loads that drawing as a block in the current drawing and then inserts that block. Inserting a drawing into the current drawing does not alter the drawing that is being inserted. Once inserted into the current drawing, the drawing does not have to be available on the disk for you to insert the block again. The block is now a part of the current drawing.

You can also tell AutoCAD specifically to look on the disk for the specified block by including a drive designation and/or a subdirectory name with the block name. Or you can instruct the INSERT command to assign a different name to the inserted block by specifying a block name equal to a drawing name. Once again, you can specify a drive designation and subdirectory name.

If you load the drawing that was inserted and alter it, the changes are not automatically incorporated into the drawing in which it was inserted. In fact, if you insert the block again, you actually read the description of the block that was loaded into the current drawing rather than reread the drawing file from the disk. To explicitly read the drawing file or to update the current drawing with any changes that were made to the block/drawing file, respond to the prompt for the block name with the name of the block, an equal sign, and the name of the drawing file (if the block name and the drawing file name are exactly the same, you can include just the equal sign).

```
Command: INSERT
Block name (or ?): block=drawing
```

Release 9 Only   AutoCAD release 9 has several additional options for use with the INSERT command. With release 9, you have the option of presetting the scale and/or rotation angle of a block prior to selecting the insertion point. Once the scale and/or rotation angle is provided, the block can be dragged to its insertion point. The block will be scaled and/or rotated to the values provided by the preset values. These preset values can be made to preempt the normal insertion scale and/or rotation prompts or to apply only while the block is being dragged into place, after which you will be prompted to supply the insertion scale factors and rotation angles normally.

Ten different preset options are available in release 9. In each case, you enter the preset option in response to the prompt for the insertion point. If you respond by entering *Scale*, the INSERT command prompts for a scale factor. This scale factor is applied to the X-, Y-, and Z-axes. The block named in the INSERT command is first scaled by the preset scale factor, and the command then prompts for the insertion point and rotation angle. The block can be dragged into place. No additional prompts for scale factors are issued.

```
Command: INSERT
Block name (or ?): (enter block name)
 Insertion point: Scale
 Scale factor: .5
 Insertion point: (select insertion point)
 Rotation angle <0>: (enter rotation angle)
```

You can provide the X-, Y-, or Z-scale factors individually as preset by entering *Xscale*, *Yscale*, or *Zscale* in response to the prompt for the insertion point. (See the following examples.) In this situation, only the preset scale factor is applied. The block being inserted will be scaled by the factors provided. After selecting the insertion point, the INSERT command prompts for the remaining scale factors that were not preset. This method also allows you to preset differing scale factors for the X-, Y-, and Z-axes (the Scale option presets the same scale factor to all three axes).

```
Command: INSERT
Block name (or ?): (enter block name)
 Insertion point: Xscale
 X scale factor: .5
 Insertion point: (select insertion point)
 Rotation angle <0>: (enter rotation angle)
```

```
Command: INSERT
Block name (or ?): (enter block name)
 Insertion point: Xscale
 X scale factor: .5
 Insertion point: Yscale
 Y scale factor: .75
 Insertion point: (select insertion point)
 Rotation angle <0>: (enter rotation angle)
```

You can preset the rotation angle at which the block will be inserted by entering *Rotate* in response to the prompt for the insertion point. The INSERT command then prompts for the rotation angle. You can enter this angle from the keyboard or specify it by selecting two points. Once the angle is supplied, the command prompts again for the insertion point. The block is shown at the preset rotation angle and can be dragged into place. This is illustrated in the following example.

```
Command: INSERT
Block name (or ?): (enter block name)
 Insertion point: Rotate
 Rotation angle: 45
 Insertion point: (select insertion point)
 X scale factor <1> / Corner / XYZ: (enter X scale factor or Return)
 Y scale factor (default=X): (enter Y scale factor or Return)
```

The preset scale factor and preset rotation angle can be combined, as in the following example.

```
Command: INSERT
Block name (or ?): (enter block name)
 Insertion point: Scale
 Scale factor: .5
 Insertion point: Rotate
 Rotation angle: 45
 Insertion point: (select insertion point)
```

You can make the preset values apply simply while the block is being dragged into place by preceding the preset option with a *P*, for example, PScale, PXscale, PYscale, PZscale, and PRotate. The method (see the following example) and the result while dragging the block into place are identical to those described previously. Once the insertion point has been selected, however, you are prompted to provide the final scale factors and/or rotation angle normally.

```
Command: INSERT
Block name (or ?): (enter block name)
 Insertion point: PXscale
 X scale factor: .5
 Insertion point: PYscale
 Y scale factor: .75
 Insertion point: PRotate
 Rotation angle: 45
 Insertion point: (select insertion point)
 X scale factor <1> / Corner / XYZ: (enter actual X scale factor or Return)
 Y scale factor (default=X): (enter actual Y scale factor or Return)
 Rotation angle <0>: (enter actual rotation angle or Return)
```

These temporary preset methods can be combined with the other preset method described previously. You will be prompted to supply the final value for any temporary preset values as shown in the next example.

```
Command: INSERT
Block name (or ?): (enter block name)
 Insertion point: Scale
 Scale factor: .5
 Insertion point: PRotate
 Rotation angle: 45
 Insertion point: (select insertion point)
 Rotation angle <0>: (enter actual rotation angle)
```

Presets are useful in conjunction with menu macros. The preset values can also be passed as AutoLISP variables.

Examples   You can use the *name = f* feature of the INSERT command to create a composite sheet of several separate details. This is a particularly useful feature because it is easier and faster to edit the individual details and then insert them than it is to draw several details on one drawing. In addition, when you are composing a sheet containing details that need to be plotted at different scales, it becomes confusing trying to decide what the insertion scale factor or plotting scale should be. If you first create the individual details as separate drawings and then insert them onto a composite sheet that will be plotted at a scale factor of 1:1, your task of determining the insertion scale factor is greatly simplified.

In this example, the individual details were first created as separate drawings. They were named DOOR, WHEEL, and RAKE. When they were inserted onto this composite sheet, the blocks were simply called DETAIL-1, DETAIL-2, and so on. The INSERT command looked like this:

```
Command: INSERT
Block name (or ?): DETAIL-1=DOOR
```

A change to the door drawing could be made a part of the composite drawing by entering the same INSERT command again. AutoCAD indicates that the corrections have been made by displaying a message that the block has been redefined. If you don't need to relocate the block or to insert another copy of it, when AutoCAD prompts for the insertion point, press Ctrl-C. You will have to regenerate the drawing before you can see the changes made to the block.

If the block was called DOOR when it was inserted, you could redefine the block by entering the insert command as

```
Command: INSERT
Block name (or ?): DOOR=
```

This is the same thing as saying DOOR = DOOR.

AutoCAD's INSERT command has many uses. Blocks (with associated attributes) representing the bore holes were inserted in the geological map in Appendix D (see Figure D-3). The attributes were later extracted to an analysis program.

By varying the insertion scale factor, the two-dimensional drawing of a small airplane in Appendix D (see Figure D-1) has the appearance of a true perspective three-dimensional drawing. The internal, structural parts of the plane were first saved as blocks.

Warnings

Before you can insert an object into a drawing, you must first create it as a block within the current drawing or as a separate drawing file.

When you are using the corner method or pointing to indicate the scaling factor for the insertion of a block, it is difficult to be sure that the X- and Y-scales are precisely what you intend. Use a specific snap spacing or object snap or disregard this method and enter the scale factors explicitly.

When you are dragging large blocks during insertion, the cursor may move very slowly. If this condition occurs, turn DRAGMODE off before inserting the block.

You cannot explode a block that was inserted with different X- and Y-scale factors.

Inserting blocks using *name has the same effect as exploding a block. Entities drawn on the 0 layer remain on the 0 layer when they are inserted. Entities drawn with color or linetype BYBLOCK will be white and continuous.

Tips

If Ortho mode is on, the rotation angle for the insertion of a block will be forced to a 90-degree angle, unless you specify the angle explicitly.

If DRAGMODE is on, you can visually drag the block to the insertion point, drag the scaling factors, and drag the rotation angle by entering the DRAG command in response to the INSERT command prompts. If DRAG-MODE is set to auto, AutoCAD will always drag the block.

You can get a listing of every defined block already in the drawing by responding to the prompt for the Block name with a question mark (?).

When supplying the insertion point, if you supply a Z-coordinate, the block will be inserted at an elevation equal to the Z-coordinate (only in AutoCAD version 2.6 and release 9 when ADE-3 is included).

If you insert a block containing attributes and the system variable ATTDIA has a value of 0, normal attribute prompts are issued (subject to the value of ATTREQ. If ATTDIA equals 1, a dialog box is displayed for the entry of the attribute information (only with AutoCAD release 9; see Advanced User Interface, ATTDIA, ATTREQ, and DDATTE).

If you insert a block containing attributes and the system variable ATTREQ has a value of 1, normal attribute prompting occurs. If ATTREQ has a value of 0, no attribute prompts are issued. You have to use one of AutoCAD's attribute editing commands to alter the default attribute values (only with AutoCAD release 9; see Advanced User Interface, ATTREQ, ATTDIA, ATTEDIT, and DDATTE).

If you are using AutoCAD release 9 (with the standard menu ACAD.MNU), you can select the INSERT command from the Draw pull-down menu. In this case, the ATTDIA system variable is set to 1. A dialog box is displayed if the block being inserted contains any attributes (only with AutoCAD release 9; see Advanced User Interface and ATTDIA).

The easiest way to prepare a composite drawing made up of several separate drawings is to create each drawing separately and then use the INSERT command to create the composite sheet. Since this can often become confusing when the individual drawings need to be presented at different scales, most users create their composite sheets at full size (1 = 1) and insert the details at varying scales. A scale-factor chart, perfect for this application, is provided in Appendix B.

ISOPLANE

[ADE-2] [all versions]

Options

L	Left plane
R	Right plane
T	Top plane
Blank	Toggles to next plane

Overview   AutoCAD enables you to create isometric drawings, that is, two-dimensional representations of three-dimensional objects. Isometric drawings should not be confused with AutoCAD's three-dimensional drawing mode. In the three-dimensional mode, you are actually describing an object in three-dimensional space. In Isometric mode, you are simply drawing a three-dimensional view on a two-dimensional plane, much the same as you might draw on a piece of paper. AutoCAD's isometric mode always uses three preset planes, which are denoted as left, right, and top, as shown in the following drawing. There is no facility for altering the angle or orientation of these planes.

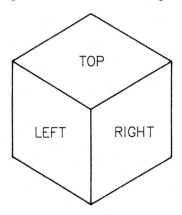

The ISOPLANE command allows you to switch between the three planes of AutoCAD's isometric view. You must first select the Isometric view mode with the SNAP command (see SNAP), choosing the isometric snap style. The ISOPLANE command then enables you to choose the current isometric plane in which to draw. When you first select the isometric snap style, the left plane is set as the current plane. The ISOPLANE command allows you to specifically select a different isometric plane or to simply toggle between the three planes.

Procedure   The ISOPLANE command allows you to indicate three specific settings or to toggle between the three planes. When using the ISOPLANE command, you can tell AutoCAD which plane to draw in. Alternatively, you can simply toggle between each plane, first left, then top, then right, then left again, and so on. Pressing Return in response to the plane/toggle prompt toggles the cursor between the planes. When a particular plane is current, the orientation of the screen cursor changes accordingly.

```
Command: ISOPLANE
Left/Top/Right/<Toggle>:
```

Release 9 Only   If you are using AutoCAD Release 9, all of the functions of the ISOPLANE command are available from within the dialog box presented by the DDRMODES command. In addition, Isometric mode can be turned on and off from within this dialog box (only with AutoCAD release 9; see Advanced User Interface and DDRMODES).

Examples   In the following example, the ISOPLANE command is entered after the Snap mode has been set first to isometric. Return is pressed repeatedly to toggle through the three isometric planes. The following drawing illustrates the appearance of the screen cursor for each isometric plane.

```
Command: ISOPLANE
ISOPLANE Left/Top/Right/<Toggle>: ( RETURN )
Current Isometric plane is: Left
Command: ( RETURN )
ISOPLANE Left/Top/Right/<Toggle>: ( RETURN )
Current Isometric plane is: Top
Command: ( RETURN )
ISOPLANE Left/Top/Right/<Toggle>: ( RETURN )
Current Isometric plane is: Right
```

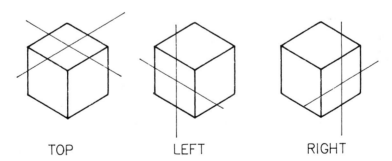

TOP   LEFT   RIGHT

Tips   In addition to the ISOPLANE command, you can toggle between planes transparently (while another command is active) by using the isoplane toggle key (Ctrl-E). This transparent toggle can be included in menus as ^E.

321

The current isometric plane is stored in the system variable SNAPISO-PAIR. You can toggle between isometric planes using the SETVAR command or AutoLISP. The following settings select the current isoplane: 0 = left, 1 = top, 2 = right.

When the snap style is set to isometric, you can draw orthagonal lines within each isometric plane by turning Ortho mode on, even if Snap mode is off.

If AutoCAD's GRID is turned on while you are drawing in the Isometric mode, an isometric grid will display.

While in Isometric mode, you can use the Isometric option of the ELLIPSE command to draw a representation of a circle projected onto any of the three isometric planes.

Last Entity Selection

[all versions]

Overview   During entity selection, AutoCAD remembers the last object added to a drawing. Thus, you can select this last object for editing.

Procedure   You can select the last entity added to a drawing and visible on the screen by responding with *L* (for last) to the prompt to select an object (see Entity Selection).

Examples   Being able to recall the last object added to a drawing is particularly useful when you want to erase the last object because it was drawn incorrectly. For example, if you have just used the HATCH command to cross-hatch an object but find that the pattern is drawn incorrectly, you can simply erase the last object.

Warnings   The last object is actually the most recent object that is visible on the screen.

Tips   If you create your own menu, build a last entity selection into editing commands, especially the ERASE command. You may actually find it advantageous to have a menu selection labeled ERASE Last, consisting of a menu macro that does just that. The macro would be written ERASE L.

Last Point

[all versions]

Overview   AutoCAD remembers the last coordinate point specified in a command. Being able to immediately retrieve and use this coordinate enables you to enter coordinates relative to a previous coordinate point.

Procedure   To use the last point specified within a relative coordinate, precede your coordinate entry with an at symbol @ (see At Symbol).

Examples   If the last point entered was (3,3), preceding your next entry with the at symbol would instruct AutoCAD to place the next point at a coordinate in relation to that (3,3) point. For example, @4,4 would place the cursor at the coordinate point (7,7), which is 4 units to the right and 4 units above the (3,3) coordinate. Entering @3 < 45 would instruct AutoCAD to place the cursor at the coordinate that is 3 units away from the (3,3) point and at a 45-degree angle from it.

Tips   If used in response to the LINE command prompt for the first point, preceding your coordinate with the @ (at) symbol would actually start your line at a relative coordinate from the last specified point. This can be seen in the following example and in Figure 61. The dotted line indicates how the starting point of the new line (7,7) is measured 4 units to the right and 4 units above the last point (3,3). The dotted line would not actually be drawn.

```
Command: LINE
From point: @4,4
To point: @3,0
To point: ( RETURN )
```

LASTANGLE

[v2.5, v2.6, r9]

Overview   The LASTANGLE system variable stores the ending angle of the last arc drawn using the ARC command. It is a read-only variable.

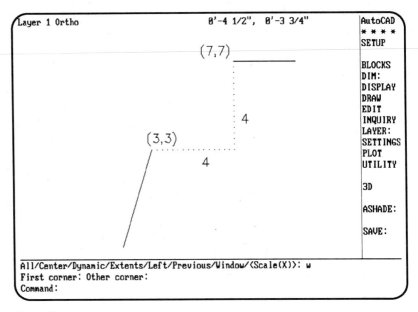

Layer 1 Ortho 0'-4 1/2", 0'-3 3/4" AutoCAD
* * * *
SETUP

BLOCKS
DIM:
DISPLAY
DRAW
EDIT
INQUIRY
LAYER:
SETTINGS
PLOT
UTILITY

3D

ASHADE:

SAVE:

(7,7)

4

(3,3)

4

All/Center/Dynamic/Extents/Left/Previous/Window/(Scale(X)): w
First corner: Other corner:
Command:

Figure 61

Procedure

The LASTANGLE value can be returned using the SETVAR command or (getvar) AutoLISP function. Its only real use is in AutoLISP routines that need to use this angle value in further calculations.

Examples

The first example uses the AutoLISP (getvar) function to return the LASTANGLE value and save it to the variable LA.

```
(setq LA (getvar "LASTANGLE"))
```

The second example uses the SETVAR command to observe the LAST-ANGLE value.

```
Command: SETVAR
Variable name or ?: LASTANGLE
LASTANGLE = 297 (read only)
```

Tips

The LASTANGLE value changes every time an arc is drawn. The current value is discarded at the end of the editing session.

324

LASTPOINT

[v2.5, v2.6, r9]

Overview   The LASTPOINT system variable stores the point coordinate of the last point specified. The value is exactly the same as that referenced by using the symbol @ (see At Symbol and Last Point).

Procedure   The LASTPOINT value can be returned using the SETVAR command. More often it is obtained within AutoLISP routines through the use of the (getvar) function. This allows the AutoLISP routine to retrieve the last coordinate for use in specifying relative coordinates. You can also use AutoLISP to change the LASTPOINT value by using the (setvar) function, although a real use for this feature is more obscure.

Examples   The first example uses the AutoLISP (getvar) function to return the LAST-POINT value and to save it to the variable *LP*.

```
(setq LP (getvar "LASTPOINT"))
```

The second example uses the SETVAR command to observe the LAST-POINT value.

```
Command: SETVAR
Variable name or ?: LASTPOINT
New value for LASTPOINT <7.7994,5.3306>: ( RETURN )
```

Tips   The LASTPOINT value changes every time an object is drawn. The current value is discarded at the end of the editing session.

LASTPT3D

[ADE-3] [v2.6, r9]

Overview   The LASTPT3D system variable stores the X-, Y-, and Z-coordinate points of the last point specified. The Z-coordinate is the current elevation if the last point was a two-dimensional point.

Procedure   The LASTPT3D value can be returned using the SETVAR command. More often it is obtained within AutoLISP routines through the use of the (getvar) function. This allows the AutoLISP routine to retrieve the last coordinate for use in specifying relative coordinates. You can also use AutoLISP to change the LASTPT3D value using the (setvar) function, although a real use for this feature is more obscure.

Examples
The first example uses the AutoLISP (getvar) function to return the LASTPT3D value and to save it to the variable *LP3D*.

```
(setq LP3D (getvar "LASTPT3D"))
```

The second example uses the SETVAR command to observe the LASTPT3D value.

```
Command: SETVAR
Variable name or ?: LASTPT3D
New value for LASTPT3D <8.6076,7.9152,6.0000>:
```

Warnings
LASTPT3D is not available in AutoCAD version 2.5.

Tips
The LASTPT3D value changes every time an object is drawn. The current value is discarded at the end of the editing session.

LAYER

[all versions]

Options

M	Makes a layer the current layer, creating it if necessary
S	Sets the current layer to an existing layer
N	Creates new layers
On	Turns existing layers on
Off	Turns existing layers off
F	Freezes existing layers [ADE-3]
T	Thaws existing layers [ADE-3]
C	Sets specific layers to a color
L	Sets specific layers to a linetype
?	Lists layers and their associated colors, linetypes, and settings

Overview
The LAYER command gives you complete control over the visibility of layers and the color and linetype values associated with them.

326

One of the most basic features of AutoCAD is that you can create layers and then select the current layer on which to draw. AutoCAD allows you to create an infinite number of layers, giving them any name, up to thirty-one characters in length. Think of layers as transparent overlays that are always in perfect registration. At any given time, these layers can possess properties of color and any valid linetype. Each layer can also be on, off, frozen (except the current layer, which can never be frozen), or thawed at any given time.

When you first enter a drawing, AutoCAD always creates a 0 layer, also known as the universal layer. The 0 layer possesses certain special properties involved in the creation and use of blocks (see BLOCK). It can never be deleted or renamed. You can, however, rename other layers with the RENAME command or delete them with the PURGE command (see RENAME and PURGE).

When you start a drawing, the 0 layer has the continuous linetype and the color number 7 (white), but you can change these properties. If you are using AutoCAD's associative dimensioning feature, AutoCAD automatically creates a layer named DEFPOINTS. Depending on changes that you may have made to the prototype drawing, other layers, or different color and linetype settings, may exist on your system. When you edit an existing drawing, the current layer will be the layer that was current when that drawing was last saved.

When you select a current layer on which to draw, color and linetype are initially set to BYLAYER. This means that whatever you draw will be assigned the color and linetype set for that layer. The COLOR command and the LINETYPE command permit you to select specific settings regardless of the current layer (see COLOR and LINETYPE). Only lines, arcs, circles, and polylines are affected by the linetype setting. Text is always drawn with solid lines. Blocks are inserted with whatever linetype properties they were created with.

Any layer can be visible (on) or invisible (off) at any given time. The current layer can even be turned off. Regardless of the visibility setting, objects on these layers are always part of the drawing. They just are not displayed or plotted. Freezing a layer, on the other hand, removes that layer from consideration when the drawing is regenerated. In addition, frozen layers are not included in a WBLOCK. One advantage of freezing layers when they are not needed is that the drawing's regeneration time will improve. A frozen layer cannot be made visible until it is "thawed".

Thus, AutoCAD refers to these two options as the Freeze and Thaw options. While it is possible to turn the current layer off, you cannot freeze the current layer.

Procedure   All layering operations begin with the LAYER command. Whenever you enter the LAYER command, AutoCAD displays all the options (Freeze and Thaw are available only if the ADE-3 package is present). You can select any layering option at this point, answer the appropriate prompts, and then select another layering option. You can also change any layer setting you want while in the LAYER command. None of the settings takes effect until you press Return a final time to return to the command prompt.

```
Command: LAYER
?/Make/Set/New/ON/OFF/Color/Ltype/Freeze/Thaw: (select an option)
```

You can select any of the displayed options. Only those letters capitalized need to be entered to tell AutoCAD which option you have chosen. Thus, entering *F* and pressing Return is sufficient to inform the program that you have chosen to freeze selected layers.

To display a list of layers, select the ? option. AutoCAD then asks you to define which layer names you want listed. The program displays only the layers that match a user-supplied pattern. You can list every layer name that exists in the drawing by pressing RETURN to accept the default of *. The asterisk is a wildcard that tells AutoCAD to display every layer name. The asterisk matches any character from its position to the end of the name. Thus, EL* would match Elevation and Elcctrical but not Entity.

The question mark (?) is also a wildcard. It matches any character that occurs in its position but does not affect those characters that follow it. Thus, ?-DOORS would match 1-DOORS and 2-DOORS but would not match NEW-DOORS. You can use the question mark and the asterisk in combination. In any case, responding to the LAYER command's option prompt with ? allows you to display a list of selected layers, along with their current state (on, off, frozen, or thawed), their color, and their linetype. Colors 1 through 7 have the same meaning on all computer systems, so their names are displayed along with their numbers. Other colors are listed as a number value only. Figure 62 shows a typical screen in response to requesting a layer list.

```
Command: LAYER
?/Make/Set/New/ON/OFF/Color/Ltype/Freeze/Thaw: ?
Layer name(s) for listing <*>: (enter a pattern or press Return to list all)
```

You can create a layer and set it as the current layer in one step by using the Make option. When you make a layer, its initial setting is on, its linetype continuous, and its color 7 (white). You can use one of the other LAYER options to change these settings before drawing on this new layer. If the layer already exists, it is simply made the current layer, the same as if you used the Set option. Basically the Make option combines the steps of the New and the Set options into one step. The current layer is given as the default selection. Pressing Return without entering a selection keeps the current layer default and places AutoCAD back at the LAYER option prompt.

```
Command: LAYER
?/Make/Set/New/ON/OFF/Color/Ltype/Freeze/Thaw: M
New current layer <current layer>: (enter layer name)
```

The Set option is similar to the Make option, except that only a layer that already exists can be selected as the current layer. Pressing Return without entering a selection keeps the current layer default and places AutoCAD back at the LAYER option prompt.

```
     Layer name      State      Color      Linetype
    ------------    --------   ----------  ------------
    0                 On        7 (white)   CONTINUOUS
    1                 On        7 (white)   DOT
    1-WALL            On        3 (green)   CONTINUOUS
    1-DOOR            On        7 (white)   DASHED
    1-WINDOW          On        2 (yellow)  CONTINUOUS

    2-WALL            Off       3 (green)   CONTINUOUS
    2-DOOR            Off       7 (white)   DASHED
    2-WINDOW          Off       2 (yellow)  CONTINUOUS

    Current layer: 1-WALL

    ?/Make/Set/New/ON/OFF/Color/Ltype/Freeze/Thaw:
```

Figure 62

```
Command: LAYER
?/Make/Set/New/ON/OFF/Color/Ltype/Freeze/Thaw: S
New current layer <current layer>: (enter layer name)
```

The New option creates one or more layers without selecting one as the new current layer. It is handy to create several new layers at one time. When creating more than one layer, separate each layer name by a comma, and do not enter a space between layer names. Pressing Return without entering a selection places AutoCAD back at the LAYER option prompt without creating any new layers.

```
Command: LAYER
?/Make/Set/New/ON/OFF/Color/Ltype/Freeze/Thaw: N
New layer name(s): (enter new names)
```

For example,

```
New layer name(s): LAYER1,LAYER2,LAYER3
```

You can turn on any layers that have been turned off by using the On option.

```
Command: LAYER
?/Make/Set/New/ON/OFF/Color/Ltype/Freeze/Thaw: ON
Layer name(s) to turn On: (enter layer names)
```

You can enter a list of layer names or use wildcards to select layers. Frozen layers cannot be turned on until they are thawed. If you use wildcards or specify a layer that is currently frozen, AutoCAD displays a message that a particular layer is frozen.

Any layers can be turned off with the Off option. Again, you can enter specific layer names or use wildcards. In the case of both the On and the Off option, if you press Return without entering any layer names, Auto-CAD returns to the LAYER option prompt without changing the state of any layers.

```
Command: LAYER
?/Make/Set/New/ON/OFF/Color/Ltype/Freeze/Thaw: OFF
Layer name(s) to turn Off: (enter layer names)
```

You can change the color of any layer or layers with the Color option. AutoCAD first asks for the color. You can enter any color number (1 to 255) or color name. AutoCAD then asks for the layer name or names to be set to the selected color. You can enter any layer names specifically, separating the names with commas but no spaces, or you can use wildcards.

```
Command: LAYER
?/Make/Set/New/ON/OFF/Color/Ltype/Freeze/Thaw: C
Color: (enter color name or number)
Layer name(s) for color x <current>: (enter layer names)
```

If you press Return when asked to select the layers to be changed to the selected color, only the current layer will be changed. If you press Return when prompted to select the color, AutoCAD will return to the LAYER option prompt.

You can select the linetype associated with any layer or layers in the same fashion you use to select a color. The Ltype option is used. The linetype name that you enter must be a valid linetype. If the selected linetype has not yet been used in the current drawing, AutoCAD loads it automatically from the ACAD.LIN file. Again, you can provide a single layer name, a list of names separated by commas, or wildcard names.

```
Command: LAYER
?/Make/Set/New/ON/OFF/Color/Ltype/Freeze/Thaw: L
Linetype (or ?) <CONTINUOUS>: (enter linetype name) (or ? for listing)
Layer name(s) for linetype xxx <current>: (enter layer names)
```

Pressing Return to the linetype prompt selects the continuous linetype. Pressing Return to the Layer name prompt selects only the current layer. You can obtain a list of all the linetypes already loaded in the current drawing by entering *?* in response to the linetype prompt. A list of the linetype names and a diagram of each linetype will be displayed on the text screen.

You can use the Freeze option to tell AutoCAD to ignore any entities on the specified layers when regenerating the drawing. This is helpful when a drawing gets large and not all the layers are needed all the time. Auto-CAD asks for the names of the layers to freeze. You can enter any layer names, with the name separated by commas (but no spaces), or you can use wildcards. The current layer can never be frozen. Pressing Return without entering any layer names causes no other action than to place AutoCAD back at the LAYER option prompt.

```
Command: LAYER
?/Make/Set/New/ON/OFF/Color/Ltype/Freeze/Thaw: F
Layer name(s) to Freeze: (enter layer names)
```

Frozen layers must be thawed before they can be displayed or plotted. The Thaw option is provided for this purpose. AutoCAD prompts for the names of the layers you want to thaw. Again, you can enter any name or names, separated by commas, or use wildcards. If you press Return without entering any names, AutoCAD reverts to the LAYER option prompt.

```
Command: LAYER
?/Make/Set/New/ON/OFF/Color/Ltype/Freeze/Thaw: T
Layer name(s) to Thaw: (enter layer names)
```

Release 9 Only   In AutoCAD release 9, the functions of the LAYER command are available from within the Modify Layer dialog box presented by the DDLMODES command. The Entity Creation Modes dialog box, activated by the DDEMODES command, also allows you to select the layer name requestor button, with which you choose a new current layer from a list of existing layer names. These features are available only with AutoCAD release 9. (See Advanced User Interface, DDEMODES, and DDLMODES.)

Examples   This lengthy example uses the LAYER command to create a number of layers, assign colors and linetypes, turn off the layers that begin with the number 2, and establish the layer 1-WALL as the current layer.

```
Command: LAYER
?/Make/Set/New/ON/OFF/Color/Ltype/Freeze/Thaw: N
New layer name(s): 1-WALL,1-DOOR,1-WINDOW,2-WALL,2-DOOR,2-WINDOW
?/Make/Set/New/ON/OFF/Color/Ltype/Freeze/Thaw: C
Color: 3
Layer name(s) for color 3 (green) <0>: ?-WALL
?/Make/Set/New/ON/OFF/Color/Ltype/Freeze/Thaw: C
Color: 2
Layer name(s) for color 2 (yellow) <0>: ?-WINDOW
?/Make/Set/New/ON/OFF/Color/Ltype/Freeze/Thaw: LT
Linetype (or ?) <CONTINUOUS>: DASHED
Layer name(s) for linetype DASHED <0>: ?-DOOR
?/Make/Set/New/ON/OFF/Color/Ltype/Freeze/Thaw: OFF
Layer name(s) to turn Off: 2-*
?/Make/Set/New/ON/OFF/Color/Ltype/Freeze/Thaw: S
New current layer <0>: 1-WALL
?/Make/Set/New/ON/OFF/Color/Ltype/Freeze/Thaw: [RETURN]
```

Careful use of AutoCAD's LAYER command permits one base drawing to be used by several construction trades. In the electrical power and lighting plan in Appendix D (see Figure D-7), the light fixtures, receptacles, ceiling grid, wiring, walls, furniture, and notes were all drawn on separate layers. This ensured precise alignment of each object. The power plan and lighting plan were then created from the same AutoCAD drawing file by selectively turning the proper layers on and off.

Warnings   If the linetype you want to use is a valid linetype but is not one of Auto-CAD's standard linetypes (that is, is user-supplied), you must first load the linetype into the drawing using the LINETYPE command. Otherwise, the LAYER command displays an error message.

If a layer was off when it was frozen, that layer will still be off when it is thawed. To display such a layer, you must turn it on.

Layers that are FROZEN do not display when they are thawed until the drawing is regenerated. Normally, AutoCAD automatically regenerates the drawing whenever you thaw a layer. But if REGENAUTO is off, you must use the REGEN command to make AutoCAD regenerate the drawing. There is, however, an exception to this rule. If REGENAUTO is off and you freeze a layer, that layer will display again when you thaw it, without your having to regenerate the drawing, if you have not regenerated the drawing since freezing that layer.

If you freeze a layer that has blocks inserted onto it, the block also becomes frozen, even if some of the entities of the block were drawn on a different layer that is not frozen. In other words, inserted blocks are referenced to the layer that was current when they were inserted, and that reference layer must be thawed for the block to be displayed. The layer may be turned off, however. In this case, only entities that were drawn on that layer will be invisible. Any other parts of the block drawn on visible layers will still display and plot.

Because all entities are placed on the current layer and it is possible to turn the current layer off, it is equally possible to draw objects that will not be visible until you turn the current layer back on again. If you are drawing but nothing shows up, check to see if the current layer is on or off.

Tips   The special associative dimensioning definition points are placed on the DEFPOINTS layer and display even though the layer is normally off. The definition points do not plot, however, unless the DEFPOINTS layer is turned on.

Normally, selecting a layer to be set to a particular color with the Color option will also turn that layer on. If you prefer to have that layer turned off or remain off, precede the color name with a minus sign (–).

Freezing layers when they are not needed can save a great deal of time when you regenerate a drawing. It also makes AutoCAD respond faster during entity selection and when using the PAN, ZOOM, and VPOINT commands.

LIMCHECK

Overview The LIMITS command lets you turn AutoCAD's limits checking facility on and off. When that facility is enabled, AutoCAD does not allow you to draw outside the drawing's limits (see LIMITS). The current setting of limits checking (on or off) is stored in the LIMCHECK system variable.

Procedure You can enable or disable limit checking outside the LIMITS command by varying the value of the LIMCHECK variable. A value of 1 turns limits checking on, and a value of 0 turns it off. No other values are valid. You can alter the LIMCHECK variable during the operation of another command by using the SETVAR command. Alternatively, you can use an AutoLISP routine to change the value by using the (getvar) and (setvar) functions.

Examples This example uses the 'SETVAR command to turn limits checking off while in the middle of the LINE command.

```
To point: 'SETVAR
>>Variable name or ?: LIMCHECK
>>New value for LIMCHECK <1>: 0
Resuming LINE command.
To point:
```

Tips The initial LIMCHECK value is determined from the prototype drawing (default value is 0, limits checking is off). The current value is saved with the drawing.

LIMITS

Options

Point   Sets lower left drawing limit, prompts for upper right
On   Turns limits checking on
Off   Turns limits checking off

Overview   The limits of a drawing define its boundaries and also control the area of the drawing covered by the visible grid of dots displayed by the GRID command. At one time, AutoCAD could not be made to draw outside the defined limits. To make the drawing bigger, one first had to expand the limits, which proved to be very limiting (pardon the pun). To get around this limitation, it is now possible to turn limits checking off. When limits checking is on, it is still not possible to draw outside the drawing limits. When limits checking is off, the only effects of the LIMITS command are to set the area in which the grid is displayed and to determine how much of the drawing is displayed by the ZOOM—All command.

     The limits of a drawing, whether limits checking is on or off, are still retained as part of a drawing. The limits are defined by absolute coordinates for the lower left and upper right corners of the drawing.

Procedure   The LIMITS command has three options turn limits checking on, turn limits checking off, and select or change the current drawing limits. When limits checking is on, you cannot enter points outside the limits, although some entities, such as circles, can extend beyond the limits. If you attempt to specify a point beyond the limits, AutoCAD responds with an error message that you are outside the limits. If limits checking is off, you can enter points anywhere. The limits are remembered, but no error condition will exist.

     You can change the drawing limits at any time by specifying coordinates for the lower left and the upper right limit corners. AutoCAD prompts first for the lower left corner, displaying the current value. You can enter a new coordinate or press Return to accept the current limit. The display then prompts for the upper right corner. Again, you can enter a new coordinate or press Return to accept the current value.

Examples   This example establishes the drawing's limits as − 10′, − 10′ for the lower left corner and 200′,150′ as the upper right corner.

```
Command: LIMITS
ON/OFF/<Lower left corner> <0'-0",0'-0">: -10',-10'
Upper right corner <1'-0",0'-9">: 200',150'
```

Tips   If you are concerned with drawing outside the "paper" (the area that will fit on your plotted drawing at a given scale), you may still want to turn limits checking on.

You can set the limits to any coordinate, including negative coordinates.

You can control limits checking by altering the system variable LIM-CHECK, using either the SETVAR command or AutoLISP. A LIMCHECK value of 1 means that limits checking is on; a value of 0 turns it off.

The current drawing limits are stored in the system variables LIMMAX, which represents the upper right drawing limit, and LIMMIN, which represents the lower left drawing limit. You can change these coordinates at any time by using either the SETVAR command or AutoLISP.

LIMMAX

[all versions]

Overview   The upper right coordinate set by the LIMITS command as the drawing limit is stored in the system variable LIMMAX (Limits Maximum). The LIMMAX value is a coordinate point.

Procedure   Normally the LIMITS command sets a drawing's limits. But you can use the SETVAR command or the AutoLISP functions (getvar) and (setvar) to alter the drawing's upper right limit by changing the LIMMAX coordinate, even while another AutoCAD command is active.

Examples   In this example, the 'SETVAR command changes the upper right limit of the drawing while the LINE command is active.

```
To point: 'SETVAR
>>Variable name or ?: LIMMAX
>>New value for LIMMAX <12.0000,9.0000>: 100,75
Resuming LINE command.
To point:
```

Tips   The initial LIMMAX value is determined from the prototype drawing (default value is 12.0000,9.0000). The current value is saved with the drawing file.

LIMMIN

Overview   The LIMMIN (Limits Minimum) system variable complements the LIMMAX variable. LIMMIN stores the lower left coordinate set by the LIMITS command as the drawing limit.

Procedure   Normally the LIMITS command sets a drawing's limits. But you can use the SETVAR command or the AutoLISP functions (getvar) and (setvar) to alter the drawing's lower left limit by changing the LIMMIN coordinate, even while another AutoCAD command is active.

Examples   In this example, the 'SETVAR command changes the lower left limit of the drawing while the LINE command is active.

```
To point: 'SETVAR
>>Variable name or ?: LIMMIN
>>New value for LIMMAX <0.0000,0.0000>: -10,-10
Resuming LINE command.
To point:
```

Tips   The initial LIMMIN value is determined from the prototype drawing (default value is 0.0000,0.0000). The current value is saved with the drawing file.

LINE

Options

Blank	As a reply to From point:, starts a new line at the end of the previous line or arc
C	As a reply to To point:, closes a polygon
U	As a reply to To point:, undoes a segment

Overview

A line is one of AutoCAD's basic entities, and the LINE command is probably the most basic of all of AutoCAD's drawing commands. LINE is usually the first command a new user tries. In its simplest form, the LINE command draws a line from point A to point B. It permits you to draw continuous segments of lines (each segment is actually an individual line entity) until you end the command. But the LINE command also allows lines to form a closed polygon, returning to the first point specified when the command was activated. You can draw lines that start out tangent to an arc that was just completed. And you can remove each segment of a line, one at a time, while the LINE command is active.

Procedure

Activate the LINE command by entering LINE at the command prompt. AutoCAD then asks for the first point of a line by prompting From point:. To this prompt, you can enter a coordinate, use the screen cursor to select a point on the screen, select a point relative to a known point (the last point entered by a previous command), or press Return to continue a line from the last line or tangent to the last arc drawn (if the arc was the last entity drawn).

After you make your first selection, the command continues to prompt To point: after you select each successive point, until you either press Return to end the command or enter C to draw a final line segment back to the point originally specified as the starting point. As each line segment is drawn, a rubberband line attaches itself from the last point you selected to the screen cursor, so you can see where the next line will be drawn.

```
Command: LINE
From point: (select point)
To point: (select next point)
To point: (select next point)
  .
  .
  .
To point: ( RETURN ) (to end the command)
```

If you incorrectly enter a line segment, you can remove the segment by entering *U* (for Undo), at the next prompt. The Undo option works in reverse order to that in which the line segments were entered. Each successive *U* removes the previous line segment. It is possible to undo a series of line segement right back to the original point specified at the LINE command's first prompt.

```
Command: LINE
From point: (select point)
To point: (select next point)
To point: U (removes the previous segment)
```

Examples   To create a closed polygon, you do not have to specify the starting point as the final point when AutoCAD displays the `To point:` prompt. Simply enter C (for Close) at the `To point:` prompt. The LINE command will end, closing the polygon automatically.

```
Command: LINE
From point: (select point A)
To point: (select point B)
To point: (select point C)
To point: C (closes the following polygon back to point A)
```

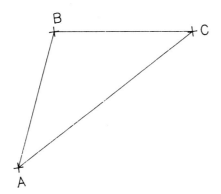

If you press Return in response to the `From point:` prompt, AutoCAD starts the line segment from the end of the most recently drawn line or arc. If the new line segment starts from the end of an arc (this happens if the arc was the most recently drawn entity), the line segment is drawn tangent to the arc, and AutoCAD prompts for the length of the line rather than for a coordinate, as illustrated in the following example and drawing.

```
Command: LINE
From point: ( RETURN )
Length of line: (enter value)
```

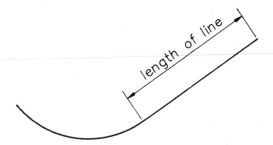

Tips
You can specify the starting point of a line segment as a point relative to the last point drawn by using the at symbol (@) to specify a point. The at symbol saves the last point entered in response to a drawing command. The coordinate @2,3 tells AutoCAD to draw from a point two units to the right and three units above the last point entered.

In AutoCAD release 9 (with the standard menu ACAD.MNU), you can select the LINE command from the pull-down menu under the Draw menu bar. The menu item repetition of release 9 causes the command to repeat automatically until you cancel it by pressing Ctrl-C or selecting another command from the menu (only with AutoCAD release 9; see Advanced User Interface and Repeating Commands).

LINETYPE
[all versions]

Options

?   Lists all linetypes in a linetype library file
C   Creates a linetype definition
L   Loads a linetype definition
S   Sets the current entity linetype
      name   Sets entity linetype name
      BYBLOCK   Sets floating entity linetype
      BYLAYER   Uses the current layer's linetype for entities
      ?   Lists loaded linetypes

Overview
The LINETYPE command allows you to select the linetype to be used for subsequently drawn entities. Normally the linetype is set BYLAYER, meaning entities are drawn with whatever linetype was specified for the current layer. The LINETYPE command permits you to override this by specifying a particular linetype or a floating linetype (BYBLOCK). Whichever method

you select, the current linetype affects all subsequently drawn lines, arcs, circles, and polylines (text is always drawn with solid lines) until you choose a different linetype.

AutoCAD has eight predefined linetypes (in addition to the default continuous linetype), the definitions of which are stored in the ACAD.LIN library file. You can display the names of these linetypes along with a description of each by using the LINETYPE command. The LINETYPE command also allows you to create new linetypes or to load linetypes from other linetype library files.

Procedure   Activate the LINETYPE command by entering the command name in response to AutoCAD's command prompt. You then have numerous options. Typing a question mark ? and pressing Return will display a list of available linetypes in any user-specified file. You can create linetypes by typing C, for create. You can load linetypes from an existing linetype library file using the Load option. Finally, you can override the linetype assigned to the current layer by using the Set option.

Each option is described fully in the following examples.

Release 9 Only   In AutoCAD release 9, you can select a current linetype from the Select Linetype menu, which is activated when you select the linetype requestor button from the Entity Creation Modes dialog box activated by the DDEMODES command. (See Advanced User Interface and DDEMODES.)

Examples   **Listing Linetypes in a Library**   To view a list of linetypes available for loading, enter the LINETYPE command and select the ? option. AutoCAD asks for the name of the linetype library file you want to view. A list of linetype names and a graphic description of each are displayed on the text screen.

```
Command: LINETYPE
?/Create/Load/Set: ?
File to list <default>: (enter file name) (or press Return)
```

To view the default file, simply press Return. To view a different linetype library file, enter the file name. You can include a drive designation and a subdirectory name. Do not include a file extension name; .LIN is assumed.

In the following example and in Figure 63, all of the linetypes in Auto-CAD's default library file are displayed.

341

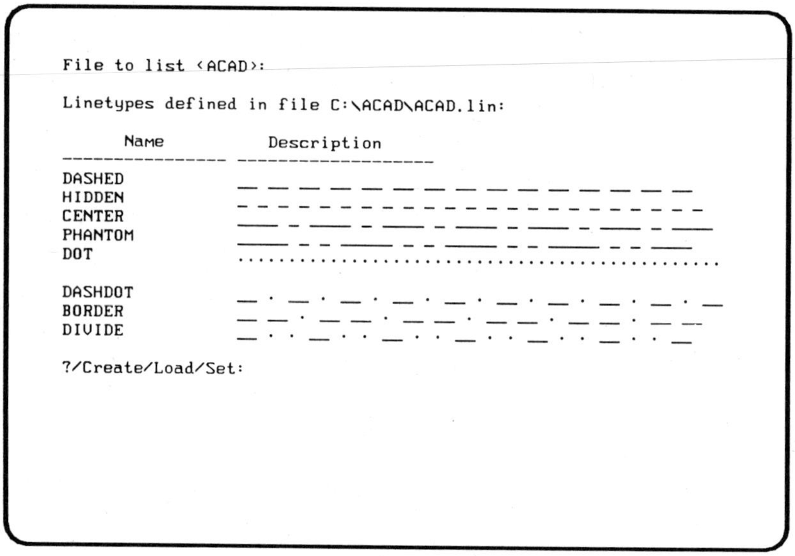

```
File to list <ACAD>:

Linetypes defined in file C:\ACAD\ACAD.lin:

        Name              Description
-----------------    ---------------------
DASHED               — — — — — — — — — — — — — —
HIDDEN               - - - - - - - - - - - - - - - - - -
CENTER               —— - —— - —— - —— - —— - —— -
PHANTOM              —— - - —— - - —— - - —— - - ——
DOT                  .................................

DASHDOT              — · — · — · — · — · — · — · — ·
BORDER               — — · — — · — — · — — · — — ·
DIVIDE               — · · — · · — · · — · · — · · —

?/Create/Load/Set:
```

Figure 63

Command: **LINETYPE**
?/Create/Load/Set: **?**
File to list <ACAD>: (RETURN)

Creating Linetypes   To create a new linetype and store it in a library file, use the Create option. AutoCAD prompts for the name of the linetype you want to create and for the name of the file where the linetype will be stored, with the current linetype library file given as the default. If you enter a different library file name, do not include a file extension. The extension .LIN is assumed. You can also provide a drive designation and/ or subdirectory name, if you want.

Command: **LINETYPE**
?/Create/Load/Set: **C**
Name of linetype to create: **(enter linetype name)**
File for storage of linetype <*default*>: **(enter file name)**

If a linetype with the same name already exists in the specified file, Auto-CAD asks if you want to redefine it. In this case, the current definition of the linetype is also displayed.

```
Name already exists in this file.  Current definition is:
  *linetype-name [, description]
  alignment, dash-1,dash-2,...
Overwrite (Y/N)<n>?:
```

After you answer these prompts, AutoCAD asks for descriptive text and the linetype pattern.

```
Descriptive text: (enter descriptive text)
Enter pattern (on next line):
A,(enter pattern)
```

All linetypes are stored on disk in a linetype library file. You can create new linetypes using the LINETYPE command's Create option or simply by using a text editor to add to or alter the linetype library file. The descriptive text is nothing more than a series of periods and underscores in the approximate pattern of the linetype. This is the graphic description that is provided whenever you use the LINETYPE ? option or the LAYER Ltype ? option to display linetype names.

The linetype pattern follows on the next line, after the letter A (which is an alignment field and is the only type of alignment currently permitted by AutoCAD). The linetype pattern is simply a description of pen-down and pen-up sequences. A positive number indicates a pen-down condition, a negative number a pen-up; a zero indicates a dot. The values are separated by commas with no intervening spaces. The values for these numbers are in drawing units. The actual length of the pen-up and pen-down sequences depends on these values being multiplied by the LTSCALE factor (see LTSCALE). The A alignment specification simply tells AutoCAD that the linetype must begin and end with a dash.

The following example is of a linetype library file description of Auto-CAD's standard dot and dashdot linetypes:

```
*DOT,.............................................
A,0,-.25
*DASHDOT,__ . __ . __ . __ . __ . __ . __ . __
A,.5,-.25,0,-.25
```

Each linetype definition in the library file is preceded by an asterisk. Next comes the name that AutoCAD uses to define the linetype, followed by a comma, then the descriptive text. The next line, the actual linetype definition, starts with the alignment A, followed by a comma, then the pen-up and pen-down description. The dot linetype is simply a series of dots (the zero)

343

with a .25-drawing-unit space between each ($-$ .25, since negative numbers mean a pen-up condition). The dashdot linetype consists of pen down for .5 drawing unit, pen up for .25 unit, a dot, then pen up for .25 units again.

Loading a Linetype   Before you can load a linetype, its definition must already exist in a linetype library file. If the linetype is one of AutoCAD's standard linetypes, you do not have to load it. AutoCAD does this automatically when needed. To load a linetype from a different library file, use the LINETYPE Load option. AutoCAD asks first for the name of the linetype and then for the name of the library file in which to search for it. Do not include a file extension. The file extension .LIN is assumed for all linetype library files. You can also include a drive designation and/or a subdirectory name.

```
Command: LINETYPE
?/Create/Load/Set: L
Name of linetype to load: (enter linetype name)
File to search <default>: (enter the library file name)
```

If the linetype you specify is not found in the library file, AutoCAD displays an error message and cancels the Load option. If the linetype was previously loaded, it is possible that you have changed the definition in the library file since it was loaded. For this reason, AutoCAD prompts you and provides you with the option of not reloading the linetype.

```
Linetype was loaded before.  Reload it? <Y>: (enter Y or N) (or press Return)
```

If the linetype has been redefined and you choose to reload it, all entities drawn using the previous linetype description will be changed to the new description the next time the drawing is regenerated. (If REGENAUTO is on, AutoCAD immediately regenerates the drawing.)

Setting a Current Linetype   The Set option allows you to set a new linetype as the current linetype. Every arc, circle, line, or polyline entity subsequently drawn will be drawn with the current linetype regardless of the linetype of the current layer.

```
Command: LINETYPE
?/Create/Load/Set: S
New entity linetype (or ?) <current>:
```

Initially the current entity linetype is BYLAYER, which causes entities to adopt the linetype of the current layer. You can always return to this setting by entering BYLAYER to the entity linetype prompt. If you supply a different name in response to the prompt, every subsequent entity will be drawn with that linetype. If the linetype name you specify is not yet loaded (or not in the ACAD.LIN library file), AutoCAD responds with the error message

```
Linetype (name) not found.  Use the LOAD option to load it.
```

You can also specify the linetype BYBLOCK. This causes entities drawn with this linetype to appear with the continuous linetype until they are grouped into a block. Later, when the block is inserted into the drawing, the entities drawn with the BYBLOCK linetype will adopt whatever linetype is set at the time of the block's insertion.

One final choice available with the LINETYPE Set option is to list the linetypes currently loaded, using the ?. This produces a list on the text screen identical to the list provided by the LAYER command's Ltype ? option.

Warnings The ability to define and set linetypes within a drawing should not be confused with the ability of some plotters to provide linetypes assigned to pen colors at the time of plotting. Mixing linetypes other than continuous with plotter linetypes will produce unpredictable results.

If you redefine a linetype within a library file, the new definition replaces the old one. If you are experimenting, you should maintain a backup copy of the original linetype library file.

Mixing linetypes on the same layer can become confusing. Blocks with entities drawn with different linetypes and linetypes defined BYBLOCK may produce confusing results when those blocks are later inserted.

LIST
[all versions]

Overview The LIST command allows you to view all of the database information stored for a selected entity or entities in a drawing. The information displayed depends on the type of entity selected but always contains the layer on which the entity was drawn, its position in the drawing, and its entity type. If printer echoing is turned on, the list is also printed.

If you select a line entity, the LIST command displays the coordinates of its end points, its length, the angle at which it was drawn, and its distance along the X- and Y-axes (delta X and delta Y). If you select a circle, Auto-CAD responds with its center point, radius, circumference, and area. An arc displays the coordinates of its center point, its radius, and its start and end angles. Besides listing the layer on which it was drawn, a text entity displays its style, font, text insertion point, height, text, rotation angle, width scale factor, obliquing angle, and generation mode. Blocks list the block name, insertion point, the scaling factors, rotation angle, and any attributes. If attributes are present, they include their layer, text style, font, insertion point, height, value, tag, rotation angle, width scale factor, obliquing angle, display mode (flags), and generation mode. Similar types of listings are provided for every entity type in the drawing.

Of particular interest is the listing provided for polylines. The LIST command causes a polyline to display the coordinate of each vertex, the polyline width, and arc information for polyline arc segments. Polyline listings may be quite long. At the end of the listing, the LIST command displays an area and a perimeter. If the polyline is a closed polyline, the area is the precise area enclosed by the polyline. If the polyline is open, the LIST command displays the actual length of the drawn portions of the polyline and lists the area as if a straight line were drawn from the ending point of the polyline back to its starting point.

Procedure

You can display a list of any entity or entities with the LIST command. Once activated, the LIST command prompts you to select objects to list. You can select the objects using any valid AutoCAD method (pointing, last, previous, window, and so on).

```
Command: LIST
Select objects: (select objects to list)
```

Long lists may scroll off the text screen. To pause the display, press either Ctrl-S or Shift and Scroll Lock. Pressing any key resumes the listing. To cancel the listing at any point, press Ctrl-C.

Examples

The LIST command is used to display the AutoCAD database information (see Figure 64) about the circle in the illustration that follows.

```
Command: LIST
Select objects: (select point on circle)
Select objects: ( RETURN )
```

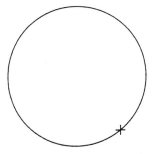

```
        CIRCLE    Layer: 0
center point, X=   4.0000  Y=   4.0000
radius    3.0000
Circumference =  18.8496,  Area =   28.2743
```

Figure 64

Tips

The LIST command is particularly handy for determining the layer an object was drawn on or the height of some existing text. It is also faster than the DISTANCE command for measuring the length of a line.

When doing survey or site plan work, using the LIST command to display a listing of a closed polyline, which represents the boundary of a site, will display the perimeter and the area enclosed by the polyline. From this information, you can easily calculate the area of an irregular site (see AREA).

The LIST command displays the area and the circumference of a circle. It is usually faster to draw a circle and then use the LIST command than it is to calculate the perimeter and area by hand (see AREA).

The system variables AREA and PERIMETER store the values of the last circle or polyline listed. You can access them by using the SETVAR command or AutoLISP.

LOAD

Overview   Shapes are special entities that are defined within a shape definition file. While the inclusion of shapes in a drawing is very simple, the defining of shapes is not. Before a shape can be included in a drawing, its shape definition file must be loaded. The LOAD command provides this facility. After the definition file is loaded, the SHAPE command is used to place a shape from the definition file into the drawing.

Procedure   The LOAD command places a shape definition file into a drawing prior to actually placing a shape entity into the drawing. When you activate the LOAD command, AutoCAD prompts for the name of the shape file to load. You must enter a valid shape file name. You can include a drive designation and/or directory name. Do not include a file extension. The extension .SHX is assumed. If you respond to the LOAD command's prompt with a question mark (?), AutoCAD lists the names of shape files that are currently loaded.

```
Command: LOAD
Name of shape file to load (or ?): (enter shape file name)
```

Examples   The following command sequence loads the shape file ES.SHX from the subdirectory \SHAPES on the D drive.

```
Command: LOAD
Name of shape file to load (or ?): D:\SHAPE\ES
```

Warnings   Before a shape file can be loaded, it must be compiled from a .SHP file (the form in which a shape file is initially created) into a .SHX file (a compiled shape file). You accomplish this by using task 7 from AutoCAD's main menu:

```
7. Compile shape/font description file
```

Tips   Most users never define their own shapes. AutoCAD's blocks provide a much easier means for defining symbols. But although shapes are more difficult to define, they take up much less memory than blocks. For objects such as simple forms (electrical circuit symbols, for example), you may want to create or buy a predesigned shape library file. AutoCAD's text fonts are a similar type of shape file. Here again, rather than design your own, you may want simply to purchase additional text fonts to augment those provided with the AutoCAD program.

The AutoCAD program comes with a sample shape file called PC.SHP.

LTSCALE

[all versions]

Overview   Each linetype definition describes the lengths of dashed or dotted (pen-down) and open (pen-up) sequences. The lengths of these sequences are defined in drawing units (see LINETYPE). The LTSCALE command applies a global multiplication factor to these basic definitions. Thus, if a description of a dashed linetype defines the linetype initially as a sequence of dashed lines and open spaces, each 0.25 drawing unit long, and if the LTSCALE factor is set to 4, then the dashed line will appear in the drawing with dashed lines and open spaces each 1 drawing unit long.

The LTSCALE factor affects every linetype in the drawing globally. Change the LTSCALE and every line is changed.

The current LTSCALE factor is stored in the LTSCALE system variable as a real number.

Procedure   The LTSCALE command simply prompts the user for a new scale factor. Respond with any positive number greater than zero. The current value is always the default. If REGENAUTO is off, the drawing first has to be regenerated before existing lines are affected. If REGENAUTO is on, Auto-CAD automatically regenerates the drawing any time you change the LTSCALE.

```
Command: LTSCALE
New scale factor <default>: (enter value greater than 0)
```

You can change the of the LTSCALE system variable value by using the LTSCALE command, the SETVAR command, or AutoLISP (see LTSCALE).

349

Examples

The following drawing illustrates the change in appearance of a dashed linetype when the LTSCALE factor is changed. Notice that the drawing must be regenerated before the change is visible.

```
Command: LTSCALE
New scale factor <1.0000>: 2
Regenerating drawing.
```

LTSCALE = 2

LTSCALE = 1

The following example illustrates a method for returning the original LTSCALE value, which is saved to a variable called LS. Then, an LTSCALE value of 1000 is set. The original value can be reestablished later by resetting LTSCALE to the LS value.

```
(setq LS (getvar "LTSCALE"))
(setvar "LTSCALE" 1000.0)
```

Tips

Since dashed lines take longer for AutoCAD to draw when regenerating or redrawing the display, you can save time by setting the LTSCALE factor to a very large number. If the LTSCALE is too large to display one series of dashes and open spaces, AutoCAD displays a given line as a solid line. In this way, the speed with which the drawing is displayed will be greatly increased. When you are ready to plot your drawing, reset LTSCALE to the correct setting and regenerate the drawing just before you plot.

The initial LTSCALE value is determined from the prototype drawing (default value is 1.0000). The current value is saved with the drawing.

LUNITS

[ADE-1] [all versions]

Overview

The type of units you select using the UNITS command is stored as an integer value in the LUNITS (Linear Units) system variable. Values of 1 to 4 are valid with the following meanings:

VALUE	MEANING
1	Scientific units
2	Decimal units
3	Engineering units
4	Architectural units

Procedure   Normally you change unit types by using the UNITS command. But you can use the SETVAR command or the AutoLISP functions (getvar) and (setvar) to change the type of units in effect in a drawing. You can actually change the linear unit mode using this method even while another command is active.

Examples   This example uses the 'SETVAR command to change the linear unit type while the LINE command is active. The method used to display linear units on AutoCAD's status line (at the top of the screen) is immediately altered.

```
To point: 'SETVAR
>>Variable name or ?: LUNITS
>>New value for LUNITS <2>: 4
Resuming LINE command.
To point:
```

Tips   The initial LUNITS value is determined from the prototype drawing (default value is 2, decimal units). The current value is saved with the drawing.

LUPREC

[ADE-1] [all versions]

Overview   The number of decimal places or the denominator of a fractional representation of linear dimension is stored in the LUPREC (Linear Units Places Record) system variable. LUPREC can be any integer.

Procedure   The UNITS command controls the LUPREC value. You can also change the number of places to the right of the decimal place or the denominator of the smallest fraction to be displayed by using the SETVAR command or AutoLISP's (getvar) and (setvar) functions to change the LUPREC value.

Examples   This example uses the 'SETVAR command to change the number of displayed decimal places while the LINE command is active. The number of decimal places used to display linear units on AutoCAD's status line (at the top of the screen) is immediately altered.

```
To point: 'SETVAR
>>Variable name or ?: LUPREC
>>New value for LUNITS <4>: 2
Resuming LINE command.
To point:
```

Tips   The initial LUPREC value is determined from the prototype drawing (default value is four decimal places). The current value is saved with the drawing.

MEASURE   [ADE-3] [v2.5, v2.6, r9]

Overview   The MEASURE command visually measures an arc, circle, line, or polyline into segments of a specified length. Markers are placed along the selected object at the measured points. The markers can be AutoCAD points (the default condition) or any predefined block. The selected object is not altered in any way. The markers are simply added to the drawing.

Procedure   Enter the MEASURE command and select the object by pointing. You may not use window, last, or crossing to select the object to measure. AutoCAD then prompts for the segment length that you want the object to be measured in or for the name of a block to use as a marker. You can use the screen cursor to select two points, the distance between which AutoCAD will use as the measured distance, or you can simply enter the distance.

```
Command: MEASURE
Select object to measure: (select object)
<Segment length>/Block: (select distance)
```

AutoCAD measures the object selected, starting at the end point closest to the point at which you selected the object. Point entities are placed at the measured points. You can later use the object snap NODE to select these points.

If you enter *B*, for block, AutoCAD prompts for the name of the block that you want to use as a marker. The block that you name must have already been defined in the drawing. Next, AutoCAD asks whether you want each insertion of the block to be aligned with the object. If you answer yes, each insertion of the block will be aligned with its X-axis tangent to the object being measured. If you respond no, each insertion of the block will be aligned as originally defined. Finally, AutoCAD asks for the length of the segments that you want the object measured in.

```
Command: MEASURE
Select object to measure: (select object)
<Segment length>/Block: B
Block name to insert: (enter name of existing block)
Align block with object? <Y>: (enter Y or N or press Return)
Segment length: (select distance)
```

AutoCAD places the block at the measured points. Each occurrence of the block is treated as an individual insertion; later you can use object snap INSERT to select the measured points.

Examples   The MEASURE command has many powerful uses. You will probably be able to come up with many more. One use is to place station points along the center line of a road. It is general practice in roadway design to measure the length of a road by placing markers, or station points, every 100 feet along the road's centerline. If the centerline is drawn using a polyline, the MEASURE command can insert a block representing these station points. Simply select the polyline by picking a point near the end point from which you want to begin measuring and then specify a measure segment distance of 100 feet. AutoCAD does the rest, placing either a point or a block marker every 100 feet. When the roadway is curved, this is the only precise method available within AutoCAD to measure distances along that curve.

An example of this application follows. A block named STAT, which is used as the station point marker, was saved previously.

```
Command: MEASURE
Select object to measure: (select point A near end of polyline)
```

```
<Segment length>/Block: B
Block name to insert: STAT
Align block with object? <Y>: N
Segment length: 100'
```

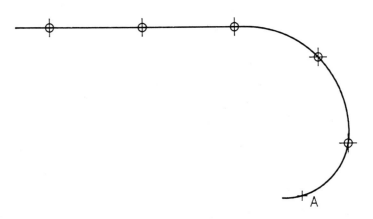

Another useful application of the MEASURE command is to quickly lay out parking spaces in a parking lot. Typically, parking spaces are denoted by a series of lines, 18 feet long and 9 feet apart. This is easy to do when the parking lot layout is rectilinear. But what about when the curbs of the lot are curved? You might calculate the angle between each stripe around the curve by drawing two of them. Then you would use the ARRAY—Polar command. But if the curb is drawn as a polyline and the parking lot stripe is saved as a block, the MEASURE command can place the stripe block, aligned with the polyline curb, every 9 feet.

This example is illustrated in the following drawing. Prior to the command sequence, two blocks were created, named STRIPE1 and STRIPE2, that consist of a vertical line, 18'-0" long. STRIPE1 saved the line with the lower end point as the insertion point, and STRIPE2 saved it with the upper end point as the insertion point.

```
Command: MEASURE
Select object to measure: (select point A on polyline)
<Segment length>/Block: B
Block name to insert: STRIPE1
Align block with object? <Y> ( RETURN )
Segment length: 9'
```

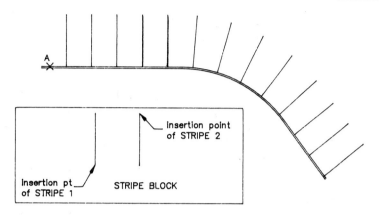

insertion point
of STRIPE 2

Insertion pt
of STRIPE 1

STRIPE BLOCK

A

Warnings

If the object you select to measure is not an arc, a circle, a line, or a polyline, AutoCAD responds:

```
Cannot measure that entity.
```

and the MEASURE command will terminate.

If you select to insert a block at the measured points, the block must already exist in the drawing. The MEASURE command will not load a block from a disk file.

If you use simple points as markers, you may not be able to see these points in the drawing. AutoCAD, however, has placed the points into the drawing.

The MEASURE command, when used with spline-curved polylines, "sees" only the actual curve, not the frame (see PEDIT; release 9 only).

Tips

When the points used for the MEASURE command are not readily visible in the drawing, you can change the way points are displayed by using the SETVAR command to alter the style of the displayed points (PDMODE) and/or the size at which they are displayed (PDSIZE). Remember, however, that when you plot, AutoCAD plots points exactly the way they are displayed on the screen. So unless you want points plotted as other than simple dots, change the PDMODE and PDSIZE back to zero.

It is often useful to define a special block, nothing more than two lines in an X, that you use with the MEASURE command. Then when you use the MEASURE command, you can specify your special block, possibly called X, as the marker to denote the measured points.

Any points or blocks you place into the drawing with the MEASURE command can be removed from the drawing with ERASE previous (until you select a new object in response to an AutoCAD command).

The MEASURE command is similar to the DIVIDE command, except that DIVIDE marks an object in the specified number of equal parts and MEASURE marks the object in segments of the specified length.

MENU   [all versions]

Overview   The MENU command loads a menu file into the menu area. One of Auto-CAD's most powerful features is its open architecture. A particular aspect of this open architecture is that you can create and load different menus of AutoCAD commands. These menus allow you to select commands from the screen menu area, a tablet menu, the buttons on your cursor, or all three. You can even add an auxiliary function-box menu similar to those available on many mainframe CAD systems. Menus are relatively simple to create, yet they can be made to custom-tailor and automate virtually any application.

Whenever you begin a new drawing, the prototype drawing determines which menu will be used. At any time during the creation and editing of a drawing, however, you can load a different menu using the MENU command.

Procedure   Activate the MENU command in response to the command prompt. Auto-CAD prompts for the name of the new menu file that you want to load. The current menu is listed as the default. The new menu file must exist or AutoCAD will respond with an error message and prompt you again. You can include a drive designation and/or a directory name. Do not include a file extension. The extension .MNX is assumed. The actual menu text file has the extension .MNU, but AutoCAD compiles the menu into an MNX file type whenever a menu is first loaded or after it has been edited.

Command: **MENU**
Menu file name or . for none <*current*>: (**menu file name**)

Examples   This example loads a menu file named MYMENU, which is stored in the subdirectory \CUSTOM on drive D.

```
Command: MENU
Menu file name or . for none <current>: D:\CUSTOM\MYMENU
```

Tips　　　If you want to disable the menu feature, for example, to blank the screen menu area, respond to the prompt for the menu file name with a period (.).

AutoCAD release 9 has the additional menu facilities of a menu bar, pull-down menus, and icon menus, all of which can be user-customized (see Advanced User Interface).

All versions of AutoCAD come complete with a predesigned menu file, which contains every AutoCAD command and the appropriate options. This supplied menu utilizes all the menu features available in a particular version of AutoCAD and illustrates structure of a typical menu file.

Menu Bar

[ADE-3] [r9]

Overview　　The menu bar is a feature of the user-customizable menu area, which appears across the top of the graphics screen. When you move the screen cursor to the top of the screen, if the graphics device supports the Advanced User Interface, the menu bar replaces the status line, as shown in Figure 65. The menu bar is an extension of AutoCAD's standard menu facilities.

Procedure　　You can access the menu bar only by using a pointing device. You cannot use the keyboard arrow keys. To display the menu bar, move the pointing device so that the screen cursor moves into the status line area. The status line area is then replaced by the menu bar. When you move the pointing device back into the graphics area, the status line reappears.

A menu bar appears only if the menu contains sections named ∗∗∗POP*n*, where *n* is a number from 1 to 10. If no such sections exist or if the graphics display device doesn't support the Advanced User Interface features, the menu bar section is ignored. If any of the sections ∗∗∗POP1 through ∗∗∗POP10 are present and supported, the menu bar and pull-down menu feature can be used.

The menu bar is actually made up of the first line of each of up to ten pull-down menus (see Pull-Down Menu). When a menu bar item is highlighted, you access its associated pull-down menu by pressing the pick button.

```
 Tools  Draw  Edit  Display  Modes  Options  File        AutoCAD
                                                         * * * *
                                                         SETUP

                                                         BLOCKS
                                                         DIM:
                                                         DISPLAY
                                                         DRAW
                                                         EDIT
                                                         INQUIRY
                                                         LAYER:
                                                         SETTINGS
                                                         PLOT
                                                         UTILITY

                                                         3D

                                                         ASHADE:

                                                         SAVE:


 Loaded menu D:\ACAD9\ACAD.mnx
 Command:
```

Figure 65

Examples   The second menu bar and pull-down menu from the ACAD.MNU file, the
             Draw menu, follow, with limited annotations (for more information, see
             Pull-Down Menu).

PROGRAM	COMMENT
***POP2	Menu section name
[Draw]	Menu bar title
[Line]*^C^C$S=X $s=line line	
[Arc]*^C^C$S=X $s=poparc arc	
[Circle]*^C^C$S=X $s=popcircl circle	
[Polyline]*^C^C$S=X $s=pline pline	
[Insert]^C^Csetvar attdia 1 $s=insert insert	
[Dtext]*^C^C$S=X $s=Dtext Dtext	
[Hatch]^C^C$i=hatch1 $i=*	

Warnings   The menu bar feature is available only in AutoCAD release 9 when used
             with graphics devices that support the Advanced User Interface. (See
             Advanced User Interface.)

358

If the first character of a menu bar title is a blank space, the title does not appear in the menu bar. AutoCAD left-justifies all menu bar items, which means menu bar items to the right of a blank item shift left. You might find this confusing because a pull-down menu would appear below a seemingly empty space on the menu bar (or since the space on the menu bar is empty, it is not apparent as a selection).

Tips

Each title in the menu bar can be up to fourteen characters in length. But since most graphics screens provide a screen width of only 80 characters, you should limit your menu bar titles to eight characters if you intend to use all ten menu bars and pull-down menus.

The standard menu file ACAD.MNU supplied with AutoCAD release 9 contains a menu bar/pull-down menu section. Study this file before you design your own menu. If you use the ACAD.MNU as a template, be sure you keep a copy of the original.

MENUECHO [v2.5, v2.6, r9]

Overview

The MENUECHO system variable controls the echoing of menu items to the command line and/or to the printer. It is an integer value.

Procedure

You can change the MENUECHO system variable by using the SETVAR command or AutoLISP.

Normally your response when you select from the AutoCAD screen menu appears on the command line. For example, if you select the Object Snap mode MIDPOINT from the screen menu, the word MIDPOINT appears on the command line. But you can change the value of MENUECHO to suppress echoing to the command line, a printer, or both.

Menu echoing is controlled by the sum of the possible values of the MENUECHO variable. If the value of MENUECHO is changed from its default of 0 to a value of 1, menu echoing on the command line is suppressed. The word MIDPOINT would not appear on AutoCAD's command line if you selected it from the screen menu. However, any menu item preceded by ^P still echoes to the command line, even if you have suppressed menu echoing. For example, if the screen menu file actually reads ^PMIDPOINT, the word MIDPOINT would still be echoed to the command line, even if the value of MENUECHO had been changed to 1. A MENU-ECHO value of 2 allows menu items to be echoed to the command line but prevents them from being echoed to a printer if printer echoing is turned on. If the value of the MENUECHO variable is set to 3 (the sum condition of 1 and 2), both command line echoing and printer echoing are suppressed; however, a ^P still allows command line echoing. A value of 4 disables the ^P toggling of command-line echoing entirely.

Examples   The following example illustrates the appearance of the command line when you select the LINE command from a menu. This example is true only when you select the command from a menu, not when you type it in on the command line. In the first part of the example, MENUECHO is set to 0. Notice that the LINE command appears following the command prompt. In the second part of the example, MENUECHO is set to 1. This time the LINE command does not appear after the command prompt, but the LINE command prompt From point: still appears.

```
Command: SETVAR
Variable name or ?: MENUECHO
New value for MENUECHO <0>: ( RETURN )
Command: LINE From point:
Command: SETVAR
Variable name or ? <MENUECHO>: ( RETURN )
New value for MENUECHO <0>: 1
Command:  From point:
```

Warnings   In AutoCAD version 2.5, MENUECHO could have values of only 0, 1, 2, or 3.
     In AutoCAD release 9, the *n* selected, *n* found messages during entity selection and the INSERT command's prompt to enter attribute values do not display when the MENUECHO is 2 and input is coming from a menu item.

Tips   The initial value of MENUECHO is 0 when you first start AutoCAD. The current MENUECHO value is saved only while AutoCAD is running. It is carried over from one drawing to the next within the same editing session but is discarded when you exit AutoCAD. The next time you start the AutoCAD program, the initial value of MENUECHO will once again be 0.

MENUNAME

Overview　　The MENUNAME system variable stores the name of the currently loaded menu file, including the drive designation and path if you entered them. MENUNAME is a read-only string variable.

Procedure　　The MENU command loads a new AutoCAD menu file. You can use the SETVAR command or the (getvar) AutoLISP function to observe the name of the currently loaded menu.

Examples　　This example shows how to use the SETVAR command to return the MENUNAME value.

```
Command: SETVAR
Variable name or ?: MENUNAME
MENUNAME = "acad" (read only)
```

Warnings　　The MENUNAME variable is available only in AutoCAD release 9.

Tips　　The initial value of MENUNAME is determined from the prototype drawing (default value is "acad," the menu supplied with the AutoCAD program). The current value is saved with the drawing.

MINSERT

[v2.5, v2.6, r9]

Options

name	Loads file *name* and forms a rectangular array of the resulting block
name = f	Creates block *name* from file *f* and forms a rectangular array
?	Lists names of defined blocks
C	As reply to X-scale prompt, specifies scale via two points (corner specification of scale)
XYZ	As replay to X-scale prompt, readies MINSERT for X-, Y-, and Z-scales [ADE-3]

361

Overview   The MINSERT command creates multiple insertions of a block in a rectangularly arrayed pattern of rows and columns. The pattern is saved as a single object. The pattern, once created, cannot be exploded.

Procedure   The initial prompts are identical to those of the INSERT command (see INSERT). AutoCAD asks for the name of the block, an insertion point, an X- and a Y-scaling factor, and a rotation angle. You can choose the name for inserting the block by using options similar to those available with the INSERT command. You can select a Block that currently exists in the drawing by simply entering its name; you can enter the name of a block that is actually a separate drawing on the disk; or you can give a block name equal to a different drawing file name. Once you have satisfied those prompts, the MINSERT command issues four prompts very similar to those of the ARRAY command. You are asked for the number of rows, the number of columns, the distance between rows, and the distance between columns. You must enter a column and row count of 1 or more. You can either enter the actual distance between rows and columns or use the screen cursor to show AutoCAD the spacing.

```
Command: MINSERT
Block name (or ?): (enter block name) (or block-name=drawing-file)
Insertion point: (select point)
X Scale factor <1>/Corner/XYZ: (enter selection)
Y Scale factor <default=X>: (enter number)
Rotation angle <0>: (enter number or select point)
Number of rows (---) <1>: (enter number) (1 or more)
Number of columns (|||) <1>: (enter number) (1 or more)
Unit cell or distance between rows (---): (enter distance or point)
Distance between columns: (enter distance)
```

You can enter a C in response to the prompt for the X-scale factor and then use two corner points to specify the scale factors, just as with the INSERT command. You can either enter or show the rotation angle.

Once you have responded to all the prompts, the inserted block will be arrayed in the rectangular pattern specified. If you have entered a rotation angle, the individual blocks will appear at the specified angle, and the entire array will be rotated at the same angle. The result appears the same as if you had created a rectangular array at a standard orientation and then rotated the entire array.

Examples
The MINSERT command is useful for creating a repetitious pattern, such as for a printed circuit board drawing or to represent rows of chairs in an auditorium. Since the entire object is treated as one entity, it takes up very little room in the drawing file. However, because objects inserted with the MINSERT command cannot be edited individually, it is not a flexible command for objects that are subject to individual changes.

The following example and drawing illustrates the use of the MINSERT command to represent rows of chairs.

```
Command: MINSERT
Block name (or ?): CHAIR
Insertion point: (select point A)
X scale factor <1> / Corner / XYZ: ( RETURN )
Y scale factor (default=X): ( RETURN )
Rotation angle <0>: 20
Number of rows (---) <1>: 3
Number of columns (¦¦¦) <1>: 4
Unit cell or distance between rows (---): 3'
Distance between columns (¦¦¦): 2'
```

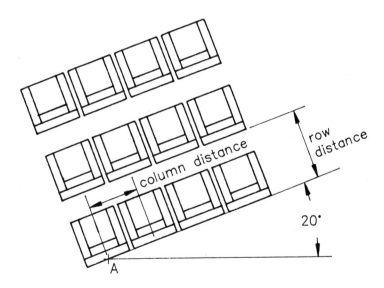

Warnings
Before you can insert an object into a drawing for use with the MINSERT command, it must be created, either as a block within the current drawing or as a separate drawing file.

You cannot explode an object you have inserted using MINSERT. The MINSERT command does not allow a block to be inserted with an asterisk prefix (*blockname*).

The warnings that apply to the INSERT command also apply to MINSERT.

Tips You can get a listing of every defined block already in the drawing by responding to the prompt for a block name with a question mark (?).

Although you cannot explode an object inserted with the MINSERT command, you can redefine the block that the MINSERT command used. All occurrences of the redefined block, including the definition referenced by the MINSERT command, will be changed.

MIRROR

[ADE-2] [all versions]

Overview The MIRROR command makes mirror images of objects that are already part of a drawing. You can retain or delete the original objects. The selected objects are mirrored about an axis that you select.

Procedure The MIRROR command first prompts you to select the objects to be mirrored. You can use any of AutoCAD's methods of selecting objects (window, pointing to individual objects, last, previous selection set, and so on). Objects can be added to or deleted from the selection set as required (see Entity Selection). When you have selected all of the desired objects, press Return. AutoCAD then prompts you to specify the axis about which to mirror the objects. You can enter coordinates or use the screen cursor to point out the line on the screen. Last, you must tell AutoCAD if you want to delete the original objects, keeping only their mirrored version, or let the originals remain. If Ortho mode is on, the objects are mirrored about a horizontal or vertical axis (unless the snap style is isometric).

```
Command: MIRROR
Select objects: (select objects) (press Return when done)
First point of mirror line: (select point)
Second point: (select point)
Delete old objects? <N>: (enter N or Y)
```

Examples

The following example illustrates use of the MIRROR command to complete a drawing quickly. Rather than the entire wide-flange beam being drawn, one quarter of the beam was drawn and the remainder simply added using the MIRROR command.

```
Command: MIRROR
Select objects: W
First corner: (select point A)
Other corner: (select point B)
Select objects: [ RETURN ]
First point of mirror line: <Ortho on> ENDPOINT
of: (select point C)
Second point: (select point D)
Delete old objects? <N> [ RETURN ]

Command: MIRROR
Select objects: W
First corner: (select point E)
Other corner: (select point F)
Select objects: [ RETURN ]
First point of mirror line: ENDPOINT
of: (select point G)
Second point: (select point H)
Delete old objects? <N> [ RETURN ]
```

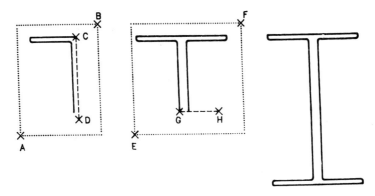

Tips

You can control the mirroring of text and attributes by using the MIRRTEXT system variable, which is accessible with the SETVAR command or AutoLISP. If the value of MIRRTEXT equals 1, text is mirrored along with the objects. If MIRRTEXT equals 0, the text remains right-reading when the objects are mirrored, subject to the two following qualifications:

365

1. Multiple lines of text will remain right-reading, but if you mirror about a horizontal line, the order of the lines will be inverted, as shown in this example:

```
This is the first line!        This is the first line!
The second line of text.       The second line of text.
Who's on third?                Who's on third?
                               --------------------------
                               Who's on third?
                               The second line of text.
                               This is the first line!
```

2. Text that is part of an inserted block will always be mirrored when the block is mirrored.

MIRRTEXT

[ADE-2] [v2.5, v2.6, r9]

Overview   The MIRRTEXT (Mirror Text) system variable controls whether AutoCAD text will be mirrored along with other objects affected by the MIRROR command or if text will remain right-reading (read normally from left to right).

Procedure   You can use the SETVAR command or AutoLISP to change the value of the MIRRTEXT variable. A value of 1 causes text to be mirrored. A value of 0 causes text to maintain its original direction even when mirrored (see MIRROR).

Examples   In the first example, the text is mirrored because the MIRRTEXT value is 1. The text in the second example remains right-reading because the MIRR-TEXT value is 0.

```
Command: MIRROR
Select objects: L
Select objects: ( RETURN )
First point of mirror line: (select point)
Second point: (select point)
Delete old objects? <N> ( RETURN )
```

```
Command: SETVAR
Variable name or ?: MIRRTEXT
New value for MIRRTEXT <1>: 0

Command: MIRROR
Select objects: P
Select objects: ( RETURN )
First point of mirror line: (select point)
Second point: (select point)
Delete old objects? <N> ( RETURN )
```

This is the first line!   ¡ǝuᴉl ʇsɹᴉɟ ǝɥʇ sᴉ sᴉɥꓕ
The second line of text.   .ʇxǝʇ ɟo ǝuᴉl puoɔǝs ǝɥꓕ
Who's on third?   ¿pɹᴉɥʇ uo s,oɥM

MIRRTEXT = 1

This is the first line!   This is the first line!
The second line of text.   The second line of text.
Who's on third?   Who's on third?

MIRRTEXT = 0

Tips   The initial MIRRTEXT value is determined from the prototype drawing (default value is 1, text is mirrored). The current setting is saved along with the drawing file.

MOVE

[all versions]

Overview   The MOVE command moves objects already in the drawing from one location to another. The command does not change the rotation angle or the size of the objects selected.

Procedure   The MOVE command first prompts you to select the objects you want to move. You can use any valid method for selecting the objects, adding or deleting objects from the selection set (see Entity Selection). As you select objects, AutoCAD highlights them. When you have selected all the objects you want to move, press Return. AutoCAD prompts for the base point or displacement. You can enter two points to indicate a "from point" and a "to point" (AutoCAD moves the objects based on the distance and the angle between the two selected points), or you can simply enter a displacement and press Return in response to the prompt for the second point.

```
Command: MOVE
Select objects: (make selection) (press Return when done)
Base point or displacement: (enter first point or x,y distance)
Second point of displacement: (enter second point or press Return)
```

Examples   In this example, the MOVE command is used to move the selected objects to the right 10 units. The objects are selected by window, and the distance is provided as a displacement.

```
Command: MOVE
Select objects: W
First corner: (select point A)
Other corner: (select point B)
Select objects: [ RETURN ]
Base point or displacement: 10,0
Second point of displacement: [ RETURN ]
```

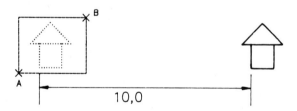

Tips   If DRAGMODE is on or auto, you can use the DRAG feature to visually position the objects while using the MOVE command.

If you make a mistake, you can restore the drawing to its condition prior to the command by entering U (for Undo) at the next command prompt.

If you supply a Z-coordinate when specifying the displacement, you can move an object to a new elevation (only with AutoCAD version 2.6 and release 9 if ADE-3 is included).

In AutoCAD release 9 (with the standard menu ACAD.MNU), you can select the MOVE command from the Edit pull-down menu. In this situation, the Single and Auto selection modes are used. The menu item repetition in release 9 causes the command to repeat automatically until you cancel it by pressing Ctrl-C or selecting another command from the menu (only with AutoCAD release 9; see Advanced User Interface and Entity Selection).

MSLIDE

[ADE-2] [all versions]

Overview   The MSLIDE (Make Slide) command captures an image of the current AutoCAD drawing screen to a slide file. The slide file is an image file, not a drawing. It simply shows what the screen looked like at the instant the slide was created. It cannot be edited. You can view slides by using the VSLIDE command.

Procedure   Use AutoCAD's other commands to create a drawing and to display it the way you want to capture it. When you are ready, enter the MSLIDE command. AutoCAD prompts you to supply the name of the slide file you want to create. The current drawing name is the default, but you can give it any name (up to the DOS limit of eight characters). You can include a drive designation and subdirectory as required, but do not include a file extension name. The extension .SLD is assumed. Only what is actually on the screen at that instant will be captured as part of the slide file.

```
Command: MSLIDE
Slide file <current>: (enter name)
```

Examples   In this example, the current image on the screen is saved as a slide file named SLIDE01. AutoCAD is directed to save it to the directory named \SLIDES on drive D.

```
Command: MSLIDE
Slide file <EXAMPLE>: D:\SLIDES\SLIDE01
```

Warnings   If you supply a slide file name that already exists, the old slide file will be replaced with the new one.

Tips

Use slides to save a drawing that you will need to refer to again later, when you are working on a different drawing or a different view of a drawing. You can view a slide without affecting the current drawing (see VSLIDE).

Slide files are the basis of the images displayed by icon menus in Auto-CAD release 9. To facilitate the management of slide files, release 9 is provided with the utility program SLIDELIB.EXE, which you can use to combine multiple slide files into a slide library (available only in AutoCAD release 9; see Slides, Icon Menus, and Advanced User Interface).

MULTIPLE

[r9]

Overview

The MULTIPLE command is a command modifier that can be combined with other AutoCAD commands to enable them to repeat continuously.

Procedure

Enter the MULTIPLE command in response to AutoCAD's command prompt. The MULTIPLE command itself does not issue a prompt. However, any command entered immediately after the MULTIPLE command will repeat indefinitely until you press Ctrl-C.

Examples

In this example, the MULTIPLE command modifier is entered. No prompt is issued. Then the CIRCLE command is entered. The CIRCLE command then prompts continuously until canceled. The result is shown in the following drawing.

```
Command: MULTIPLE
CIRCLE
3P/2P/TTR/<Center point>: (select point A)
Diameter/<Radius>: (select point B)
CIRCLE 3P/2P/TTR/<Center point>: (select point C)
Diameter/<Radius>: (select point D)
CIRCLE 3P/2P/TTR/<Center point>: (select point E)
Diameter/<Radius>: (select point F)
CIRCLE 3P/2P/TTR/<Center point>: (select point G)
Diameter/<Radius>: (select point H)
CIRCLE 3P/2P/TTR/<Center point>: (Ctrl-C)
```

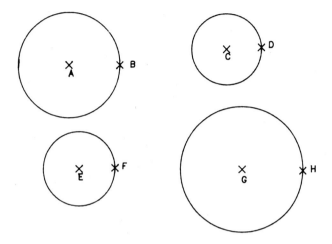

Warnings   The MULTIPLE command is available only with AutoCAD release 9.

You cannot use the MULTIPLE command inside an AutoLISP (command) function.

You cannot use the MULTIPLE command with the PLOT and PRPLOT commands.

Tips   The MULTIPLE command is most useful with drawing, editing, and inquiry commands.

Object Snap

Overview   The first inclination of most users, when they are trying to attach a new line to the end point of an existing line in a drawing, is simply to place the cursor as close as possible to that point and press the pick button. Sometimes the user will even zoom in somewhat on the existing end point and try to start the new line from the existing point. Depending on the scale at which you ultimately plot the drawing, this method may give a suitable appearance. But if you zoomed in very close to the two lines, you would immediately see that you were not successful, as shown in the following drawing. It is virtually impossible to latch on to the end point of an existing object by visually manipulating the cursor. Visually, there may be nothing wrong with this approximate method of attaching one line to another. But mathematically, it is not accurate, and accuracy is one of the basic reasons for using a computer to draw. It is also unreasonable to have to remember the coordinates of objects in the drawing so you can enter the coordinates from the keyboard every time you want to connect one line to another.

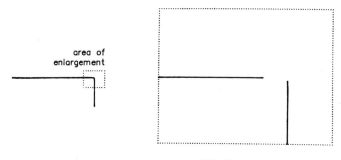

area of
enlargement

ZOOMed view

AutoCAD's Object Snap feature enables you to lock the cursor onto an existing drawing object. With this feature, you can attach to the end point of a line or arc, the center point of a circle or arc, the midpoint of a line or arc, a point, the intersection of any two entities, any of the four quadrants of a circle, the insertion point of an entity, or the nearest point of an object to the cursor. In addition, you can use object snap to draw entities that are tangent or perpendicular to an existing object.

AutoCAD actually has two Object Snap modes. The OSNAP command turns on a running object snap. This running Object Snap mode is initially turned on through the use of the OSNAP command and remains in effect until you turn it off by executing the OSNAP command again.

The second Object Snap mode is turned on for a single selection. This one-time object snap can be used any time AutoCAD prompts for selection of an object or a point. Once you place the aperture box over an entity and make a selection (by pressing the pick button), this one-time object snap is turned off. This method can also be used to override an existing running object snap.

Procedure   During any drawing command, inquiry command, or editing command, you can turn on a one-time object snap simply by entering the type of object snap that you want to use. For example, to start a line from the end point of an existing line, activate AutoCAD's LINE command. When the program prompts you for the "from point," type in *ENDP* (for end point). The drawing cursor immediately changes to include an aperture box. This lets you know that an Object Snap mode has been activated. Move the cursor so that this aperture box is near the end point of the line on to which you want to lock. Press the pick button. AutoCAD attaches your new line to the end point you have just selected. The LINE command continues by prompting you for the "to point." Notice that an aperture box is no longer attached to the cursor. This is because you activated a one-time object snap. That object snap locked the new line onto the end point of the existing line and was then canceled.

If you had pressed the pick button while the aperture was not over an entity that had an end point, AutoCAD would have displayed the error message

`No Endpoint found for specified point.`

You can activate a running object snap by using the OSNAP command (see OSNAP). AutoCAD provides an object snap mode for locking onto most drawing entities using geometric constructions. Each Object Snap mode can be abbreviated down to as few as its first three letters. Descriptions of the Object Snap modes follow.

Center   This mode snaps to the center of a circle or an arc. You must position the aperture over the circle or arc, not over the center point.

Endpoint   This mode snaps to the closest end point of an arc, line, or polyline vertex. You do not need to have the aperture box physically over the end point. AutoCAD automatically snaps to the end point closest to the point at which you picked the entity.

Insert   This mode snaps to the insertion point of a text, block, or shape entity. In the case of a block or a shape, this insertion point is obviously the insertion point that was referenced by the INSERT command or the start point specified with the SHAPE command. For a text entity, the insertion point is the first point referenced by the TEXT or DTEXT command when you placed the text into the drawing. For example, the insertion point for left-justified text is the left start point; for right-justified text, it would be the right-ending point.

Intersection   This mode locks onto the intersection of two drawing entities. These entities can be two lines, a line and an arc or a circle, or two arcs or circles. The intersection object snap can also be used to snap to the corners of a trace or a solid. The intersection object snap does not find the intersection of entities with arcs or circles that are part of a block but does locate the intersection of lines that are part of a block. When using the intersection mode, the intersection must be within the aperture box.

Midpoint   This mode locks onto the midpoint of an arc or a line. Again, it is not necessary to place the aperture directly over the midpoint. If you place the aperture anywhere over the entity, AutoCAD will snap to the midpoint.

Nearest   This mode locks onto the point on an arc, a circle, or a line, or onto the point entity that is closest to the cross-hairs in the middle of the aperture box. However, arcs or circles that are part of blocks are not selected by this Object Snap mode.

Node   This mode locks onto a point entity. The point to which you want to snap must be within the aperture box.

Perpendicular   The perpendicular object snap is used to snap to a point on an arc, a circle, or a line so that a line drawn from the the last point to the referenced entity forms a 90-degree angle. Arcs or circles that are part of blocks are not selected by this Object Snap mode.

Quadrant   This object snap locks onto the nearest visible quadrant of an arc or a circle. The quadrants are the 0-, 90-, 180-, and 270-degree points. If the arc or circle is part of a block that has been rotated, the quadrant points are also rotated.

Tangent   The tangent object snap locks onto the point on a circle or an arc so that a line drawn from the last point to the referenced arc or circle is drawn tangent to the arc or circle. Tangent object snap does not reference a circle or an arc that is part of a block.

You can combine any two or more Object Snap modes by entering the modes, separated by commas. For example, you could have AutoCAD locate the end point or intersection of two objects by entering *ENDP,INT*. In this case, AutoCAD scans the entities in the aperture box and locks onto the end point or intersection of two entities, depending on which condition it finds closest to the cross-hairs.

When selecting a point on an entity by using object snap, AutoCAD scans all of the entities that are located within the aperture box. The point that satisfies the active Object Snap mode that is closest to the cross-hairs within the aperture box is the point selected. Only visible objects are scanned. As your drawing gets larger, this scanning process may get progressively longer. You can force AutoCAD to use a faster method to locate the object snap. The Quick option causes AutoCAD to select the first point it finds within the aperture box that satisfies the active Object Snap mode. This first point is usually the corresponding point on the entity that was most recently added to the drawing. The Quick option must be used in combination with one of the other Object Snap modes. For example, to snap to the first end point that AutoCAD locates within the aperture box, specify the Object Snap mode as *QUI,ENDP*. The Quick mode does not work when you are snapping to an intersection. In that instance, a complete scan of the entities within the aperture box is done whether you have specified a quick snap or not.

Examples   The following drawing and command sequence illustrate some of AutoCAD's object snaps. Not every object snap has been included, but you should be able to see the general method with which they are used. After entering the desired Object Snap mode, press the Return key. AutoCAD then issues the prompt to or of indicating you should now select an object.

```
Command: LINE
From point: ENDPOINT of (select point A)
To point: TANGENT to (select point B)
To point: PERPEND to (select point C)
To point: CENTER of (select point D)
To point: QUADRANT of (select point E)
To point: PERPEND to (select point F)
```

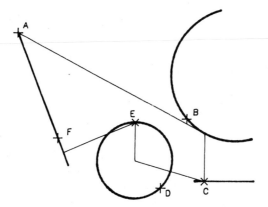

Warnings   Remember that arcs and circles that are part of blocks will not be recognized by nearest, intersection, perpendicular, or tangent object snaps.

When using the Quick Snap mode, AutoCAD snaps to the first point on an entity that satisfies the object snap selected. This sometimes results in the wrong point being selected. The point selected is usually on the entity within the aperture box that was most recently added to the drawing.

When you are selecting objects using an object snap, only visible entities or the visible portions of noncontinuous linetypes will be seen by AutoCAD.

In running Object Snap mode, if no point satisfying the object snap is found, the cross-hair position is selected. In one-time Object Snap mode (or Override mode), if no satisfactory point is found, AutoCAD displays an error message to that effect and usually reissues the preceding prompt.

You can abbreviate object snaps down to their first three letters, but avoid entering *END* for *Endpoint*. Use *ENDP* instead. Many users find themselves out of the drawing editor and back at AutoCAD's main menu when they enter *END* in response to AutoCAD's command prompt, thinking they are entering an object snap while in the middle of a command. No real harm is done (since the END command also saves the drawing before exiting), but the time lost while ending the drawing and then reloading it can be frustrating.

Object snap, when used to lock onto a spline curved polyline, sees only the actual curve, not the frame (see PEDIT; release 9 only).

Tips   You can override a running object snap (previously set using the OSNAP command) by issuing a one-time object snap. You can even turn the Object Snap feature off for a one-time selection by entering *off* or *none* instead of an Object Snap mode.

When you are using the LINE command, you can select the perpendicular or tangent object snap in response to the prompt for the "from point," after which you can point to an existing object. When you next pick a point in response to the prompt for the "to point," AutoCAD constructs a line from that "to point" to the existing object such that the line is perpendicular or tangent to it. The LINE command then continues normally. In this case the rubberband line normally displayed during the LINE command is not drawn.

When using the CIRCLE command's three-point construction method, you can issue tangent and perpendicular object snaps to construct a circle tangent to three other circles or lines.

You can use the APERTURE command to change the size of the aperture box displayed when using the object snap feature.

Most good menu systems, including AutoCAD's standard menu that comes with the program, provide all of the Object Snap modes on screen menus, tablet menus, or both. In particular, look for menu systems that automatically place an object snap menu onto the screen menu area each time a drawing, editing, or inquiry command is activated. As you become more proficient in AutoCAD, you will find yourself using the Object Snap feature more often. In addition, some CAD operators, using multibutton pointing devices, program some of the object snaps to the available buttons. AutoCAD's standard menu causes an object snap menu to be displayed on the screen menu area whenever you click on the three asterisks on the screen menu or press the third button on a multibutton cursor.

When working with AutoCAD's three-dimensional entities (in a three-dimensional view), you can use the endpoint and midpoint object snaps to snap to a 3DLine. In addition, you can use the Intersection object snap to snap to the corners of a 3DFace. The Z-coordinate will be taken from the three-dimensional object. When you are viewing the objects in a two-dimensional view, the current elevation is used rather than the Z-coordinate. (This feature is not available with AutoCAD version 2.5; see 3DLINE and 3DFACE.)

If you are working with AutoCAD in a three-dimensional view and input of a three-dimensional coordinate is accepted by AutoCAD (for example, when you use the 3DLINE or 3DFACE command), you can use object snaps to snap to the top edges of existing extruded objects. (This feature is available only with AutoCAD release 9; see 3DLINE and 3DFACE.)

OFFSET

[ADE-3] [v2.5, v2.6, r9]

Options

number	Specifies offset distance
T	(Through) Allows specification of a point through which the offset curve is to pass

Overview

The OFFSET command draws a new entity at a specified distance from the first or so that the new entity passes through a specified point. The new entity will be parallel to the original. You can use the OFFSET command to construct parallel lines, arcs, and polylines and concentric circles and ellipses.

Procedure

The command first prompts for the distance you want to offset any selected objects. You can enter a distance explicitly or use the screen cursor to indicate or measure a distance within the drawing. If you enter or indicate a distance, you are next prompted to select the objects to offset. Finally, AutoCAD asks you to indicate, by pointing to a side of the selected object, the direction to offset the new entity.

```
Command: OFFSET
Offset distance or Through <last distance>: (enter distance)
Select object: (select object by pointing)
Side to offset: (point to one side or the other)
```

The last distance used by the command is the default. The OFFSET command runs continuously. Once you have specified a distance, you can offset any number of objects, one at a time, until you press Return to end the command.

As an alternative, you can instruct AutoCAD to calculate the offset distance so that the new entity passes through a given point. To do this, respond to the first prompt with *T*, for Through. The command waits until you select the through point. AutoCAD then constructs the new entity so that it passes through the selected point, if possible.

```
Command: OFFSET
Offset distance or Through <last distance>: T
Through point: (select point)
```

Examples   The OFFSET command is useful for drawing double lines to indicate walls. Because the command runs continuously until you cancel it, it is much more useful than the COPY command as long as the distance you want to copy objects doesn't vary. For example, you can use OFFSET for small areas of reflected-ceiling plan grids. It is not as powerful as the ARRAY command when you are doing much repetitive copying, but is easier to use for small copying jobs. It is the only command that automatically constructs concentric circles and parallel curves.

Several drawings created using the OFFSET command follow.

```
Command: OFFSET
Offset distance or Through <last distance>: 4.5
Select object to offset: (select point A)
Side to offset? (select point B)
Select object to offset: (select point C)
Side to offset? (select point D)
Select object to offset: ( RETURN )
```

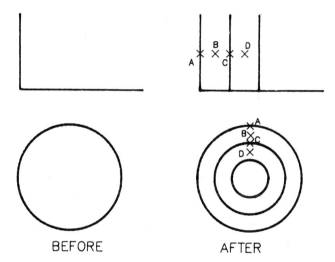

BEFORE   AFTER

Warnings   If you are offsetting a line and the through point would construct a new line directly over the original, AutoCAD displays the message

```
Invalid through point.
```

If you are offsetting a circle or an arc and have specified an offset distance or "through point" that would result in a nonexistent curve (one with a negative radius), AutoCAD displays the message

```
No parallel at that offset.
```

If you try to offset an entity other than an arc, a circle, a line, or a polyline, AutoCAD displays the error message

```
Cannot offset that entity.
```

Tips

If you make a mistake, you can erase individual entities that were created with the OFFSET command. If you want, you can remove all the objects added to the drawing by the OFFSET command by entering *U*, for Undo, at the next command prompt. In one step, this removes every entity just added.

OOPS

[all versions]

Overview

The OOPS command immediately restores the most recent entities removed from a drawing by the ERASE command. OOPS restores only the last entity or group of entities removed by the most recent ERASE command. If you use the ERASE command to remove a group of objects and then use it again to remove a second group of objects, only the second, most recent group can be restored with the OOPS command.

When you use the BLOCK or WBLOCK command to combine a group of entities into a block, AutoCAD initially removes those entities from the drawing. If you want to have the original entities remain in the drawing as individual entities, the OOPS command, used immediately after the BLOCK or WBLOCK command, restores them to the drawing.

Procedure

To restore an erased object or objects, enter the command OOPS.

```
Command: OOPS
```

Examples

You might use the following command sequence to create a block named WINDOW, and still allow the original entities to remain in the drawing.

380

First, the BLOCK command is used with the window object selection method to create the block.

```
Command: BLOCK
Block name (or ?): WINDOW
Insertion base point: MIDPOINT of (select point A)
Select objects: W
First corner: (select point B)
Other corner: (select point C)
Select objects: (RETURN)
```

At this point, the objects are erased from the drawing. To restore the objects to the drawing, the OOPS command is issued.

```
Command: OOPS
```

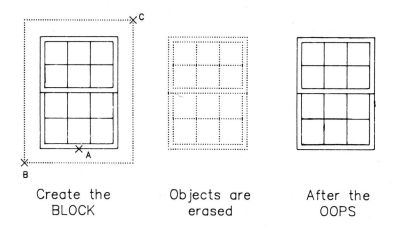

Create the   Objects are   After the
BLOCK   erased   OOPS

Warnings   Each time you use the ERASE command, AutoCAD maintains a list of the objects that were erased. The OOPS command restores that list. If you use the ERASE command again, the first list of erased objects is discarded and the new list replaces it. For this reason, OOPS restores only the most re-cently erased objects. Use the U, or UNDO, command to restore earlier objects. The OOPS and UNDO commands work only while in the current editing session. If you end the drawing and come back to it at another time, previously erased objects are gone forever.

ORTHO

Options

On   Forces lines to horizontal or vertical
Off   Does not constrain lines

Overview   Often you need to draw lines that are always at right angles. The ORTHO command turns AutoCAD's Ortho mode on or off. When the Ortho mode is on, all lines drawn will be at right angles to the current cursor orientation, also called the snap grid orientation. In most cases, this will mean that lines will be drawn at angles of 0, 90, 180, or 270 degrees. It is possible to use the SNAP command to rotate the cursor and the grid orientation. When the grid is rotated and the Ortho mode is turned on, lines are drawn orthogonal to the current grid orientation. When the Ortho mode is off, lines are not forced to any particular orientation.

    There is one special case to remember when you are using AutoCAD's isometric construction facilities. If you are drawing in one of the isometric planes and the Ortho mode is on, lines will be constrained to the equivalent of an orthogonal line within the current isometric plane. (See ISOPLANE.)

Procedure   To turn the Ortho mode on or off, use the ORTHO command.

```
Command: ORTHO
ON/OFF: (enter selection)
```

Release 9 Only   In AutoCAD release 9, the action of the ORTHO command can be duplicated by the Ortho checkbox available within the dialog box presented by the DDRMODES command (only with AutoCAD release 9; see Advanced User Interface and DDRMODES).

Examples   The two drawings that follow illustrate the difference between when the Ortho mode is on and when it is off. The exact same command sequence was used to create both drawings. The drawing on the left was created with the Ortho mode on, the one on the right with the Ortho mode off.

```
Command: LINE
From point: (select point A)
To point: (select point B)
To point: (select point C)
To point: (select point D)
To point: ( RETURN )
```

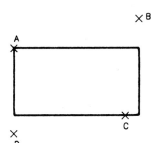
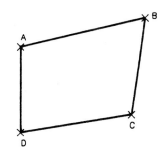

Tips
You can also control the current setting of the Ortho mode by using the Ortho toggle key. On all systems, this is controlled by Ctrl-O. On DOS systems, the F8 key is also an Ortho toggle key. In addition, you can build the Ortho toggle into menus by including ^O at a menu location.

You can control the current setting of the system variable ORTHOMODE by using the SETVAR command or AutoLISP.

ORTHOMODE   [all versions]

Overview
The ORTHOMODE system variable duplicates the ORTHO command and the Ortho toggle. Both the ORTHO command and the Ortho toggle control the ORTHOMODE value. This value can be either 0 or 1.

Procedure
An ORTHOMODE value of 0 indicates that the Ortho mode is off; a value of 1 means that ORTHO is on. The ORTHO command and the Ortho toggle actually change the value from 0 to 1 when they are used to turn ORTHO on. In addition, you can use the SETVAR command and the (setvar) Auto-LISP function to change the ORTHOMODE value, turning ORTHO on or off.

Examples   The following sequence uses AutoLISP to return the current ORTHO-MODE value and save it to the variable OM. Then the current value is set to 0, regardless of the previous setting. This routine could be included in AutoLISP routines that need to establish a particular setting. You could then restore the original setting at the end of the routine by resetting the ORTHOMODE value equal to OM.

PROGRAM	COMMENTS
(setq OM (getvar "ORTHOMODE"))	
(setvar "ORTHOMODE" 0)	
.	The body of the AutoLISP routine
.	would occur here
.	
(setvar "ORTHOMODE" OM)	

Tips   The initial ORTHOMODE value is determined from the prototype drawing (default value is 0, Ortho mode is off). The current value is saved with the drawing.

You can also change the ORTHOMODE value by selecting the Ortho checkbox from within the dialog box displayed when the DRRMODES command is activated (available only with AutoCAD release 9; see Advanced User Interface and DDRMODES).

OSMODE
[ADE-2] [all versions]

Overview   The OSMODE (Object Snap Mode) system variable controls the current object snap mode setting. What the OSNAP command actually does is change the value of the OSMODE system variable.

Procedure   The current Object Snap mode is determined by the integer value of the OSMODE variable. Each Object Snap mode has a specific integer value:

VALUE	MODE
1	Endpoint
2	Midpoint
4	Center
8	Node
16	Quadrant
32	Intersection
64	Insert
128	Perpendicular
256	Nearest
512	Quick

Since multiple Object Snap modes can be active at the same time, AutoCAD determines which snap modes are active by looking at the value as the sum of the active modes. Thus, an OSMODE value of 6 indicates that both the Midpoint and the Center Object Snap modes are active. A value of 161 means that the Perpendicular, Intersection, and Endpoint Object Snap modes are all currently active.

You can use AutoCAD's SETVAR command to observe or change the current OSMODE value. Of greater importance, however, is that you can check the current value while within an AutoLISP routine through the use of the (getvar) function and change the Object Snap mode while in the middle of a routine through the use of the (setvar) function.

Examples The following sequence uses AutoLISP to return the current OSMODE value and save it to the variable OS. A new current value could then be set in the body of the routine. The original setting would be restored at the end of the routine by resetting the OSMODE value equal to OS.

PROGRAM	COMMENTS

```
(setq OS (getvar "OSMODE"))
.
.                           The body of the AutoLISP routine would
.                           occur here
(setvar "OSMODE" OS)
```

385

Tips The initial OSMODE setting is determined from the prototype drawing (default value is 0, no OSNAP). The current value is saved along with the drawing file whenever you save or end the drawing.

OSNAP

Options

CENT	Center of arc or circle
ENDP	Closest endpoint of arc or line
INSERT	Insertion point of text/block/shape
MIDP	Midpoint of arc or line
NEAR	Nearest point of arc/circle/line/point
NODE	Node (point)
PERP	Perpendicular to arc/line/circle
QUAD	Quadrant point of arc or circle
QUICK	Quick mode (first find, not closest)
TANG	Tangent to arc or circle

Overview The OSNAP command turns on a running object snap that stays in effect until you turn it off.

AutoCAD has a virtually limitless magnification ratio. With it you can see that no matter how precisely you draw one line near another, without a way of mathematically ensuring that the new line starts exactly where the old one ends, the two lines won't touch.

The OSNAP command ensures that objects really correspond. By using the OSNAP command, you know that if you start drawing a new line from the end of a previous one, the new line starts precisely at the end of the old one.

Besides being able to lock onto the end point of a line, or an arc for that matter, you can use OSNAP to lock onto the geometric center of an arc or a circle; the insertion point of a block or a line of text; the midpoint of an arc or a line; the nearest point of an object, a node, or a point entity; or the nearest of one of the four quadrants of an arc or a circle. In addition, you can draw entities that are tangent or perpendicular to an existing object by using OSNAP with the Tangent or Perpendicular option.

Procedure

The OSNAP command is similar to AutoCAD's Object snap, except that the OSNAP command turns on a continuous object snap that remains in effect until you turn it off. The same Object Snap options, or modes, are used for every selection. When you enter the OSNAP command, AutoCAD prompts for you to enter the Object Snap modes you desire. You can enter one or more Object Snap modes, separating them with commas. The first three letters of a snap mode are sufficient. (For a complete explanation of the different modes, see Object Snap.)

```
Command: OSNAP
Object snap modes: (enter desired modes)
```

You always know when OSNAP is active. When you activate any command that accepts screen pointing, the cursor has a small box, called an aperture, around it. When you respond to any command by selecting a point, AutoCAD evaluates everything in the aperture box in relation to the current OSNAP mode. The object that matches the current OSNAP mode that is closest to the center of the aperture box (where the cursor crosses) is selected. If no object matches the current OSNAP mode, the actual point is selected.

Examples

In the drawing that follows, the OSNAP command was used to establish a running Quadrant object snap to draw the square inside the circle. In this case, notice that there was no need to enter *QUADRANT* each time the LINE command prompted for a point selection. Using the OSNAP command in this fashion cuts down on the number of keystrokes.

```
Command: OSNAP (activates OSNAP command)
Object snap modes: QUADRANT (selects object snap mode)
Command: LINE
From point: (select point A)
To point: (select point B)
To point: (select point C)
To point: (select point D)
To point: C (closes the line back to start point)
Command: OSNAP (activates OSNAP command again)
Object snap modes: NONE (turns off running object snap mode)
```

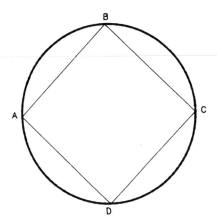

Warnings   If more than one OSNAP mode is active, it is possible that more than one matching condition will be satisfied for a selected point. In this case, the order in which the modes were specified will control.

Tips   You can override any running OSNAP modes by entering a single Object Snap mode in response to any AutoCAD prompt that expects a point selection. For example, if you currently have a running Endpoint OSNAP mode, but want to snap to the nearest point to draw a line this time, you could respond to the To point: prompt by entering *ENDP*. This will override the current OSNAP for this one selection. The running OSNAP will regain control on the next selection (see Object Snap).

You can disable the current OSNAP for a single override by entering *NONE* in response to a point selection prompt.

You can also set the running OSNAP modes using the SETVAR command or AutoLISP to control the value of the system variable OSMODE.

In using AutoCAD release 9 (with the standard menu ACAD.MNU), you can select the OSNAP command from the Tools pull-down menu by pressing the object snap button on a multibutton pointing device (only with AutoCAD release 9; see Advanced User Interface).

'PAN

[all versions]

Overview   The PAN command moves the current display window. One description of the image you see on AutoCAD's graphic screen is an analogy to being in a blimp. There's a whole world beneath you, seen through a viewport of a limited size. To see more of the world, you can increase your altitude, but you can't see as much detail. If you are at the correct altitude to see enough detail but aren't looking in the spot you want to see, you can simply move the blimp. That's the way the PAN command operates. PAN lets you move your viewport of the drawing to a different location without changing the current magnification.

Procedure   To use the PAN command, you must first know where you are going. If the area that you want to move to is off in a corner of the current screen, you can simply show AutoCAD where to pan to by pointing with the screen cursor. If the area you want to view is off the screen, you can enter coordinates or distances when you instruct the PAN command.

PAN prompts for a displacement and a second point. You can give the command a pair of coordinates ("move from this first coordinate to the second") or just enter a displacement in response to the first prompt and press Return in response to the prompt for the second point. When you use the first method, the command behaves exactly as it would if you were pointing to distinct points on the screen. With the displacement method, if you want to move to a new viewport that is 8 units to the right and 6 units up, you must key in the displacement as $-8,-6$. Then you press Return in response to the prompt for the second point.

```
Command: PAN
Displacement: -8,-6
Second point: ( RETURN )
```

You are actually telling AutoCAD that you want to move the image under your viewport to the left 8 units (-8) and down 6 units (-6). You could have entered this same sequence as

```
Command: PAN
Displacement: 0,0
Second point: -8,-6
```

but that takes more keystrokes. What you told AutoCAD, in effect, was the same as if you had used the screen cursor to first select the 0,0 point and then to select the −8,−6 point.

You can use the PAN command transparently (while another command is active) by preceding the command with an apostrophe, as in 'PAN. You can do this as long as the command does not require the drawing to be regenerated.

Examples

In most cases, you respond to the PAN command's prompts simply by selecting points on the screen with your pointing device, as shown in the following example and in Figure 66.

```
Command: PAN
Displacement: (select point A)
Second point: (select point B)
```

Warnings

You cannot use the transparent PAN command while the VPOINT command or a PAN, ZOOM, or VIEW command is active or if the operation would require a regeneration of the drawing. Fast Zoom mode, as set by the VIEWRES command, must be on for transparent operation.

The transparent PAN command is not available in AutoCAD version 2.5.

Tips

You can pan so far that the drawing will need to be regenerated. Avoid unwanted regeneration by setting REGENAUTO to off. Then AutoCAD will prompt you if a regeneration is required. You can also use the ZOOM−Dynamic command to pan within the drawing (see ZOOM).

In AutoCAD release 9 (with the standard menu ACAD.MNU), you can select the PAN command from the Display pull-down menu. A transparent PAN will then be executed. (Only with AutoCAD release 9; see Advanced User Interface.)

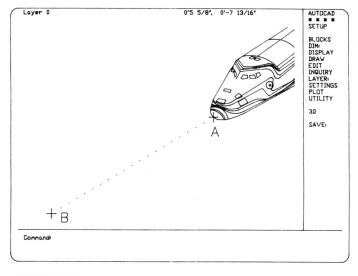

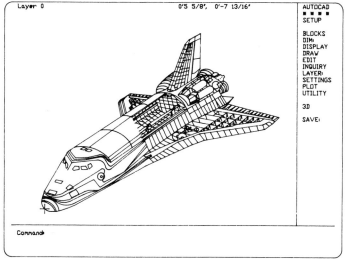

Figure 66

PDMODE

[v2.5, v2.6, r9]

Overview   The POINT command places point entities into a drawing. Under normal circumstances, these points appear in the drawing as simple dots. No matter what zoom magnification you use, each point always appears as a single pixel. Under some circumstances, however, you may want the points to display as something other than single pixel. For example, when used to mark locations along an object in conjunction with the DIVIDE or MEASURE command, a simple pixel point would be difficult to locate (see DIVIDE and MEASURE). The PDMODE (Point Display Mode) system variable allows you to change the appearance of point entities.

Procedure   The appearance of point entities is controlled by the PDMODE variable through the use of numerical values. The basic single pixel point is represented by a value of 0; a point entity is invisible if the PDMODE value is 1. A value of 2 causes a point to be represented as a cross; a value of 3 causes the point to be displayed as an X; and a value of 4 displays a point as a short vertical line.

In addition to these basic settings, a value of 32, 64, or 96 can be added to the initial PDMODE value to add a circle, a square, or a circle and a square around the basic point. You can change the PDMODE value using Auto-CAD's SETVAR command or AutoLISP.

Examples   The twenty available point descriptions follow.

0	1	2	3	4
32	33	34	35	36
64	65	66	67	68
96	97	98	99	100

Warnings   Changing the PDMODE value affects all points in the drawing. All point entities already in the drawing change to the new description the next time the drawing is regenerated. A point added to the drawing after changing the PDMODE value has the new appearance.

Tips   The initial value of PDMODE is determined from the prototype drawing (default value is 0, single pixel points). The current value is saved along with the drawing file.

PDSIZE   [v 2.5, v2.6, r9]

Overview   Point entities are normally displayed as single pixels regardless of the zoom magnification. The PDSIZE (Point Display Size) system variable controls the real size, in drawing units, of point entities whenever the PDMODE value is not equal to 0 or 1.

Procedure   You can use the SETVAR command to manipulate the PDSIZE variable, as well as the AutoLISP functions (getvar) and (setvar). When the PDSIZE value is a positive number, the variable represents the actual size, in drawing units, of the point entity. Changing the zoom magnification factor results in a change of the displayed point entities. If the PDSIZE value is a negative number, AutoCAD uses the value as a relative percentage of the screen size. In that case, point entities appear the same size regardless of a change in the zoom magnification.

Examples   The following sequence illustrates the use of the SETVAR command to change the PDSIZE value; the drawing shows the result. Notice that it was necessary to regenerate the drawing for the change to take effect. Changing the PDSIZE does not cause an automatic regeneration.

```
Command: SETVAR
Variable name or ?: PDSIZE
New value for PDSIZE <0.0000>: 3
Command: REGEN
Regenerating drawing.
```

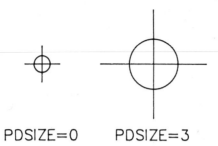

PDSIZE=0   PDSIZE=3

Warnings   Changing the PDSIZE value affects all points in the drawing. Every point entity already in the drawing changes to the new size description the next time the drawing is regenerated. A point entity added to the drawing after changing the PDSIZE value is drawn at the new size.

Tips   The initial PDSIZE value is determined from the prototype drawing (default value is 0.0000). The current value is saved along with the drawing file.

PDSIZE has no effect on points when the PDMODE value is 0 (normal single pixel point display) or 1 (invisible points).

PEDIT

[ADE-3] [all versions]

Options

C	Closes an open polyline
D	Decurves a polyline
E	Edits vertex (see suboptions on following page)
F	Fits curve to polyline
J	Joins to polyline
O	Opens a closed polyline
S	Splines curve fitting [r9]
U	Undoes one editing operation
W	Sets uniform width for polyline
X	Exits from PEDIT command

SUBOPTIONS FOR VERTEX EDITING

B   Sets first vertex for Break
G   Goes (performs Break or Straighten operation)
I   Inserts new vertex after current one
M   Moves current vertex
N   Makes next vertex current
P   Makes previous vertex current
R   Regenerates the polyline
S   Sets first vertex for Straighten
T   Sets tangent direction for current vertex
W   Sets new width for segment following current vertex
X   Exits from vertex editing or cancels Break/Straighten request

Overview   The PEDIT (Polyline Edit) command provides features for editing polylines. Polylines, being complex entities made up of multiple segments, are
powerful drawing entities that demand additional editing features not
available with AutoCAD's standard editing commands. PEDIT allows you
to add a segment that joins the first vertex of a polyline to its last vertex (C,
close), remove that segment (O, open), change a jagged polyline to one
with smooth curves connecting each vertex (F, fit curve), remove those
curves (D, decurve), join a polyline to another or to other lines and arcs (J,
join), change the width of all the segments of a polyline (W, width), undo
something you have just done to the polyline (U, undo), or edit any individual vertex (E, edit vertex).

When you are editing vertices, PEDIT presents an entirely separate set of
suboptions. These suboptions allow you to break a polyline into two separate polylines (B, break), insert a new vertex into the polyline (I, insert),
move the current vertex (M, move), move the current vertex to a different
vertex (N, next; P, previous), change the width of an individual segment
(W, width), set a tangent direction for an individual segment (T, tangent),
straighten a curve segment (S, straighten), or regenerate the polyline so
you can see changes without having to exit from PEDIT (R, regenerate).

Release 9 Only   AutoCAD Release 9 has an additional option to the PEDIT
command. Besides the regular Fit option, cubic B-spline curves can also be
created using the new Spline option. A spline curve is created by using the
vertices of the selected polyline as the control points, or frame, of the curve.
The spline curve is constructed so that it passes through the first and last

control points (except for closed polylines) and is pulled toward the other points but does not pass through them. This is different from the regular Fit Curve option, where the resulting curve passes through all the vertices.

Procedure

When you enter the PEDIT command, you must first select the polyline you want to edit. You can edit only one polyline at a time. You can use any of AutoCAD's entity selection methods to choose the polyline (pick box, window, last, and so on).

```
Command: PEDIT
Select polyline: (use any method to select)
```

If the entity you select is actually a line or an arc, PEDIT can turn it into a polyline. AutoCAD will prompt

```
Entity selected is not a polyline.
Do you want it to turn into one? <Y>:
```

Once you have changed the line or arc into a polyline, you can edit it with the PEDIT command. This is a useful way to combine arcs and lines into one polyline using the Join option.

Once you have selected the polyline, you are given a choice of options. The default option is to exit from the PEDIT command. The options are not in alphabetical order. Because one option is generally the opposite of another (Fit curve/Decurve), the options are paired. If the polyline you select is an open polyline, the Close option is presented. If the polyline is already closed, the Open option appears in the list. You need enter only the capitalized letter of the option and press Return to select that option. Each option is discussed individually in the next section.

```
Close/Join/Width/Edit vertex/Fit curve/Decurve/Undo/eXit <X>:
```

Examples

Closing (C) and Opening (O) Polylines Closing a polyline causes AutoCAD to draw a straight polyline segment from the first vertex of the polyline to the last vertex. (See the following drawing.) Opening a polyline causes AutoCAD to remove this segment. To open a closed polyline, enter *O* in response to the PEDIT prompt. To close an open polyline, enter *C*. You will know whether the polyline is open or closed because the PEDIT command will display the only available option. There is no further prompting if you choose the Open or Close option. PEDIT returns to its option prompt.

```
Command: PEDIT
Select polyline: (select point A)
Close/Join/Width/Edit vertex/Fit curve/Decurve/Undo/eXit <X>: C
Open/Join/Width/Edit vertex/Fit curve/Decurve/Undo/eXit <X>:  RETURN
```

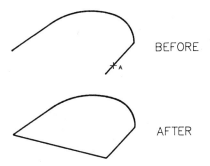

BEFORE

AFTER

Join (J)   With the Join option, you can join arcs and lines to a polyline (see the following drawing) or join one polyline to another, thus creating one continuous polyline. To join any entity to a polyline, that entity must already share an endpoint with an end vertex of the selected polyline. The entity and the polyline must end exactly at the same point or they will not be joined. This option is available only if the selected polyline is open.

The PEDIT command prompts you to select the objects to join to the polyline. You can pick one or more entities. Each entity is highlighted when you select it. You can include the polyline if you want.

```
Close/Join/Width/Edit vertex/Fit curve/Decurve/Undo/eXit <X>: J
Select objects: (select)
```

When the Join option is completed, AutoCAD tells you how many segments were added to the polyline. The command then returns to the PEDIT option prompt.

```
Command: PEDIT
Select polyline: (select point A)
Close/Join/Width/Edit vertex/Fit curve/Decurve/Undo/eXit <X>: J
Select objects: (select point B)
Select objects:  RETURN
1 segments added to polyline
Close/Join/Width/Edit vertex/Fit curve/Decurve/Undo/eXit <X>:  RETURN
```

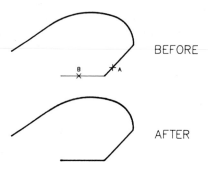

Width (W) The Width option lets you change the widths of all of the segments of the selected polyline to one constant width. (See the following drawing.) AutoCAD prompts you for the new width. You can enter a width or use the cursor to indicate a width, the width being the distance between two selected points. The polyline changes to indicate the new width, and the command then returns to the option prompt.

```
Close/Join/Width/Edit vertex/Fit curve/Decurve/Undo/eXit <X>: W
Enter new width for all segments: (enter width)
```

```
Command: PEDIT
Select polyline: (select point A)
Close/Join/Width/Edit vertex/Fit curve/Decurve/Undo/eXit <X>: W
Enter new width for all segments: 1
Close/Join/Width/Edit vertex/Fit curve/Decurve/Undo/eXit <X>: ( RETURN )
```

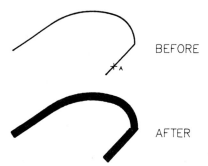

Fit Curve (F)/Decurve (D) The Fit Curve option computes a smooth curve that connects all of the vertices of the polyline. (See the following drawing.) A curve consists of a pair of arcs joining each pair of vertices. To accomplish this, AutoCAD adds additional vertices into the polyline. If you have used the Fit Curve option and don't like the result, you can return the polyline to its previous state by using the Decurve option. There are no additional prompts for these options.

```
Command: PEDIT
Select polyline: (select point A)
Close/Join/Width/Edit vertex/Fit curve/Decurve/Undo/eXit <X>: F
Close/Join/Width/Edit vertex/Fit curve/Decurve/Undo/eXit <X>: RETURN
```

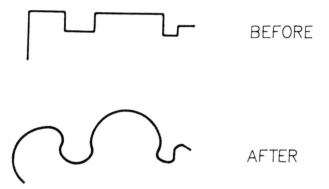

BEFORE

AFTER

Spline Curve (S) [Release 9 only]   The Spline Curve option computes a smooth curve that is pulled toward the vertices but passes through only the first and last vertices in the case of open polylines. The more vertices in a particular area of a polyline, the more pull that is exerted.

The original polyline is used as a frame. This frame is remembered and can be displayed along with the spline curved polyline by setting the system variable SPLFRAME to 1 (see SPLFRAME). If displayed, the frame always has a continuous linetype and a zero width. To restore the polyline to its original frame appearance, use the Decurve option.

If the original polyline contains arc segments, they are straightened when forming the spline frame. A spline curved polyline with varying segment widths simply tapers continuously from the width of the first vertex to the width of the last.

The actual number of segments used to approximate the spline curve is controlled by the system variable SPLINESEGS (see SPLINESEGS).

The following drawing illustrates the differences between a regular polyline, a fit curve polyline, and a spline curve polyline.

```
Command: PEDIT
Select polyline: (select point)
Close/Join/Width/Edit vertex/Fit curve/Spline curve/
Decurve/Undo/eXit <X>: S
Close/Join/Width/Edit vertex/Fit curve/Spline curve/
Decurve/Undo/eXit <X>: ( RETURN )
```

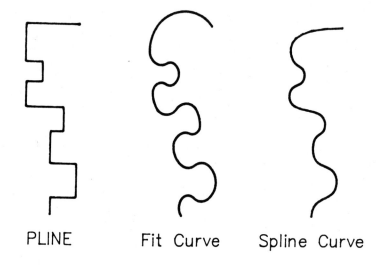

PLINE   Fit Curve   Spline Curve

Undo (U)   The Undo option reverses the previous PEDIT operation. By entering U several times, you can step back through the previous edits to the first thing you did in the current PEDIT command. You cannot use the PEDIT Undo option to reverse something that you did in a previous PEDIT session.

Exit (X)   To end the PEDIT command, you must exit back to the command prompt by entering X in response to the PEDIT option prompt (or by pressing Return when X is the default).

```
Close/Join/Width/Edit vertex/Fit curve/Decurve/Undo/eXit <X>: ( RETURN )
```

Edit Vertex (E)   All of the options mentioned so far act on the entire polyline at once. The Edit Vertex option has ten suboptions that allow you to edit individual segments of the selected polyline, one at a time.

```
Next/Previous/Break/Insert/Move/Regen/Straighten/Tangent/Width/eXit <N>:
```

When you elect to edit the individual vertices of the polyline, this option places an X on the first vertex of the polyline actually displayed on the screen.

(See the following drawing.) This indicates the current vertex. The vertex editing commands operate on one vertex at a time, whichever one is current.

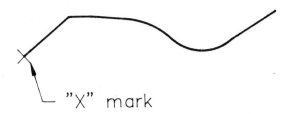

Next(N)/Previous (P)   The Next and Previous suboptions move the X marker from one vertex to another. When the Edit Vertex option is initially selected, the default setting for the suboption prompt is N (Next). Simply pressing Return moves the X to each successive vertex. If you go past the vertex you want to edit or need to go in the other direction, enter *P* (for Previous) to start moving the other way. Once you have entered *P*, it becomes the new default and pressing Return continues to move you from vertex to vertex in that direction. Entering *N* changes the default again and moves you back in the original direction. You cannot move continuously around a closed polyline. When you reach one end, you must reverse directions.

```
Command: PEDIT
Select polyline: (select point)
Close/Join/Width/Edit vertex/Fit curve/Decurve/Undo/eXit <X>: E
Next/Previous/Break/Insert/Move/Regen/Straighten/Tangent/Width/
eXit <N>: ( RETURN )
Next/Previous/Break/Insert/Move/Regen/Straighten/Tangent/Width/
eXit <N>: ( RETURN )
Next/Previous/Break/Insert/Move/Regen/Straighten/Tangent/Width/eXit <N>: P
Next/Previous/Break/Insert/Move/Regen/Straighten/Tangent/Width/
eXit <P>: ( RETURN )
Next/Previous/Break/Insert/Move/Regen/Straighten/Tangent/Width/eXit <P>: X
Close/Join/Width/Edit vertex/Fit curve/Decurve/Undo/eXit <X>: ( RETURN )
```

Break (B)   The Break suboption marks a vertex to be used as a break point. You can then break the polyline into two pieces or move to another vertex, mark it as a second break point, and then remove the segments between the two vertices. (See the following drawing.) When you enter *B* to select the current vertex as a break point, you are given four options:

```
Next/Previous/Go/eXit <N>:
```

401

Next and Previous work normally to move you to successive or preceding vertices. Go tells AutoCAD to complete the break operation, using the vertex or vertices you have marked. If you mark a vertex in error or change your mind, X cancels the Break suboption and returns you to the Edit Vertex suboptions prompt.

```
Command: PEDIT
Select polyline: (select point A)
Close/Join/Width/Edit vertex/Fit curve/Decurve/Undo/eXit <X>: E
Next/Previous/Break/Insert/Move/Regen/Straighten/Tangent/Width/
eXit <N>: (RETURN)
Next/Previous/Break/Insert/Move/Regen/Straighten/Tangent/Width/
eXit <N>: (RETURN)
Next/Previous/Break/Insert/Move/Regen/Straighten/Tangent/Width/eXit <N>: B
Next/Previous/Go/eXit <N>: (RETURN)
Next/Previous/Go/eXit <N>: (RETURN)
Next/Previous/Go/eXit <N>: G
Next/Previous/Break/Insert/Move/Regen/Straighten/Tangent/Width/eXit <N>: X
Close/Join/Width/Edit vertex/Fit curve/Decurve/Undo/eXit <X>: (RETURN)
```

BEFORE

AFTER

Insert(I)   The Insert option adds a new vertex to the polyline at a chosen location. The new vertex is added after the current vertex, in the direction that you would move when proceeding to the Next vertex. You are prompted to select the location for the new vertex, after which the poly-line is immediately regenerated. (See the following drawing.) You can select the point by entering absolute coordinates or coordinates relative to the current vertex or by pointing on the screen.

```
Command: PEDIT
Select polyline: (select point A)
Close/Join/Width/Edit vertex/Fit curve/Decurve/Undo/eXit <X>: E
Next/Previous/Break/Insert/Move/Regen/Straighten/Tangent/Width/
eXit <N>: (RETURN)
```

```
Next/Previous/Break/Insert/Move/Regen/Straighten/Tangent/Width/eXit <N>: I
Enter location of new vertex: (select point B)
Next/Previous/Break/Insert/Move/Regen/Straighten/Tangent/Width/eXit <N>: X
Close/Join/Width/Edit vertex/Fit curve/Decurve/Undo/eXit <X>: ( RETURN )
```

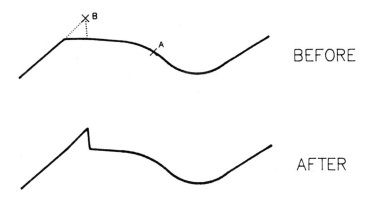

BEFORE

AFTER

Move (M) The Move option moves the current vertex to a different location. (See the following drawing.) AutoCAD prompts for the new location, which you can enter as absolute coordinates, as coordinates relative to the current vertex, or by pointing. As soon as you select the new location, the polyline regenerates to reflect the change.

```
Command: PEDIT
Select polyline: (select point A)
Close/Join/Width/Edit vertex/Fit curve/Decurve/Undo/eXit <X>: E
Next/Previous/Break/Insert/Move/Regen/Straighten/Tangent/Width/
eXit <N>: ( RETURN )
Next/Previous/Break/Insert/Move/Regen/Straighten/Tangent/Width/eXit <N>: M
Enter new location: (select point B)
Next/Previous/Break/Insert/Move/Regen/Straighten/Tangent/Width/eXit <N>: X
Close/Join/Width/Edit vertex/Fit curve/Decurve/Undo/eXit <X>: ( RETURN )
```

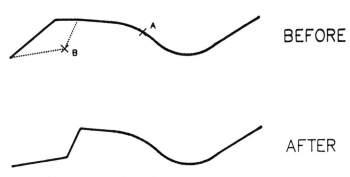

BEFORE

AFTER

Regen (R)   The Regen option regenerates the polyline after the width of an individual segment is changed. This option affects only the selected polyline. It does not regenerate the entire drawing.

Straighten (S)   The Straighten suboption removes all vertices between two specified vertices and substitutes one straight segment instead. When you enter *S* to select this suboption, AutoCAD uses the current vertex (marked with an X) as the first vertex. You are then given four options:

```
Next/Previous/Go/eXit <N>:
```

Next and Previous move you along from one vertex to another. When you are at the second desired vertex, entering *G* causes all of the vertices between the two that were selected to be removed and replaced with one straight polyline segment. (See the following drawing.) If you selected the wrong vertex or simply change your mind, enter *X* to cancel the Straighten command and to return to the Edit Vertex suboptions prompt.

```
Command: PEDIT
Select polyline: (select point A)
Close/Join/Width/Edit vertex/Fit curve/Decurve/Undo/eXit <X>: E
Next/Previous/Break/Insert/Move/Regen/Straighten/Tangent/Width/
eXit <N>: ( RETURN )
Next/Previous/Break/Insert/Move/Regen/Straighten/Tangent/Width/
eXit <N>: ( RETURN )
Next/Previous/Break/Insert/Move/Regen/Straighten/Tangent/Width/
eXit <N>: ( RETURN )
Next/Previous/Break/Insert/Move/Regen/Straighten/Tangent/Width/eXit <N>: S
Next/Previous/Go/eXit <N>: ( RETURN )
Next/Previous/Go/eXit <N>: ( RETURN )
Next/Previous/Go/eXit <N>: G
Next/Previous/Break/Insert/Move/Regen/Straighten/Tangent/Width/eXit <N>: X
Close/Join/Width/Edit vertex/Fit curve/Decurve/Undo/eXit <X>: ( RETURN )
```

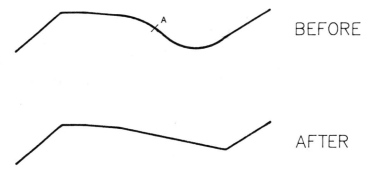

BEFORE

AFTER

Tangent (T)   The Tangent suboption lets you specify a tangent direction for the current vertex. Later, when you use the Fit Curve option, the tangent direction determines the curve that is generated. To use the Tangent suboption, move to the desired vertex and enter *T*. When AutoCAD prompts for the tangent direction, type in an angle or use the cursor to indicate an angle direction from the current vertex.

```
Command: PEDIT
Select polyline: (select point A)
Close/Join/Width/Edit vertex/Fit curve/Decurve/Undo/eXit <X>: E
Next/Previous/Break/Insert/Move/Regen/Straighten/Tangent/Width/
eXit <N>: ( RETURN )
Next/Previous/Break/Insert/Move/Regen/Straighten/Tangent/Width/eXit <N>: T
Direction of tangent: (select point B) (or enter angle)
Next/Previous/Break/Insert/Move/Regen/Straighten/Tangent/Width/eXit <N>: X
Close/Join/Width/Edit vertex/Fit curve/Decurve/Undo/eXit <X>: F
Close/Join/Width/Edit vertex/Fit curve/Decurve/Undo/eXit <X>: ( RETURN )
```

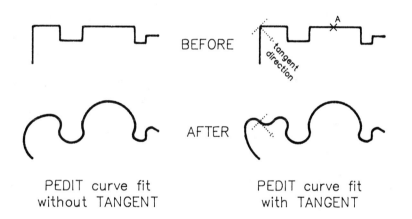

BEFORE

AFTER

PEDIT curve fit
without TANGENT

PEDIT curve fit
with TANGENT

Width (W)   The Width suboption changes the starting and ending widths of the polyline segment immediately following the current vertex. Move the X marker to the desired vertex and enter *W*. AutoCAD prompts for the starting width and the ending width. If you want to change the width of the polyline segment but desire a constant width, press Return in response to the ending width prompt. The starting width you specify is the default value for the ending width. The current width of the polyline segment is provided as the default value for the starting width.

```
Next/Previous/Break/Insert/Move/Regen/Straighten/Tangent/Width/eXit <N>: W
Enter starting width <current>: (enter width)
Enter ending width <starting width>: (enter width)
```

This option is different from the PEDIT Width option. Only one segment is affected at a time. The polyline is not immediately redrawn after you change the width of a segment. You must use the R (Regenerate) suboption to see the changes. The Width suboption affects the next segment after the current vertex. If you are not sure which is the next segment, use the Next suboption to move to the next vertex and thus determine which direction is which. The following drawings illustrate the effects of the Width suboption on a typical polyline. Only one segment of the polyline is changed.

```
Command: PEDIT
Select polyline: (select point A)
Close/Join/Width/Edit vertex/Fit curve/Decurve/Undo/eXit <X>: E
Next/Previous/Break/Insert/Move/Regen/Straighten/Tangent/Width/
eXit <N>: ( RETURN )
Next/Previous/Break/Insert/Move/Regen/Straighten/Tangent/Width/eXit <N>: W
Enter starting width <0.0000>: .25
Enter ending width <0.2500>: 1
Next/Previous/Break/Insert/Move/Regen/Straighten/Tangent/Width/eXit <N>: R
Next/Previous/Break/Insert/Move/Regen/Straighten/Tangent/Width/eXit <N>: X
Close/Join/Width/Edit vertex/Fit curve/Decurve/Undo/eXit <X>: ( RETURN )
```

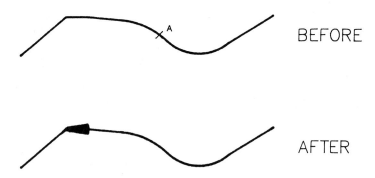

BEFORE

AFTER

Exit (X)   To end the Edit Vertex option, you must exit back to the PEDIT option prompt by entering *X* in response to the suboption prompt:

```
Next/Previous/Break/Insert/Move/Regen/Straighten/Tangent/Width/eXit <N>: X
```

This returns you to the PEDIT option prompt. You can then do additional polyline editing. To return to the AutoCAD Command prompt, you must exit from the PEDIT command by entering *X* again.

Warnings   The Spline Curve option is available only with AutoCAD release 9.
        Once you use the BREAK, TRIM, or EXPLODE command on a Fit or Spline curved polyline, just the curve remains. The curve cannot later be decurved.

Tips   If you have already drawn the polyline connecting a segment back to the first segment, AutoCAD still considers the polyline to be open. Opening or closing it, in this case, has no visible effect.
        When you join polylines, only entities that start precisely where another joined entity ends are added to the polyline. If the entities appear to meet at the same point but still don't join, use the Edit Vertex option to move the vertex to the endpoint of the desired entity, using an object snap. This is sometimes necessary when joining arcs and lines to a polyline, even though they were drawn to the same specific points.
        When you break a closed polyline, it becomes an open polyline. The closing segment is automatically removed.
        In AutoCAD release 9 (with the standard menu ACAD.MNU), you can select the PEDIT command from the Edit pull-down menu. The standard PEDIT screen menu is also displayed. The menu item repetition in release 9 causes the command to repeat automatically until you cancel it by pressing Ctrl-C or by selecting another command from the menu (only with AutoCAD release 9; see Advanced User Interface and Entity Selection).
        Stretching a spline curved polyline results in the frame being stretched and a new spline curve being generated (release 9 only).
        When you are editing the vertices of a spline curve, the X marker appears on the spline's frame (whether visible or not), not on the spline curve (release 9 only).

PERIMETER

Overview   The PERIMETER system variable contains the last perimeter value calculated by the AREA, LIST, or DBLIST command. The value is returned as a
real number. Each time the AREA, LIST, or DBLIST command is executed, a
new PERIMETER value replaces the previous value. The PERIMETER value
is read-only.

Procedure   You can use either the SETVAR command or the (getvar) AutoLISP function to obtain the current value stored in the PERIMETER variable. Using
SETVAR is useful if you have simply forgotten what the last perimeter
measured was. An AutoLISP routine that incorporates the PERIMETER
value into a calculation is of more practical use.

Examples   The SETVAR command returns the most recent PERIMETER value.

```
Command: SETVAR
Variable name or ?: PERIMETER
PERIMETER = 22.3680 (read only)
```

Tips   The PERIMETER value changes every time you use the AREA, LIST, or
DBLIST command. When you end the drawing, the current PERIMETER
value is discarded.

PICKBOX

Overview   The PICKBOX system variable controls the size of the pick box. Most
AutoCAD editing commands cause a pick box to be displayed, which is
used to select objects to be edited. Any object falling within the pick box is
a candidate for selection. Sometimes, however, the size of the pick box
causes AutoCAD to become confused and to select the wrong object. While
AutoCAD has a specific command for altering the size of the aperture box
(see APERTURE), no such command exists for changing the size of the pick
box. Instead, you must change the value of the PICKBOX system variable.

Procedure   The size of the pick box is determined by the value of the PICKBOX variable. The value is expressed as an integer that represents the height of the pick box in screen pixels. The pick box is always square. No variable exists to alter the width. You can change the PICKBOX value using either the SETVAR command or from within an AutoLISP routine using the (setvar) function.

Examples   This example illustrates the use of the SETVAR command to alter the size of AutoCAD's pick box.

```
Command: SETVAR
Variable name or ?: PICKBOX
New value for PICKBOX <3>: 5
```

Tips   The value of PICKBOX is stored in AutoCAD's configuration file (ACAD.CFG). The default value is 3 pixels.

PLINE   [ADE-3] [all versions]

Options

H	Sets new half-width
U	Undoes previous segment
W	Sets new line width
Return	Exits from PLINE command

IN LINE MODE

A	Switches to Arc mode
C	Closes with straight segment
L	Specifies segment length (continues previous segment)

IN ARC MODE

A	Included angle
CE	Center point
CL	Closes with arc segment
D	Starting direction
L	Chord length or switches to Line mode
R	Radius
S	Second point of three-point arc

Overview

The PLINE (Polyline) command draws polyline entities. Polylines are connected sequences of arcs and/or lines that are treated as one entity. They can be drawn with any linetype and can have a visible width that can vary over the length of any segment. Polylines can form complex objects. The DONUT, ELLIPSE, and POLYGON commands actually construct these objects as polylines.

You can edit polylines using AutoCAD's standard editing commands. Polylines can be moved, rotated, broken, erased, and so on. To deal with the special properties of polylines, such as width, a special editing command called PEDIT (Polyline Edit) is provided to alter these other properties (see PEDIT).

There are several other special properties of polylines that make them particularly useful. The CHAMFER and FILLET commands affect the entire polyline as one entity at one time (see CHAMFER and FILLET). The AREA and LIST commands display the area and the perimeter of a selected polyline.

Procedure

When you enter the PLINE command, AutoCAD first prompts for the start point, the same as for the LINE command. Once you select the start point, the current line width is displayed, which is initially set by the prototype drawing. Once you enter a new width, that width becomes the current line width until you change it again.

```
Command: PLINE
From point: (select start point)
Current line-width is <nnn>
```

The PLINE command assumes that you will be drawing a straight-line segment. You can draw the segment by selecting the end point of the line (just like the LINE command) or you can select one of the other PLINE options. If you select an end point, AutoCAD draws the segment and the PLINE command prompts again for another end point, displaying all of the other options. To end the command, press Return. Using the PLINE command in this fashion draws line segments exactly as in the LINE command, except that the segments are treated as one polyline entity.

```
Arc/Close/Halfwidth/Length/Undo/Width/<Endpoint of line>: (select end point)
```

To select one of PLINE's other options, enter it in response to the prompt. You don't have to enter the entire option name. The single starting letter is sufficient. Initially the PLINE command produces straight polyline segments. To draw arc segments, enter *A* in response to the PLINE prompt.

Examples   **Straight Polyline Segments**   Draw simple straight polylines by specifying the "from point" and any number of "to points," in a fashion similar
to the LINE command, as illustrated in the following example and drawing.

```
Command: PLINE
From point: (select point A)
Current line-width is 0.0000
Arc/Close/Halfwidth/Length/Undo/Width/<Endpoint of line>: (select point B)
Arc/Close/Halfwidth/Length/Undo/Width/<Endpoint of line>: (select point C)
Arc/Close/Halfwidth/Length/Undo/Width/<Endpoint of line>: (select point D)
Arc/Close/Halfwidth/Length/Undo/Width/<Endpoint of line>: ( RETURN )
```

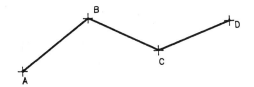

Close (C)   The Close option draws a straight polyline segment from the
end point (D in the following drawing) of the last segment drawn back to
the starting point (the "from point," which is A in the drawing) and ends
the PLINE command. The current width controls the width of this closing
segment. The polyline thus drawn is a closed polyline. The closing segment has effects on the polyline other than simply drawing a segment
back to the starting point.

```
Command: PLINE
From point: (select point A)
Current line-width is 0.0000
Arc/Close/Halfwidth/Length/Undo/Width/<Endpoint of line>: (select point B)
Arc/Close/Halfwidth/Length/Undo/Width/<Endpoint of line>: (select point C)
Arc/Close/Halfwidth/Length/Undo/Width/<Endpoint of line>: (select point D
Arc/Close/Halfwidth/Length/Undo/Width/<Endpoint of line>: C
```

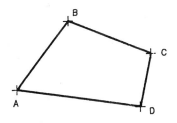

411

Length (L)   The Length option draws a polyline segment of a specified length, continuing the polyline at the same angle as that of the previous segment, as shown in the following example and drawing.

```
Command: PLINE
From point: (select point A)
Current line-width is 0.0000
Arc/Close/Halfwidth/Length/Undo/Width/<Endpoint of line>: (select point B)
Arc/Close/Halfwidth/Length/Undo/Width/<Endpoint of line>: L
Length of line: 5
Arc/Close/Halfwidth/Length/Undo/Width/<Endpoint of line>: ( RETURN )
```

If the previous segment is an arc, the line segment is drawn at an angle tangent to the end point of that arc (similar to continuing a line from the end point of an arc). This is illustrated in the following example and drawing.

```
Command: PLINE
From point: (select point C)
Current line-width is 0.0000
Arc/Close/Halfwidth/Length/Undo/Width/<Endpoint of line>: (select point D)
Arc/Close/Halfwidth/Length/Undo/Width/<Endpoint of line>: A
Angle/CEnter/CLose/Direction/Halfwidth/Line/Radius/Second pt/Undo/Width/
<Endpoint of arc>: (select point E)
Angle/CEnter/CLose/Direction/Halfwidth/Line/Radius/Second pt/Undo/Width/
<Endpoint of arc>: L
Arc/Close/Halfwidth/Length/Undo/Width/<Endpoint of line>: L
Length of line: 5
Arc/Close/Halfwidth/Length/Undo/Width/<Endpoint of line>: ( RETURN )
```

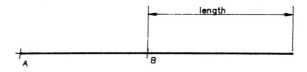

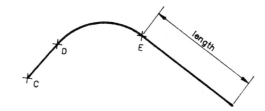

Undo (U)   The Undo option removes the last polyline segment added. As in the Undo option of the LINE command, you can enter *U* repeatedly to step back through the polyline, removing one segment at a time.

Width (W)   The Width option lets you specify the starting width and the ending width of the next polyline segment. A width of zero results in a polyline segment that looks exactly like a line or an arc. Widths greater than zero display as wide lines and arcs, which are filled in if Fill mode is on (see FILL). If the starting width is different from the ending width, the polyline can be made to taper. The current line width is the default value for the starting width. The starting width is the default value for the ending width. And the ending width is the new current line-width setting.

```
Arc/Close/Halfwidth/Length/Undo/Width/<Endpoint of line>: W
Starting width <current>: (enter value) (or use cursor to show)
Ending width <start width>: (enter value) (use cursor or press Return)
```

Changing the width enables you to draw tapered polylines. You can draw an arrow by specifying a starting width of zero and a wider ending width. The following sequence produces the arrow in the drawing.

```
Command: PLINE
From point: (select point A)
Current line-width is 0.0000
Arc/Close/Halfwidth/Length/Undo/Width/<Endpoint of line>: W
Starting width <0>: ( RETURN )
Ending width <0>: 8
Arc/Close/Halfwidth/Length/Undo/Width/<Endpoint of line>: @8,0
Arc/Close/Halfwidth/Length/Undo/Width/<Endpoint of line>: W
Starting width <8>: 3
Ending width <3>: ( RETURN )
Arc/Close/Halfwidth/Length/Undo/Width/<Endpoint of line>: @16,0
Arc/Close/Halfwidth/Length/Undo/Width/<Endpoint of line>: ( RETURN )
```

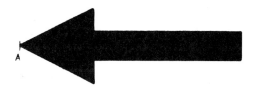

Half-width (H)   While the Width option allows you to specify the total width of a polyline segment (starting point and/or end point), the Half-width option lets you specify one-half the width. The prompts are similar to those of the Width option, except that AutoCAD prompts for the starting and ending half-widths. The half-width option may be more useful when you are using the screen cursor to show how wide the polyline segment should be, since the rubberband cursor will be attached to the center of the polyline segment at one end.

```
Arc/Close/Halfwidth/Length/Undo/Width/<Endpoint of line>: H
Starting half-width <current>: (enter value) (or use cursor to show)
Ending half-width <start width>: (enter value) (use cursor or press Return)
```

Arc Polyline Segments (A)   If you respond to the PLINE command's first set of options with *A* (Arc), the command switches to the Arc mode for the current polyline segment. You are presented with a new set of options. To return to drawing straight line segments, enter *L* in response to this Arc Polyline option prompt. The Half-width, Undo, and Width options have the same effects on arc segments as they do on straight segments.

```
Angle/CEnter/CLose/Direction/Halfwidth/Line/Radius/Second pt/Undo/Width/
<endpoint of arc>:
```

The PLINE command now assumes that you will provide the end point of an arc. The arc drawn starts at the end point of any previous polyline segment and is drawn tangent to the last segment. If this is the first segment of a new polyline, the default condition draws an arc segment tangent to the direction of the most recently drawn line, arc, or polyline segment. If this is not acceptable, you can use one of the other Arc options.

The following example shows a typical sequence of prompts and responses when using the POLYLINE command in Arc mode. The sequence creates the curved polyline in the following figure.

```
Command: PLINE
From point: (select point A)
Current line-width is 0.0000
Arc/Close/Halfwidth/Length/Undo/Width/<Endpoint of line>: A
Angle/CEnter/CLose/Direction/Halfwidth/Line/Radius/Second pt/Undo/Width/
<Endpoint of arc>: (select point B)
Angle/CEnter/CLose/Direction/Halfwidth/Line/Radius/Second pt/Undo/Width/
<Endpoint of arc>: (select point C)
Angle/CEnter/CLose/Direction/Halfwidth/Line/Radius/Second pt/Undo/Width/
<Endpoint of arc>: ( RETURN )
```

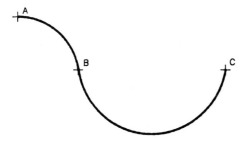

Angle (A)   The Angle option prompts for the included angle, the angle the arc is to span. This is similar to the ARC command. A positive number specifies a counterclockwise angle. A negative number results in a clockwise included angle. Once you enter the angle, the PLINE command needs to determine the end point of the arc. Providing the end point is the default condition. You can instead enter C or R to inform the PLINE command that you will provide the center point or the radius. The command then prompts for the appropriate information.

```
Angle/CEnter/CLose/Direction/Halfwidth/Line/Radius/Second pt/Undo/Width/
<Endpoint of arc>: A
Included angle: (enter angle)
Center/Radius/<Endpoint>:
```

In the following example (and the drawing on the left) point A is the start point, an angle of 90 degrees is specified, and point B is selected as the center point.

```
Arc/Close/Halfwidth/Length/Undo/Width/<Endpoint of line>: A
Angle/CEnter/CLose/Direction/Halfwidth/Line/Radius/Second pt/Undo/Width/
<Endpoint of arc>: A
Included angle: 90
Center/Radius/<Endpoint>: C
Center point: (select point B)
```

In the next example (the middle drawing in the following figure) point C is the start point, an angle of 90 degrees is specified, and a radius of 2 units entered. The command prompts for the direction of the chord (which is the chord from the start point, point C, to the end point). A chord direction can be entered or you can point on the screen. In this case, a direction of 315 degrees is entered.

```
Arc/Close/Halfwidth/Length/Undo/Width/<Endpoint of line>: A
Angle/CEnter/CLose/Direction/Halfwidth/Line/Radius/Second pt/Undo/Width/
<Endpoint of arc>: A
Included angle: 90
Center/Radius/<Endpoint>: R
Radius: 2
Direction of chord <-6>: 315
```

In the next example (drawing on the right) the start point is point D, a
radius of 90 degrees is specified, and the end point E is selected. Notice that
all three methods produce the same polyline arc.

```
Arc/Close/Halfwidth/Length/Undo/Width/<Endpoint of line>: A
Angle/CEnter/CLose/Direction/Halfwidth/Line/Radius/Second pt/Undo/Width/
<Endpoint of arc>: A
Included angle: 90
Center/Radius/<Endpoint>: (Select point E)
```

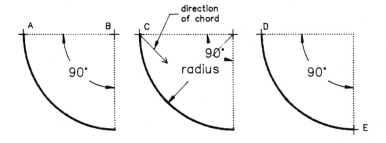

Center (CE)   Because the Center option and the Close option both start
with the letter C, you must respond with the first two letters to indicate
which of these options you want. The prompts remind you of this by
capitalizing the first two letters. The Center option allows you to specify
the center point of the arc segment. You can specify this point as an absolute or relative coordinate or use the cursor. After selecting the center
point, you must provide either the angle, the length, or the end point of
the arc. The end point is the default. The angle is the included angle (as in
the Angle option). The length is the length of the chord connecting the
start point and the end point. If you enter A or L, AutoCAD prompts for the
angle or the length, as appropriate.

```
Angle/CEnter/CLose/Direction/Halfwidth/Line/Radius/Second pt/Undo/Width/
<Endpoint of arc>: CE
Center point: (select center point)
Angle/Length/<End point>:
```

In the following example (and the drawing on the left) point A is the start
point, point B is the center point, and an included angle is specified.

```
Angle/CEnter/CLose/Direction/Halfwidth/Line/Radius/Second pt/Undo/Width/
<Endpoint of arc>: CE
Center point: (select point B)
Angle/Length/<End point>: A
Included angle: 90
```

In the next example (and the middle drawing) point C is the start point,
point D is the center point, and the length of the chord is specified.

```
Angle/CEnter/CLose/Direction/Halfwidth/Line/Radius/Second pt/Undo/Width/
<Endpoint of arc>: CE
Center point: (select point D)
Angle/Length/<End point>: L
Length of chord: 2
```

In this example (drawing on the right) point E is the start point, point F is
the center point, and point G is the end point (using the default condition).
Notice that all three methods resulted in the same drawing. Use the
method most convenient for your particular application.

```
Angle/CEnter/CLose/Direction/Halfwidth/Line/Radius/Second pt/Undo/Width/
<Endpoint of arc>: CE
Center point: (select point F)
Angle/Length/<End point>: (select point G)
```

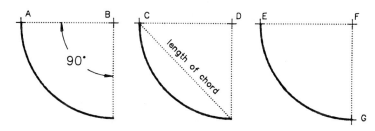

Close (CL)   The Close option in Arc mode is similar to the Close option
for straight-line polylines, but the closing segment is an arc, drawn tangent to the previous segment, as shown in the following drawing. To
indicate the Close option, type the first two letters, CL, to differentiate it
from the Center option.

```
Angle/CEnter/CLose/Direction/Halfwidth/Line/Radius/Second pt/Undo/Width/
<Endpoint of arc>: CL
```

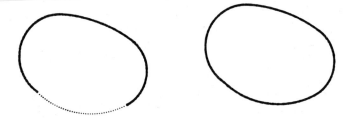

Direction (D)   The Direction option lets you specify the starting direction for the arc segment (as illustrated in the following drawing) rather than using AutoCAD's default, which draws arc segments tangent to the previous segment. AutoCAD prompts for the direction and then the end point.

```
Angle/CEnter/CLose/Direction/Halfwidth/Line/Radius/Second pt/Undo/Width/
<Endpoint of arc>: D
Direction from starting point: (enter angle) (or use cursor to indicate)
End point: (select end point)
```

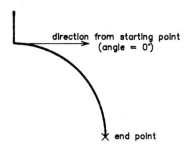

Radius (R)   The Radius option lets you first provide the radius of the arc segment. Then you can either select the end point (the default) or provide the angle that the arc subtends. The arc is drawn tangent to the previous segment.

   In the following example, point A is the start point, a radius of 4 units is specified, and point B is selected as the end point.

```
Angle/CEnter/CLose/Direction/Halfwidth/Line/Radius/Second pt/Undo/Width/
<Endpoint of arc>: R
Radius: 4
Angle/<End point>: (select point B)
```

In the second example, point C is the start point and a radius of 4 units is again specified. This time, however, an included angle of 90 degrees is entered resulting in an additional prompt requesting the direction of the chord.

```
Angle/CEnter/CLose/Direction/Halfwidth/Line/Radius/Second pt/Undo/Width/
<Endpoint of arc>: R
Radius: 4
Angle/<End point>: A
Included angle: 90
Direction of chord <43>: ( RETURN )
```

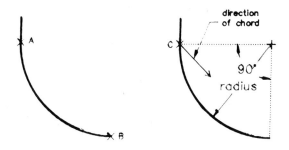

Second Point (S)   The Second Point option lets you select the second and third points of a three-point arc (see illustration), similar to the ARC command. AutoCAD prompts for the two points.

```
Angle/CEnter/CLose/Direction/Halfwidth/Line/Radius/Second pt/Undo/Width/
<Endpoint of arc>: S
Second point: (select second point)
End point: (select end point)
```

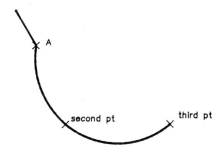

Polylines were used extensively to draw the contours and the course of the creek in the geologic survey map in Appendix D (see Figure D-3).

419

Tips   Polylines, because of their special property of being one entity, are the perfect entity to use to form the boundaries for site plans. The AREA command displays the perimeter of and the area enclosed within the polyline. If you use a polyline to draw the center line of a road, the MEASURE command will mark station points at equal distances along the polyline, without regard to vertices. If you draw the curbs of a parking area as a polyline, the DIVIDE command can quickly insert blocks that represent individual parking stalls. And, of course, AutoCAD has three specific commands, DONUT (or DOUGHNUT), ELLIPSE, and POLYGON, that actually draw these complex objects as closed polylines.

In AutoCAD release 9 (with the standard menu ACAD.MNU), you can select the PLINE command from the pull-down menu under the Draw menu bar. This selection displays the standard PLINE menu in the regular screen menu area. The menu item repetition of release 9 causes the command to repeat automatically until you cancel it by pressing Ctrl-C or selecting another command from the menu. (Only with AutoCAD release 9; see Advanced User Interface and Repeating Commands.)

PLOT

[all versions]

Overview   A drawing that you produce on the screen is not much use unless you can copy it onto paper for use by others. The PLOT command causes your drawing to be plotted on the pen plotter that AutoCAD was previously configured to use. Depending on the type of plotter you are using, the PLOT command allows you to tell AutoCAD which pens to assign to particular screen colors, how large your paper is, at what scale to plot the drawing, and where to place the actual drawing on the paper. You also must specify what part of the drawing you wish to plot. As an alternative, you can instruct the program to direct your output to a disk file rather than sending the drawing directly to the pen plotter. Using this method, you could utilize a plotting utility package to plot offline or to actually copy the plot file onto another machine for subsequent plotting.

Procedure   There are several steps you must take before any drawing can be plotted. When you start the PLOT command, you are first prompted to specify what portion of the drawing you want to plot.

Command: **PLOT**
What to plot -- Display, Extents, Limits, View, or Window <*default*>:

The selection made during the most recent execution of the PLOT command is the default. Any of the five available responses specifies a rectangular area of the drawing. As with any AutoCAD command, the available options may all be abbreviated to their first letter. The Display option (D) sends the current image on the graphics screen to the plotter. The Extents option (E) causes AutoCAD to plot every portion of the drawing that contains entities. This would be the same as executing a ZOOM—Extents command and then plotting the display. Plotting the limits (L) plots only that portion of the drawing that is within the drawing limits (see LIMITS). Specifying the View option (V) causes AutoCAD to prompt for a named view that must already have been saved using the VIEW command. Once provided, only the portion of the drawing that would be displayed on the screen when the view is restored is plotted (see VIEW). The Window option (W) allows you to indicate the lower left and upper right corners of an area of the screen. Only the entities within this window are plotted.

Once you select what part of the drawing to plot, AutoCAD displays the most recent settings for the PLOT command. The screen might appear as shown in Figure 67.

```
Plot will NOT be written to a selected file
Sizes are in Inches
Plot origin is at (0.00,0.00)
Plotting area is 44.72 wide by 35.31 high (MAX size)
Plot is NOT rotated 90 degrees
Pen width is 0.010
Area fill will NOT be adjusted for pen width
Hidden lines will NOT be removed
Plot will be scaled to fit available area

Do you want to change anything? <N>
```

Figure 67

If you answer Y, AutoCAD displays the current settings for the fifteen configurable pen settings before continuing. The available settings vary depending on the type of plotter you have configured. These settings might appear as shown in Figure 68.

Each color numbered 1 through 15 in your drawing can be assigned to a specific pen. If you are using a multiple-pen plotter, the plotter can automatically change pens as it is plotting. If you are using a single-pen plotter, you can make the PLOT command pause and prompt you when it is time to change pens. You can assign the various pens to draw with different-color inks, different line widths, or any combination of pens available on your particular plotter.

Changing Color Assignments   If you answered Y to the previous question, you can change the available settings for each individual color. AutoCAD prompts

```
Enter values. blank=Next value, Cn=Color n, S=Show current values, X=Exit
```

The program then prompts for each successive color and each configurable parameter. For example, the first prompt would appear as shown in Figure 69.

```
Area fill will NOT be adjusted for pen width
Hidden lines will NOT be removed
Plot will be scaled to fit available area

Do you want to change anything? <N> y

Entity        Pen  Line  Pen      Entity       Pen  Line  Pen
Color         No.  Type  Speed    Color        No.  Type  Speed
1 (red)       1    0     36       9            1    0     36
2 (yellow)    2    0     36       10           2    0     36
3 (green)     3    0     36       11           1    0     36
4 (cyan)      4    0     36       12           2    0     36
5 (blue)      5    0     36       13           3    0     36
6 (magenta)   2    0     36       14           4    0     36
7 (white)     4    0     36       15           3    0     36
8             1    0     36

Line types    0 = continuous line
              1 = .................................
              2 = ----     ----     ----     ----
              3 = -----    -----    -----    -----
              4 = ------.  ------.  ------.  ------.
              5 = ---- -   ---- -   ---- -   ---- -
              6 = --- - -  --- - -  --- - -  --- - -
Do you want to change any of the above parameters? <N>
```

Figure 68

422

```
Layer      Pen  Line  Pen
Color      No.  Type  Speed
1 (red)    1     0    10      Pen number <1>:
```

Figure 69

You could, at this point, assign a different pen number to all entities drawn with color number 1 in the drawing to be plotted. Entering a blank accepts the current default value and moves on to the next configurable parameter, in this case, the linetype. You can move through all of the parameters and change whatever ones you want to change (remember to press Return after any change) or accept the default. You can also go directly to the color number you want to change by entering C followed by the appropriate number. For example, to go directly to color 6 (magenta) to change the pen assignment, enter *C6* in response to any parameter prompt.

After you have made some changes, view the current settings of all of the colors by entering *S*. This causes the entire color table to display, after which you can continue to change parameters. When you are satisfied that all of the parameters are correct, enter *X* to go to the other settings of the PLOT command.

Write the plot to a file?   AutoCAD next asks whether you want to send your plotter output to a plot file. If you answer yes, when the drawing is plotted AutoCAD will create a file on the computer's hard disk of the exact output that normally would have gone to the plotter. The default answer for this option is the answer from the previous plot.

```
Write the plot to a file? <default>: (Y or N)
```

If you have configured AutoCAD for a plot spooler subdirectory and you use the default file name AUTOSPOOL, the plot file is sent to the spooler subdirectory and saved with a cryptic name that begins with the letter V (printer-plotter files begin with an R when using this facility).

Size units   Enter the type of units, either *I* (for inches) or *M* (for millimeters), that you will use to measure your plots. Again, the previous execution of the PLOT command sets the default.

```
Size units (Inches or Millimeters) <default>: (I or M)
```

Plot Origin For each method of selecting the image to plot, the lower left corner of the image is mapped to the plot origin, initially the lower left corner of the paper in the plotter. You can relocate this origin to another position on the paper. For example, if you have established that you will use inches as the unit of measure, you can instruct AutoCAD to place the lower left corner of the plotted drawing at a location 4 inches to the right and 3 inches up from the lower left corner of the paper by responding to the plot origin prompt as follows:

```
Plot origin in inches <default>: 4,3
```

As usual, the default origin location is set by the previous plot. On most plotters, the 0,0 origin is approximately 1⅝ inch to the right of the left border of the paper and ⅝ inch up from the bottom. Also, the image does not plot closer than ⅝ inch from the top and right edge of the paper. This is because the plotter must hold the paper between the pinch wheels and grit wheels and cannot plot any closer to the edges. You cannot specify a negative plot origin.

Plotting Size The sizes of paper available depend on your particular plotter, but most have a wide range of paper sizes. You can specify the size exactly, or you can select one of several standard paper sizes listed by entering the letter designation in response to the plotting size prompt. The actual size can be entered as a decimal width and decimal height, separated by a comma; for example, 11,8.5 to specify a paper size 11 inches wide and 8½ inches high. Width and height are determined in the same orientation as you are looking at the graphic screen. The maximum plotter paper size is displayed, which you can select by entering *MAX* in response to the prompt.

Plot Rotation AutoCAD allows you to rotate a two-dimensional plot 90 degrees clockwise. Doing so results in the point that would have been in the lower left corner moving to the upper left, and the point that would have been in the upper left moving to the upper right. AutoCAD asks if you want to rotate the plot by prompting

```
Rotate 2D plots 90 degrees clockwise? <N>
```

Pen Width If you are using solids, traces, or wide polylines, you can adjust for the width of your pen so that these solid filled entities are drawn completely filled. To do so, enter the pen width when AutoCAD prompts

```
Pen width <default>:
```

The width is measured in inches or millimeters, depending on the system of measurement that you selected for size units. Except in cases where solid filling is a paramount concern, the default pen width generally suffices.

Area Fill Adjustment   When solid filling is a paramount concern, you can cause AutoCAD to fill solids, traces, and wide polylines to an accuracy of half the pen width specified. This is generally not necessary but can be done by responding with Y to the prompt

```
Adjust area fill boundaries for pen width? <N>
```

Hidden Line Removal   If you are plotting a three-dimensional view, you can have AutoCAD remove hidden lines from the plot. Hidden-line removal using the HIDE command does not result in removing hidden lines from a plot. You must respond with Y to the prompt

```
Remove hidden lines? <N>
```

at the time the drawing is plotted. Removing hidden lines can add a substantial amount of time to the plotting of a drawing.

Plot Scale   The drawing that you have created in AutoCAD's drawing editor has usually been drawn at full scale. When you plot the drawing, you must therefore specify the scale at which you want AutoCAD to plot your drawing. AutoCAD prompts

```
Specify scale by entering:
Plotted units=Drawing units or Fit or ? <default>
```

The default value is the scale used for the previous plot. Typing *F* or *Fit* causes AutoCAD to scale the portion of the drawing to be plotted so that it is drawn as large as possible on the size paper specified. In this case, you do not know the actual scale at which the drawing is plotted.

You can specify an exact scale by entering the number of plotted units and the number of drawing units, separated by an equal sign (=).

Examples   The exact prompts that AutoCAD displays during the PLOT command vary from system to system, depending on the type of plotter that has been selected during AutoCAD's configuration. For example, if your plotter does not internally support different linetypes (not to be confused with

425

AutoCAD's various linetypes), linetype selection is not available as part of the PLOT command. If your plotter accepts paper only up to 18×24 inches, you will not be able to select D- or E-size paper.

Warnings

You can specify a paper size that exceeds the maximum paper size of your configured plotter. If you do so, AutoCAD displays a warning. You can cancel the plot by pressing Ctrl-C.

Hidden-line removal can add substantially to plot time. Instruct Auto-CAD to remove hidden lines only when such removal is necessary. Remember, if you want hidden lines removed from your plot, you must instruct AutoCAD during the PLOT command to do so. Hidden lines that you remove using the HIDE command will still be plotted unless you remove them again when plotting.

When you have instructed AutoCAD to plot to a file, the specified paper size is not checked against the actual area required for the plot. Make sure that you later use a piece of paper at least as large as the one specified when plotting from the plot file. You will need a separate plot spooler program to plot AutoCAD's plot files. AutoCAD itself does not provide this capability. (One such plot spooler program, called PLUMP, is available from SSC Softsystems, Memphis, TN. It is a highly recommended addition to your AutoCAD system.)

Tips

The most recent plotter configuration is saved automatically in AutoCAD's configuration file (ACAD.CFG) whenever you change the default values.

You can cancel a plot after it has begun by pressing Ctrl-C at any time. Some plotters, because of their internal buffers, may continue plotting for a short time after the plot has been canceled.

Some plotters allow you to set additional features not used by AutoCAD. The PLOT command normally sends a Reset signal to the plotter at the start of plotting. This Reset signal cancels any options entered from the plotter's control panel. After you have entered the plot scale, AutoCAD prompts

```
Position paper in plotter.
Press RETURN to continue or S to Stop for hardware setup.
```

If you respond with *S*, AutoCAD resets the plotter and then prompts

```
Do hardware setup now.
Press RETURN to continue:
```

At this point, you can enter other parameters, such as pen speed, stylus pressure, and acceleration, from the plotter's control panel. AutoCAD will not override these settings when they are entered at this point of the plot routine. When you have finished setting the parameters, press Return to proceed with plotting the drawing.

If you have a single-pen plotter but have configured AutoCAD to change pens while plotting, AutoCAD prompts you when it is time to change pens by displaying a message such as

```
Install pen number 4, color 1 (red)
Press RETURN to continue.
```

Wait for the plotter to stop, change pens accordingly, and then press Return when you are ready tc continue plotting.

If a pen skips or runs out of ink while plotting, you can usually turn off layers that have plotted correctly and simply replot only those layers that were affected. In many cases, the resulting plot will still be usable. Effective use of this technique requires care in two areas. First, don't remove the current plot from the plotter until you have checked to make sure that all the lines were plotted successfully. Second, make efficient use of layers, so you can isolate entities that may not have plotted. This second step basically requires that you avoid mixing different colors on the same layer whenever feasible.

You can plot a drawing directly from AutoCAD's main menu, without first loading it into the drawing editor, by using main menu task 3. AutoCAD prompts for the name of the drawing and the portion of the drawing to be plotted:

```
Specify the part of the drawing to be plotted by entering
Display, Extents, Limits, View, or Window <D>:
```

In this case, Display represents the drawing as it was displayed when last saved (or when the drawing was ended).

You can quickly change all the colors to a particular pen speed or linetype by entering an asterisk before the pen speed or linetype. This global change affects only the colors beginning at the current pen color. For example, when AutoCAD prompts for the pen speed for color number

4, entering *18* changes all the speeds for colors 4 through 15 to a speed of 18 but does not affect colors 1, 2, or 3.

In AutoCAD release 9 (with the standard menu ACAD.MNU), the PLOT command can be selected from the File pull-down menu (only with Auto-CAD release 9; see Advanced User Interface).

PLOTTING

[all versions]

Overview   Your drawing has been created within AutoCAD's drawing editor, but what you now need is a hard copy of your drawing on paper. AutoCAD provides two commands to accomplish this, the PLOT command and the PRPLOT (Printer Plot) command. These two commands are described more fully in other sections (see PLOT and PRPLOT).

Procedure   The PLOT command directs the drawing output to the configured pen plotter. The PRPLOT command sends the hard copy drawing to the configured printer plotter.

The operation of these two commands are also duplicated as tasks that you can execute directly from AutoCAD's main menu. Selecting task 3 executes the PLOT command; task 4 the PRPLOT command. In both cases, the operation of the command is identical to the standard command (described in the individual command description) with one exception. AutoCAD first prompts for the name of the drawing you want to plot or printer plot.

Examples   To plot or printer plot a drawing that is currently on the screen, refer to PLOT or PRPLOT.

If you are plotting or printer plotting directly from AutoCAD's main menu, AutoCAD displays slightly different prompts. In Figures 70 and 71, the extents of a drawing named SAMPLE are plotted (or printer-plotted).

As you can see, after you enter the name of the drawing and the portion of the drawing that you want to plot, the remainder of the prompts are identical to those normally displayed for the PLOT or PRPLOT command.

Tips   For a complete description of plotting or printer plotting, see PLOT or PRPLOT.

```
            A U T O C A D
Copyright (C) 1982,83,84,85,86,87 Autodesk, Inc.
Version 2.6 (4/3/87) IBM PC
Advanced Drafting Extensions 3
Serial Number:  xx-xxxxxx

Main Menu

    0.   Exit AutoCAD
    1.   Begin a NEW drawing
    2.   Edit an EXISTING drawing
    3.   Plot a drawing
    4.   Printer Plot a drawing

    5.   Configure AutoCAD
    6.   File Utilities
    7.   Compile shape/font description file
    8.   Convert old drawing file

Enter selection: 3

Enter NAME of drawing: sample
```

Figure 70

```
            A U T O C A D
Copyright (C) 1982,83,84,85,86,87 Autodesk, Inc.
Version 2.6 (4/3/87) IBM PC
Advanced Drafting Extensions 3
Serial Number:  xx-xxxxxx
Current drawing:  SAMPLE

Specify the part of the drawing to be plotted by entering:
Display, Extents, Limits, View, or Window <D>: e

Plot will NOT be written to a selected file
Sizes are in Inches
Plot origin is at (0.00,0.00)
Plotting area is 44.72 wide by 35.31 high (MAX size)
Plot is NOT rotated 90 degrees
Pen width is 0.010
Area fill will NOT be adjusted for pen width
Hidden lines will NOT be removed
Plot will be scaled to fit available area

Do you want to change anything? <N>
Effective plotting area:  44.72 wide by 33.54 high
Position paper in plotter.
Press RETURN to continue or S to Stop for hardware setup
```

Figure 71

POINT

Overview   The POINT command draws a point entity in the drawing.

Procedure   The POINT command prompts you to enter the location at which you want to draw a point entity. You can respond with an absolute coordinate or a coordinate relative to the last object drawn or by pointing with the screen cursor. A point entity is then placed at the specified location.

Examples   The following example illustrates the typical POINT command prompts.

```
Command: POINT
Point: (select location)
```

Tips   You can use AutoCAD's object snap or the OSNAP command to lock onto a point in the drawing. The Object Snap mode to snap to a point entity is the node selection.

You can alter the way in which AutoCAD displays point entities by changing the value of the system variable PDMODE. You can alter the size at which AutoCAD draws the different images of point entities by changing the system variable PDSIZE. Either of these variables can be altered with the SETVAR command or AutoLISP. Changing either variable affects all the point entities already in the drawing, as well as any new point entities drawn (see PDMODE and PDSIZE).

You can place a point anywhere in three-dimensional space by providing the X-, Y-, and Z-coordinates (only with AutoCAD version 2.6 and release 9 when ADE-3 is included).

POLYGON

Options

E   Specifies polygon by showing one edge
C   Circumscribes around circle
I   Inscribes within circle

P

Overview   The POLYGON command draws an equal-sided polygon, having anywhere from 3 to 1,024 sides. The polygon is actually drawn as a closed polyline, which you can edit with the PEDIT command and other editing commands. You can specify the polygon either by giving the length of one edge or by providing the radius of a circle. The polygon is then drawn so the midpoint of each edge is on the circle (circumscribed) or each intersection of two edges is on the circle (inscribed).

Procedure   When the POLYGON command is first activated, you are prompted to specify the number of sides (between 3 and 1,024). Then AutoCAD prompts you to select one of the available options. You can enter *E* to inform AutoCAD that you will specify the length of an edge, or you can accept the default and provide the location of the center of the polyline.

Examples   **Center Method**   When specifying the center point of the polygon, you can provide its location by entering an absolute coordinate or a relative coordinate or simply by using the screen cursor to point. Once you specify the center point, the polygon is sized based on the radius of a circle. The polygon is drawn either inscribed within this imaginary circle (the vertices of the polygon would occur on the circle) or circumscribed around the circle (the midpoints of each side would occur on the circle). (See the following drawings.) You must enter either *I* (for inscribed) or *C* (for circumscribed) to specify which method to use. You are then prompted for the radius of the circle. If you enter the radius as a value, the polygon is drawn so that the bottom edge of the polygon is drawn at the same angle as the current snap rotation angle. If you provide the radius by entering a point or by using the screen cursor, the selected point not only determines the radius, it locates either a vertex of the polygon (inscribed) or the midpoint of one edge (circumscribed).

```
Command: POLYGON
Number of sides: (enter number of sides)
Edge/<Center of polygon>: (select center point A)
Inscribed in circle/Circumscribed about circle (I/C): (enter C or I)
Radius of circle: (enter radius or select point)
```

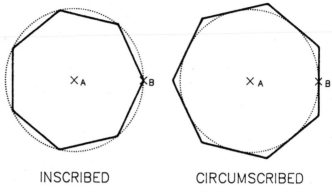

INSCRIBED   CIRCUMSCRIBED

(B = point used to select radius)

Edge Method   When you are specifying one edge of the polygon, respond to the edge/center prompt with *E*. AutoCAD then prompts for the two end points of the edge. You can give absolute or relative coordinates or use the screen cursor to indicate the points on the screen. If you use relative coordinates, specify the second end point relative to the first. The two points define the end points of one edge; the polygon will be drawn in a counterclockwise direction from the first edge. The orientation of the edge determines the orientation of the polygon. This can be seen in the following example and figure.

```
Command: POLYGON
Number of sides: (enter number of sides)
Edge/<Center of polygon>: E
First endpoint of edge: (select end point A)
Second endpoint of edge: (select end point B)
```

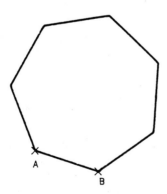

Tips If DRAGMODE is set to on or auto, you can visually drag the size and orientation of the polygon as you specify it using the screen cursor.

POPUPS

Overview The POPUPS system variable holds a value of either 0 or 1. If POPUPS equals 1, the currently configured graphics display driver supports the dialog boxes, icon menus, menu bar, and pull-down menus available under the Advanced User Interface of AutoCAD release 9. If the POPUPS value is 0, these features are not supported and any calls made to them are ignored. POPUPS is a read-only variable.

Procedure The POPUPS system variable is determined by the currently configured graphics display. The only way to control its value is through the selection of the graphics display device when configuring AutoCAD (see Configuring AutoCAD).

You can use the SETVAR command and the (getvar) AutoLISP functions to observe the POPUPS value.

Examples In this example, the SETVAR command returns the current value of POPUPS.

```
Command: SETVAR
Variable name or ?: POPUPS
POPUPS = 1 (read only)
```

Warnings The POPUPS variable is included only in AutoCAD release 9.

Tips The only graphics display drivers capable of using the Advanced User Interface features are the IBM Color/Graphics Adapter (CGA, in monochrome mode only), IBM Enhanced Graphics Adapter (EGA), IBM Video Graphics Array (VGA), Hercules Graphics Card (monochrome), and the Autodesk Device Interface (only if the individual ADI driver supplied by the graphics display manufacturer implements the new features).

Previous Selection Set

[v2.5, v2.6, r9]

Overview   Whenever AutoCAD prompts you to select objects, it is building a selection set. A selection set represents the objects most recently selected. Objects can be added to and removed from a selection set until you are satisfied with the objects you have chosen. When you respond to the prompt to select objects by pressing Return, you instruct AutoCAD to complete the operation of the active editing command. The identity of the entities chosen to be operated on by that command are now stored in AutoCAD's memory as the previous selection set.

Procedure   The previous selection set can be reused by another editing command. When an AutoCAD command prompts you to select objects, you can reuse this previous set by responding with *P*. The last group of entities selected by the most recent editing command will be highlighted to indicate their selection. You can then add or remove objects.

Examples   In the following example, a group of entities is first selected by windowing and then copied to another position in the drawing. Then the CHANGE command is activated and the original objects are selected, using the previous selection set, and changed to a different layer. These sets are illustrated in the following drawing. The original object is selected using the window method and copied from point C to point D. Then the original object is selected again, using the Previous option. This is indicated in the illustration by the object appearing with a dotted line type. On the graphics screen, the object would be highlighted instead.

```
Command: COPY
Select objects: W
First corner: (select point A)
Other corner: (select point B) 5 found.
Select objects: [ RETURN ]
<Base point or displacement>/Multiple: (select point C)
Second point of displacement: (select point D)
Command: CHANGE
Select objects: P
5 found.
Select objects: [ RETURN ]
```

```
Properties/<Change point>: P
Change what property (Color/Elev/LAyer/LType/Thickness) ? LA
New layer <0>: 1
Change what property (Color/Elev/LAyer/LType/Thickness) ? [ RETURN ]
```

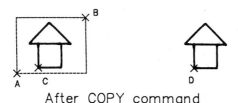

After COPY command

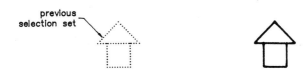

After CHANGE command

Warnings   The previous selection set is not saved across editing sessions. Once you end or plot a drawing, the Previous Selection Set is cleared from memory.

   The ERASE command clears the previous selection set from memory. After an ERASE command, there is no previous set.

Tips   Use of the previous selection set can save a substantial amount of time when you are editing large drawings, since AutoCAD does not have to scan the entire drawing database to find the selected objects. Using the previous selection set effectively requires forethought.

Printer Echo   [all versions]

Overview   You can tell AutoCAD to cause anything displayed within the command-line area of the screen (and the text screen) to be printed on your printer. In this way, layer listings, prompts, and even keyboard input can be echoed to the printer.

Procedure   You can toggle printer echoing on and off by pressing Ctrl-Q. AutoCAD's command line indicates whether the printer echo is on or off. You can include this printer toggle feature in menus by including a menu selection that sends the line ^Q.

Examples   When printer echoing is turned on, everything you type at AutoCAD's command prompt and everything AutoCAD displays there will also be printed on the printer. This is useful if you need a complete record of an editing session, to print out a list of layer names and attributes, or to debug a menu macro or AutoLISP routine.

Warnings   The printer echo feature of AutoCAD is in addition to any printer echo features included in the computer's operating system. Turning on the printer echo for both the operating system and AutoCAD could cause double echoing. AutoCAD's printer echo toggle does not affect the operating system's echo feature.

       Activating AutoCAD's printer echo when there is no printer connected could cause your computer to lock up.

Tips   The echoing of items selected from menus to the printer is disabled if the system variable MENUECHO has a value of 2 (see MENUECHO).

Prototype Drawing

[all versions]

Overview   Whenever you begin a new drawing (AutoCAD main menu task 1), certain drawing parameters are preset. Included in this set of predetermined parameters are mode settings (Ortho mode, Snap mode, Fill mode, and so on), the size of text, and dimensioning variables. All of the system variables that are saved along with the individual drawing files are predetermined. AutoCAD determines the settings for these parameters from a prototype drawing.

Procedure   Typically the prototype drawing is a drawing file called ACAD.DWG. This drawing file is supplied on the AutoCAD release disks and is normally copied into the \ACAD subdirectory. When you enter the name of the drawing you are about to create, AutoCAD creates a new file in memory with the name that you have entered and initially displays the prototype drawing on the screen.

You can create any number of additional prototype drawings that might contain your predrawn title block, predefined layer names, and even completed portions of a drawing. Then, when you begin a new drawing, you can instruct AutoCAD to use one of your own prototype drawings.

Examples　**Configuring the Prototype Drawing**　When you configure AutoCAD, you can instruct the program to use a specific drawing file as the prototype drawing. This is accomplished from the configuration menu (main menu task 5). From the configuration menu, select task 8, Configure Operating Parameters. This displays another submenu, the operating parameters menu. Task 2 in this submenu allows you to set the initial drawing setup. AutoCAD will issue the following prompt:

```
Enter name of default prototype file for new drawings
or . for none <current>:
```

AutoCAD's initial prototype file name ACAD usually appears as the default. You can enter a different drawing file name as the prototype, in which case that drawing becomes the default prototype drawing. Alternatively, you can enter a period, which indicates to AutoCAD that you do not have any prototype drawing defined. If you press Return, the existing default prototype will continue to be used as the prototype drawing. You can always override the prototype drawing each time you begin a new drawing.

Beginning a New Drawing　When you begin a new drawing, you actually tell AutoCAD which file to use as the prototype drawing. From the main menu, when you select task 1, Begin a New Drawing, AutoCAD prompts you for the name of the drawing:

```
Enter selection: 1
Enter NAME of drawing:
```

The name you enter becomes the name of the new drawing. The prototype drawing used is the drawing configured as the prototype drawing in the configuration menu (typically ACAD.DWG).

You can alternatively include the name of a drawing file to be used as the prototype drawing for this new drawing by entering the new drawing name, an equal sign, and the name of the drawing you want to use as the prototype:

```
Drawing-name=Prototype-name
```

437

In this case, all of the parameters set by the prototype drawing and any entities already drawn in the prototype drawing will be present in the new drawing. Both the drawing name and the prototype name can include drive designations and directory paths. Do not include any file extensions. The file extension .DWG is understood. If the prototype name referenced cannot be found, AutoCAD will issue the error message

```
** Prototype drawing (name) not on file.
Press RETURN to continue.
```

and will display the main menu again.

A third method is also available. You can instruct AutoCAD not to use any prototype drawing. This yields the same result as entering a period when configuring AutoCAD's prototype drawing. In this case, AutoCAD uses its default environment settings for all of the parameters that need to be predetermined. No layers are preset other than the 0 layer. This method was used to create the original ACAD.DWG drawing file that comes with the AutoCAD program. Enter the name of the new drawing you are creating and an equal sign with nothing after it:

```
Enter NAME of drawing: New-drawing=
```

No prototype drawing will be used.

Tips

Edit the ACAD.DWG file to change the settings for such parameters as units, text height, and dimensioning variables so you do not have to reset them every time you begin a new drawing.

When using the DXFIN command, use the third method to begin a new drawing to ensure that no entities are already drawn in the prototype drawing. Otherwise, if the prototype drawing has been altered, the DXFIN command may not be successful.

PRPLOT

[all versions]

Overview

The PRPLOT (Printer Plot) command causes your drawing to be plotted on the printer plotter that AutoCAD has been configured to use. Depending on the type of printer plotter you are using, the PRPLOT command allows you to tell AutoCAD to assign screen colors to particular colors available on your printer-plotter, how large your paper is, at what scale to plot the drawing, and where to place the actual drawing on the paper. Or you can instruct the program to direct your output to a disk file rather sending the drawing directly to the printer-plotter. Using this method, you can utilize a utility package to plot offline or in the backgound so you can continue working on a drawing while the printer-plotter is at work.

Procedure   The operation of the PRPLOT command is virtually the same as that of the PLOT command. Both are similar in all respects to plotting or printer plotting directly from AutoCAD's main menu. (For a complete description of AutoCAD's plotting facilities, see PLOT.)

There are several steps you must take before any drawing can be printer-plotted. When you start the PRPLOT command, you are first prompted to specify what portion of the drawing you want to plot. Once you do so, AutoCAD presents the most recent settings for the PRPLOT command (those used the last time you used the command or the default settings if this is the first time you have used the command). You will be asked if these settings are correct. If they are acceptable, you can continue with the command. If any of the parameters require changing, you must tell AutoCAD that you wish to change something. The program then takes you through a series of prompts so you can alter any of the default settings.

The only difference between the PRPLOT and PLOT commands is that PRPLOT sends the output to the configured printer-plotter rather than to the plotter. If you are generating a plot file, the output is in a form suitable for the configured printer plotter (a raster file) rather than for the plotter (a vector file). If you have configured AutoCAD for a plot spooler subdirectory and you use the default file name AUTOSPOOL, the printer-plotter file is sent to the spooler subdirectory and saved with a cryptic name that begins with the letter R (plotter files begin with a V when using this facility).

Examples   Figures 72 and 73 illustrate the prompts that would be presented when you use the PRPLOT command. Compare these prompts with the descriptions provided for the PLOT command.

```
Plot will NOT be written to a selected file
Sizes are in Inches
Plot origin is at (0.00,0.00)
Plotting area is 13.59 wide by 11.00 high (MAX size)
Plot is NOT rotated 90 degrees
Hidden lines will NOT be removed
Plot will be scaled to fit available area

Do you want to change anything? <N>
```

Figure 72

```
Write the plot to a file? <N>
Size units (Inches or Millimeters) <I>:
Plot origin in Inches <0.00,0.00>:

Standard values for plotting size

Size    Width   Height
A       10.50    8.00
MAX     13.59   11.00

Enter the Size or Width,Height (in Inches) <MAX>:
Rotate 2D plots 90 degrees clockwise? <N>
Remove hidden lines? <N>

Specify scale by entering:
Plotted Inches=Drawing Units or Fit or ? <F>:
Effective plotting area:  13.59 wide by 9.96 high
Position paper in printer.
Press RETURN to continue:
```

Figure 73

Tips   PRPLOT operates by sending horizontal lines of graphics output to the printer plotter one at a time, beginning at the top of the drawing and working toward the bottom. If there is only blank space to the right of an object, AutoCAD advances the paper one line, returns the print head to the left of the paper, and begins sending the next horizontal strip of the drawing. Thus, if you can eliminate unneccessary objects from the right side of the drawing, you can save a great deal of time when printer plotting.

Notice that the PRPLOT command does not prompt for adjustments to pen width or area fill, since these values have no meaning when printer plotting.

You can have your printer plot a drawing directly from AutoCAD's main menu, without first loading it into the drawing editor, by using main menu task 4. AutoCAD prompts for the name of the drawing and the portion of the drawing to be plotted:

```
Specify the part of the drawing to be plotted by entering
Display, Extents, Limits, View, or Window <D>:
```

In this case, Display represents the drawing as it was displayed when last saved (or when the drawing was ended).

In AutoCAD release 9 (with the standard menu ACAD.MNU), you can select the PRPLOT command from the File pull-down menu (see Advanced User Interface).

See PLOT for additional warnings and tips.

Pull-Down Menus   [ADE-3] [r9]

Overview   Pull-down menus are an extension of the standard AutoCAD menu structure. They are displayed only when you are using AutoCAD release 9 with a graphics display that supports the Advanced User Interface features. A pull-down menu is the rectangular menu that drops below a menu bar when the menu bar is selected.

Procedure   When you use the pointing device to move the screen cursor to the top of the graphics screen, the status line is replaced by a menu bar. Highlighting a menu bar item and pressing the pick button causes the associated pull-down menu to be displayed. The pull-down menu appears below the selected menu bar item. Moving the cursor down highlights the various items in the pull-down menu. Pressing the pick button when one of the pull-down menu items is highlighted executes that command or menu macro. A pull-down menu remains displayed until you pick an item from it; type on the keyboard; press the pick button while over another area of the graphics screen, menu bar, regular screen menu, or tablet menu; or make a selection using the button menu.

Pull-down menus are similar in structure to regular screen menus. A pull-down menu section is indicated by the section headers ***POP1 through ***POP10. Pull-down menus can have pull-down submenus, which you can call by using a $P= construct similar to the $S= method used with screen menus. Submenus have a header indicated by a double asterisk (**).

The first line after a header is the title that appears in the menu bar. The menu bar title occurs alone on the next line and is enclosed in square brackets. Each title can be up to fourteen characters long. However, since most graphics displays provide only eighty characters across the width of the screen, if you are going to use all ten menu bar items you should limit titles to eight characters each (allowing for blank spaces between them).

Each successive line after the header represents a pull-down menu line. The first item on each line is the actual text that appears on each line of the pull-down menu. This text is enclosed in square brackets. Many graphics cards display only twenty-one lines, so you should limit the number of lines in each pull-down menu to this number. The width of any pull-down menu is determined by the longest title in it. Because pull-down menus occur directly beneath their menu bar titles, you should keep your menu bar titles short, especially those toward the right side of the screen. Auto-CAD truncates any titles that do not fit onto the screen.

For the purposes of swapping, the ten available pull-down menus are named P1 through P10. The submenu header is used as the submenu name. Therefore, to replace pull-down menu area 3 with the submenu **ARCS, you could include a line in a menu macro as follows:

```
[Arcs]$P3=ARCS $P3=*
```

This would cause the original third pull-down menu to be replaced with the ARCS pull-down menu. The $P3=* command causes pull-down menu number 3 to be immediately displayed. Otherwise, the menu would have been swapped, but you would not see the change until you accessed the menu bar again.

Examples The following portion of a menu file is taken from the ACAD.MNU file provided with AutoCAD release 9. This section illustrates the makeup of a typical pull-down menu. Each line is annotated. The actual pull-down screen is shown in Figure 74.

MENU	EXPLANATION
***POP2	Pull-down section header.
[Draw]	Title appearing in menu bar.
[Line]*^C^C$S=X $s=line line	These lines call associated screen
[Arc]*^C^C$S=X $s=poparc arc	menus & execute one of AutoCAD's
[Circle]*^C^C$S=X $s=popcircl circle	entity draw commands. The *^C^C
[Polyline]*^C^C$S=X $s=pline pline	construct is used to cause the macro
[Insert]^C^Csetvar attdia 1 $s=insert insert	to repeat. (See Repeating
[Dtext]*^C^C$S=X $s=Dtext Dtext	Commands.)
[Hatch]^C^C$i=hatch1 $i=*	

Warnings The pull-down menu feature is available only with AutoCAD release 9 when used with a graphics card that supports the Advanced User Interface (see Advanced User Interface).

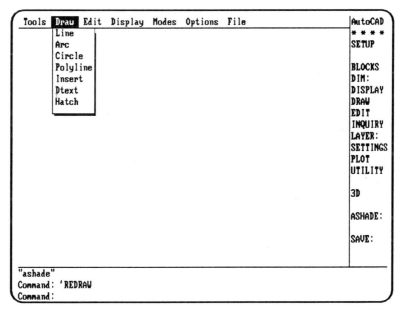

Figure 74

Tips   If a menu title begins with a tilde (˜), the title appears greyed out. This is often done to indicate that a selection is not available.

If a menu title contains two hyphens, the title expands to become a line that fills the entire width of the pull-down menu (or thirty-nine characters, whichever is less). A combination of this method with the tilde character will display a greyed-out line.

You can include control codes in menus by preceding the character with a caret (^). Thus, ^C is the same as Ctrl-C, and ^H is the same as a backspace.

You can spread out long menu macros over more than one line by placing a plus sign as the last character of the line and continuing the macro on the next line.

A semicolon (;) has the same effect as a RETURN.

A backslash (\) within a menu macro causes the macro to pause for user input. Once the input is provided, the macro continues.

If you intend to customize AutoCAD's standard menu ACAD.MNU, make the changes to a copy so you can always restore the original.

For additional information, see Icon Menus, Menu Bar, Screen Menu, and Tablet Menu.

PURGE

[all versions]

Options:

A	Purges all unused named objects
B	Purges unused blocks
LA	Purges unused layers
LT	Purges unused linetypes
SH	Purges unused shape files
ST	Purges unused text styles

Overview   Occasionally when you are working on a drawing, you may create blocks, layers, or text styles, load shape files, or assign linetypes that you end up never actually using. AutoCAD remembers all these objects once you have named them and displays them whenever you request a listing (such as with the LAYER ? option). The PURGE command allows you to remove these named objects from the drawing file, but only if they have never actually been included in the drawing.

Procedure

The PURGE command is accepted only if it is the very first command you issue when you first load a drawing from AutoCAD's main menu. Only objects that have been named but never included within the drawing can be purged. You can instruct the PURGE command to remove all unused objects or only unused blocks, layers, linetypes, shape files, or text styles. Select what you want to purge by entering the capitalized letters of the desired option (for example, type *LA* to purge a layer name). AutoCAD prompts you to verify removal of each unused object before it is actually removed.

```
Command: PURGE
Purge unused Blocks/LAyers/LTypes/SHapes/STyles/All: (select one)
```

As long as you do not issue any other command at the beginning of an editing session, you can repeat the PURGE command if necessary. If the PURGE command finds no unused objects, AutoCAD responds that it has found no unreferenced objects.

Examples

A typical use of the PURGE command might look like the following sequence. Notice that named objects are not deleted unless you enter *Y* in response to the verification prompt. In this example, a dashed linetype was assigned to the CABINET layer. Once that layer has been purged, the dashed linetype can be purged if it has not been used anywhere else in the drawing. First, however, the drawing would have to be ended and reloaded (see "Tips").

```
Command: PURGE
Purge unused Blocks/LAyers/LTypes/SHapes/STyles/All: A
Purge block TOILET? <N> Y
Purge block SINK? <N> RETURN
Purge layer ROOF? <N> RETURN
Purge layer FLOOR? <N> Y
Purge layer DOOR? <N> RETURN
Purge layer WINDOW? <N> Y
Purge layer CABINET? <N> Y

No unreferenced linetypes found.
No unreferenced text styles found.
No unreferenced shape files found.
```

445

Warnings   PURGE works only if it is the first command issued when you begin editing an existing drawing.

You cannot purge the 0 layer, the continuous linetype, or the standard text style.

The PURGE command does not remove views. Use the Delete option of the VIEW command for this purpose.

Tips   The PURGE command may not remove every unreferenced object. You may need to end the drawing, then load it and issue the PURGE command again to remove all unused objects. You may have to repeat this process several times to remove every unused object. This is because AutoCAD does not see an object as being unused if it is referenced by another object, even if that other object is itself unused. For example, if you have a dotted linetype specified for a particular layer that has never had any entities drawn on it, the first time you issue the PURGE command, the layer name will be removed. But because the linetype was not unreferenced until the layer had been removed, the PURGE command will not find the unused linetype until you end the drawing and reload it. To remove every unused object, repeat the process several times until the PURGE command reports that it finds no unreferenced objects.

QTEXT
[all versions]

Options

On   Quick text mode on
Off   Quick text mode off

Overview   It takes AutoCAD quite a bit of time to completely draw the individual letters of every text entry in a drawing. To speed the displaying of the drawing, turn on QTEXT (Quick Text) mode. When QTEXT is on, AutoCAD indicates only the location of text entities by displaying a rectangle of the same approximate height and length as the string of text it represents. When QTEXT is off, the actual text is displayed. (See the following drawing.)

Addison—Wesley
Expert Advisor
AutoCAD

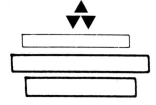

QTEXT "Off"   QTEXT "On"

Procedure   QTEXT has only two possible settings: on and off. When you issue the QTEXT command, AutoCAD prompts you to enter the desired setting and displays the current setting as the default. To accept the default, simply press Return. To change the current setting, enter either on or off and press Return.

```
Command: QTEXT
ON/OFF <current>: (enter desired setting)
```

Examples   As you enter text or attribute information, the actual text initially displays regardless of the setting of QTEXT. The next time the drawing is regenerated, however, if QTEXT is on, the text will be replaced with the rectangle representation.

Warnings   You must issue the REGEN command to force AutoCAD to regenerate the drawing before the current setting of QTEXT takes effect.

Be sure to turn QTEXT off and regenerate your drawing before plotting, or the text will be plotted as the displayed rectangles rather than as text.

Tips   Set QTEXT to on whenever your drawing contains many text or attribute entities. Text takes a long time to draw on the screen. The resulting increase in drawing speed will increase your productivity as you edit your drawing. Just remember to turn QTEXT off and to regenerate the drawing before plotting.

Even though text displays as a rectangle, the LIST command still responds with the actual text string of any selected text or attribute entities.

You can change the setting of QTEXT by changing the current setting of the system variable QTEXTMODE, using the SETVAR command or AutoLISP. QTEXT is on when this variable equals 1 and off if it equals 0.

447

QTEXTMODE

Overview   The QTEXTMODE system variable stores the integer 0 or 1. The value determines whether Qtext is on or off.

Procedure   Normally the QTEXT command turns Qtext on and off. But you can also use the SETVAR command or AutoLISP to control Qtext. Changing the setting of Qtext has no effect on text until the drawing is regenerated. A QTEXTMODE value of 1 indicates that Qtext mode is on and a value of 0 indicates that it is off.

Examples   In this AutoLISP example, the first line of the routine returns the current QTEXTMODE value and saves it to the variable QM. Then QTEXTMODE is set to 1 (Qtext On). The last line in the routine restores QTEXTMODE to its original setting. This routine could be used within an AutoLISP program that does a lot of text manipulation and that would otherwise be slowed down by continuously having to redraw the text.

PROGRAM	COMMENTS
`(setq QM (getvar "QTEXTMODE"))`	
`(setvar "QTEXTMODE" 1)`	
.	The body of the program
.	would occur here
`(setvar "QTEXTMODE" QM)`	

Tips   The initial QTEXTMODE value is determined from the prototype drawing (default value is 0, Qtext off). The current value is saved with the drawing.

QUIT

Overview   The QUIT command abandons any changes made to the drawing since the last time it was saved (if any) and returns to AutoCAD's main menu.

Procedure   To exit from the drawing without recording changes, issue the QUIT command. Because this could cause you to lose a great deal of work if done accidentally, AutoCAD asks you to verify the command.

Command: **QUIT**
Really want to discard all changes to drawing?

Answering yes or *Y* discards all changes. Any other answer cancels the QUIT command and returns you to the command prompt. If you quit the current drawing, the drawing's previous .DWG and .BAK files, if any, remain unchanged.

Examples   One particularly useful application of the QUIT command is when you are trying a what-if type of change to a drawing. First, save your drawing just prior to a potentially damaging series of alterations. If you find that you have ruined your drawing, all is not lost. Simply quit the drawing before you save the ruinous changes and reload it again. You will be back in the drawing at the same spot before you made any changes. (You could also use the UNDO command to reverse damage.)

Warnings   If you quit a drawing by mistake, you lose all changes made to that drawing since the last time it was saved. There is no way to recover those changes once you have answered *Y* to the QUIT command's verification prompt.

Tips   Build the QUIT command into any custom menus you create, but don't include an automatic *Y* in a macro. Make sure you always have the command prompt to ensure that you really want to abandon all changes.

If you enter any letter other than *Y* when the QUIT command prompts for verification, you will be returned to AutoCAD's command prompt.

In AutoCAD release 9 (with the standard menu ACAD.MNU), you can select the QUIT command from the File pull-down menu (only with Auto-CAD release 9; see Advanced User Interface).

REDEFINE   [ADE-3] [r9]

Overview   The REDEFINE command restores an AutoCAD command that was previously undefined (see UNDEFINE). When a command is undefined, it is not recognized when it is entered in response to AutoCAD's command prompt. Redefining a command restores the command so that it is once again recognized.

Procedure

To redefine a command that previously has been undefined, activate the REDEFINE command. The command prompts for a command name. Enter the AutoCAD command that had previously been undefined (using the UNDEFINE command).

Examples

If the LINE command had previously been undefined, it can be redefined using the REDEFINE command.

```
Command: REDEFINE
Command name: LINE
```

Tips

Only main AutoCAD commands (those entered directly at the command prompt) can be undefined and redefined. For example, you cannot undefine the LAYER Set option or the dimensioning subcommands).

For further information on command redefinition and its uses, see UNDEFINE.

REDO

[v2.5, v2.6, r9]

Overview

If the last command issued was either U or UNDO and you didn't really want to undo something, REDO reverses the last undo. REDO reverses only the last undo.

Procedure

Immediately after either a U or an UNDO command, issue the command REDO. The last undo will be reversed.

```
Command: REDO
```

Examples

In the following example, everything in the drawing is accidentally undone. Realizing the mistake, the REDO command immediately restores the drawing to its condition prior to the UNDO command.

```
Command: UNDO
Auto/Back/Control/End/Group/Mark/<number>: B
This will undo everything. OK? <Y> ( RETURN )
CIRCLE LINE
Everything has been undone

Command: REDO
```

Warnings   REDO reverses only the last undo. It must follow immediately after an UNDO command. Otherwise, AutoCAD responds with the message

`Last command did not undo anything.`

Tips   When you have just undone something, check the screen. If you have undone something in error, immediately execute the REDO command so you can recapture what you have accidentally undone. Then try the UNDO command again.

In AutoCAD release 9 (with the standard menu ACAD.MNU), you can select the REDO command from the Tools pull-down menu. (See Advanced User Interface.)

'REDRAW   [all versions]

Overview   The REDRAW command explicitly causes AutoCAD to refresh the screen. This removes the little blip markers left when selecting objects when BLIP-MODE is on.

Procedure   To refresh the screen, issue the REDRAW command:

`Command: REDRAW`

You can use the REDRAW command transparently (while another command is active) by preceding the command with an apostrophe, for example, 'REDRAW.

Examples   Several other commands automatically cause AutoCAD to redraw the screen, for example, when you turn a layer on or toggle the visible grid off.

Warnings   Transparent REDRAW is not available in AutoCAD version 2.5.

Tips   You can cancel the REDRAW command by pressing Ctrl-C.

     The REDRAW command also redraws the screen while in AutoCAD's dimensioning mode if you enter the REDRAW command at the dim prompt.

     In AutoCAD release 9 (with the standard menu ACAD.MNU), you can select the transparent REDRAW command from the Tools pull-down menu (see Advanced User Interface).

REGEN

Overview   The REGEN command forces AutoCAD to regenerate an entire drawing and to redraw the screen. When AutoCAD regenerates a drawing, it mathematically calculates the drawing display. This can be a lengthy process on large drawings.

Procedure   To force AutoCAD to regenerate the drawing, issue the REGEN command.

```
Command: REGEN
```

Examples   Some commands automatically force AutoCAD to regenerate the drawing under certain conditions. Since regenerating the drawing can be a lengthy process on large drawings, the command REGENAUTO allows you to control this automatic regeneration (see REGENAUTO).

Warnings   If you are using AutoCAD release 9, you can freeze the current layer using the Modify Layer dialog box activated by the DDLMODES command. If you do this, a warning dialog box is displayed informing you that any entities added to the drawing will not appear until you regenerate the drawing after thawing the current layer. You must use the REGEN command before the entities will become visible. (This applies only to Auto-CAD release 9; see Advanced User Interface and DDLMODES.)

Tips   You can cancel a REGEN command by pressing Ctrl-C.

REGENAUTO

[all versions]

Options

On   Allows automatic regeneration
Off   Prevents automatic regeneration

Overview   The REGENAUTO command controls whether the drawing will be regenerated by the actions of certain other AutoCAD commands. Normally AutoCAD automatically updates the drawing if, for example, you redefine a block, change the LTSCALE, or freeze or thaw layers. But such regenerations of a drawing can take a great deal of time. The REGENAUTO command allows you to disable this automatic regeneration feature.

Procedure   To disable automatic regeneration of a drawing, use the REGENAUTO command to turn automatic regeneration off. To turn automatic regeneration on, enter *On* in response to the prompt. The current status of REGENAUTO is displayed as the default.

```
Command: REGENAUTO
ON/OFF <current setting>: (enter ON or OFF)
```

Examples   Eight different AutoCAD commands are affected by the setting of REGENAUTO. The ATTEDIT command normally causes a drawing to be regenerated when you are doing global editing of attributes. A drawing is also normally regenerated when you redefine a block or use the INSERT command with the Block = file method. A drawing is also normally regenerated if a layer is frozen or thawed, the LTSCALE is changed, or the STYLE command is used to redefine a text style. However, if REGENAUTO is set to off, a drawing will not be regenerated until you enter the REGEN command explicitly.

Tips   When the VIEWRES command has been used to enable Fast Zoom mode and REGENAUTO is off, if you attempt to pan to an area of the drawing that

453

is off the screen or you use the ZOOM command to view an area that either is not entirely on the screen or is not shown in sufficient detail, AutoCAD will prompt

```
About to regen, proceed? <Y>
```

Pressing Return causes the drawing to be regenerated. If you respond no, the PAN or ZOOM command is canceled. This is an effective way to save time while you are working in a large drawing. If you really don't need to pan or zoom to a particular location, use this method to end the command. Then try a slightly different area of the drawing (or use the ZOOM Dynamic option).

REGENMODE

[all versions]

Overview

The REGENAUTO command determines whether certain AutoCAD commands will automatically cause the drawing to be regenerated or if the program will warn you that a regeneration is necessary to complete the current command. The REGENAUTO mode setting is stored in the REGENMODE system variable.

Procedure

Changing the current setting of REGENAUTO causes the REGENMODE value to be changed. You can also use the SETVAR command or AutoLISP to change the value of REGENMODE. If the value is 0, REGENAUTO is off; a value of 1 indicates that REGENAUTO is on.

Examples

In this example, the SETVAR command returns the current value of REGENMODE. Entering a new value of 0 turns REGENAUTO off.

```
Command: SETVAR
Variable name or ?: REGENMODE
New value for REGENMODE <1>: 0
```

Tips

The initial REGENMODE setting is determined from the prototype drawing (default value is 1, REGENAUTO is on). The current value is saved with the drawing.

RENAME

Options

B Renames block
LA Renames layer
LT Renames linetype
S Renames text style
V Renames named view

Overview
If you make a typing error when naming a block, layer, linetype, text style, or view, or if you just change your mind, the RENAME command allows you to change the name.

Procedure
To change the name of a block, layer, linetype, text style, or view that already exists in a drawing, enter the RENAME command at the command prompt. AutoCAD prompts you for the type of object whose name you want to change. You can respond with only the letters that the prompt displays in capital letters (for example, LA for layer). Next, the RENAME command asks for the old name and then the new name.

```
Command: RENAME
Block/LAyer/LType/Style/View: (enter your selection)
Old (object) name: (enter the old name)
New (object) name: (enter the new name)
```

Examples
In the following example, the layer named Door is renamed Doors.

```
Command: RENAME
Block/LAyer/LType/Style/View: LA
Old layer name: DOOR
New layer name: DOORS
```

Tips
You cannot change the name of the 0 layer or the continuous linetype. You can, however, change the name of other linetypes and the standard text style.

Repeating Commands

[all versions]

Overview   Every time you activate an AutoCAD command, the program remembers what command has been activated. When the current command is completed, you can reactivate it without having to pick it on the menu again or enter it on the keyboard.

Procedure   Simply pressing Enter, Return, or the equivalent button on your cursor causes the last active command to be repeated. Pressing the Space bar is the same as pressing Enter or Return.

Release 9 Only   In AutoCAD Release 9, two additional methods are available to repeat commands. The MULTIPLE command modifier enables commands that would normally carry out only one action (and then return to the command prompt) to continue to repeat indefinitely until you press Ctrl-C to cancel the command (see MULTIPLE).

Menu items can be made to repeat, complete with all included macros, in exactly the same way they are repeated when you use the MULTIPLE command modifier. This is accomplished by an extension to AutoCAD's standard menu macro language. If the menu macro is prefaced by the string *^C^C immediately following the label, the entire menu macro is saved as if it were a command and repeated indefinitely until you press Ctrl-C (see the following example). These two features are available only in AutoCAD release 9.

Examples   Many commands require that you first activate them to accomplish a setting and that they be reactivated to be used. One such command is the FILLET command. To set a fillet radius, you must first enter the FILLET command and then *R* to indicate that you want to change the fillet radius. After you set the radius, the command is completed. Normally you would need to reactivate the FILLET command to join two lines with an arc of the specified radius. Instead, you can just press Return to repeat the command.

Release 9 Only   If a menu macro is prefaced by the string *^C^C, the entire macro is saved exactly as a command. The entire macro repeats indefinitely until you press Ctrl-C.

The following menu macro would repeat the prompts for the ARC command, substituting the prompts for selecting the starting point, the center point, and the end point, rather than reverting to the ARC command's standard three-point method.

```
[ARC sce]*^C^CARC \C \
```

```
Command: (select menu item)
ARC Center/<Start point>: (select start point)
Center/End/<Second point>: C Center: (select center point)
Angle/Length of chord/<End point>: (select end point)
Command: ARC Center/<Start point>: (select start point)
Center/End/<Second point>: C Center: (select center point)
Angle/Length of chord/<End point>: (select end point)
Command: ARC Center/<Start point>: (select start point)
Center/End/<Second point>: C Center: (select center point)
Angle/Length of chord/<End point>: (select end point)
Command: ARC Center/<Start point>: ( Ctrl-C )  *Cancel* (to cancel)
```

Warnings The MULTIPLE command modifier and the *^C^C menu construction methods of repeating command are available only with AutoCAD release 9.

Tips If you are right-handed, keep your right hand on your digitizer cursor and your left hand over the Space bar on the keyboard. That way, you can always repeat the last command, without having to take your eyes off the screen, by simply pressing the Space bar (using it as a Return key).

'RESUME

[all versions]

Overview One of AutoCAD's powerful features is that it lets you preprogram any number of command sequences through the use of scripts. (See SCRIPT.) If an error occurs while a script is active or if you press Ctrl-C or Backspace while a script is active, the script is interrupted. You can use the RESUME command to reactivate a script at the point at which it was interrupted.

Procedure To continue a script, enter the RESUME command. If an AutoCAD command was active when the interruption occurred, precede the RESUME command with an apostrophe, as in 'RESUME, to use the RESUME command transparently.

```
Command: RESUME
```

Examples   If a script file was running and drawing lines automatically, you may have interrupted the script file and left AutoCAD sitting and waiting for a response. To make the script file pick up where it left off, use the transparent method.

```
To point: 'RESUME
```

Tips   You can interrupt command scripts by pressing Backspace. Use the RESUME command to restart command scripts at the point they were interrupted.

ROTATE

[ADE-3] [v2.5, v2.6, r9]

Overview   The ROTATE command changes the orientation of any existing object in a drawing. The object is rotated about a user-specified point. You can change the rotation by specifying the absolute angle of orientation of the object or by rotating the object to an angle relative to another angle.

Procedure   To change the orientation of an object in a drawing, activate the ROTATE command by entering it at the command prompt. The ROTATE command prompts you to select the object or objects you want to rotate. You can select objects using any acceptable method (window, last, previous, pointing, and so on). Any objects selected are rotated about the same base point. When you have finished selecting objects, press Return a final time in response to the prompt to select objects.

Next, the ROTATE command asks you for the base point. You can use the screen cursor to select a point on the screen or enter the coordinates at the keyboard. Finally, you are prompted for the rotation angle. The rotation angle is the number of degrees (or radians or grads, depending on the current UNITS) you wish to rotate the selected objects. The rotation angle you enter is only in relation to the current orientation of the objects. For example, if the object in question is already drawn at 30 degrees, responding with 45 degrees for the rotation angle causes the object to be rotated to 75 degrees. It is rotated an *additional* 45 degrees.

```
Command: ROTATE
Select objects: (select the objects to be rotated)
Base point: (select point)
<Rotation angle>/Reference:
```

You can also rotate the object in reference to another angle, by responding to the <Rotation angle>/Reference prompt with R, to indicate that the angle you are about to provide is a relative angle. The Relative option causes the ROTATE command to issue additional prompts. You are asked for a reference angle and a new angle. In this case, the object is rotated from the reference angle to the new angle. For example, if a reference angle of 75 degrees (the orientation of the object in the previous paragraph) is entered and a new angle of 30 degrees is specified, the object would now be oriented at 30 degrees. If the Relative option had not been chosen, the object would have simply been rotated an additional 30 degrees (to a 105-degree orientation).

You can enter the reference angle or use the screen cursor (usually with the help of object snap) to indicate the reference angle. Then you can enter the new angle from the keyboard or use the screen cursor again.

```
<Rotation angle>/Reference: R
Reference angle <0>: (select reference angle)
New angle: (enter new angle)
```

Examples   The following example and drawings illustrate both methods of specifying the reference angle. First, the object is selected by the window method of entity selection. The base point (about which the object will be rotated) is selected using the midpoint Object Snap (point C). The object is then rotated to an angle of 45 degrees.

The object is then selected again by the window method. Once again, the base point (point F) is selected using the midpoint Object Snap. This time the object is rotated in relation to a reference angle, in this case the angle from the base point (point F) to the intersection (point G). This angle is the current orientation angle of the object. The object is then rotated to a new orientation angle of 180 degrees. This reference angle method is particularly useful when you do not know the current orientation of an object but you do know the angle to which you want it rotated.

459

```
Command: ROTATE
Select objects: W
First corner: (select point A)
Other corner: (select point B)
Select objects: ( RETURN )
Base point: MID
of (select point C)
<Rotation angle>/Reference: 45
Command: ROTATE
Select objects: W
First corner: (select point D)
Other corner: (select point E)
Select objects: ( RETURN )
Base point: MID
of (select point F)
<Rotation angle>/Reference: R
Reference angle <0>: MID
of (select point F)
Second point: INTERSECT
of (select point G)
New angle: 180
```

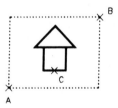

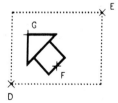

BEFORE   AFTER

Tips You can enter an angle as either a positive or a negative value. Positive values causes the rotation to occur in a counterclockwise direction. Negative values result in clockwise rotation.

RSCRIPT

Overview Script files are often used in self-running demonstrations for clients or at trade shows. If a script file is started from or interrupted within an AutoCAD drawing, the RSCRIPT command will start the script running again from the beginning. If included as the last line in the script file itself, the RSCRIPT command makes the script file run continuously.

Procedure You can use the RSCRIPT command while in a drawing (at the command prompt) to reactivate a script file after it has ended or been interrupted. The script file will restart at the beginning.

Command: **RSCRIPT**

Examples You can also include the RSCRIPT command as the last line of a script file. This causes the script to start again after it reaches the end, thus making a continuous demonstration. The following example shows a continuous slide show, which will repeat indefinitely.

PROGRAM	EXPLANATION
VSLIDE SLIDE 1	Begins slide show
VSLIDE *SLIDE2	Preloads slide 2
DELAY 2000	Pauses to allow reading of slide 1
VSLIDE	Displays slide 2
VSLIDE *SLIDE3	Preloads slide 3
DELAY 2000	Pauses to allow reading of slide 2
VSLIDE	Displays slide 3
DELAY 2000	Pauses to allow reading of slide 3
RSCRIPT	Repeats the entire cycle

Tips To use the RSCRIPT command, the script file must not contain any responses to AutoCAD's main menu. It should be created to be started initially from within a drawing using the SCRIPT command. (See SCRIPT.)

SAVE

Overview The SAVE command causes the current drawing file to be written to the disk. After the drawing has been saved, the command prompt returns and you can continue to edit the current drawing.

Procedure When you work on a drawing, the current drawing exists only in random-access memory (RAM). If your power failed, you would lose all the changes made since you loaded the drawing. As your work proceeds, you should write any changes to your disk about every 15 or 20 minutes. This ensures that you will not lose your work to an act of nature or human error. The SAVE command writes a copy of your current drawing from RAM to disk. AutoCAD provides the drawing name that was entered to load the current drawing as the default drawing name whenever you save the drawing. If you accept the default, the previous saved version of the drawing file becomes the backup file. Any previous backup is erased. In this way, you always have both the current drawing and one previous generation.

Examples This example shows the use of the SAVE command.

```
Command: SAVE
File name <current file>:
```

Do not include a file extension. The extension .DWG is assumed. If you specified a drive name or directory name when you load the drawing, it appears in the default.

Warnings If you respond with a name other than the default, AutoCAD checks to see if a drawing file with that name already exists. If it does, you will be asked if you really want to copy the current drawing over the existing file. Answering yes overwrites the existing file without creating a backup file. Answering no cancels the command.

Tips This is the single most important tip offered in this book. **Save your drawings often.** Save your drawing whenever you get up and walk away from your computer for any reason. Sometime during your use of AutoCAD,

462

something will go wrong. Your computer will crash. If you use the SAVE command often, it may mean the difference between losing 10 minutes' worth of work and losing 5 hours' worth.

You can use the SAVE command, substituting another file name for the default, to make a second copy of your drawing. This may be useful when you need to generate two different schemes for a project. Draw the information that is common to both drawings, saving one as SCHEME1 and the other as SCHEME2.

In AutoCAD release 9 (with the standard menu ACAD.MNU), you can select the SAVE command from the File pull-down menu (see Advanced User Interface).

You can use AutoCAD's time functions to create an automatic SAVE routine (see Time).

SCALE

[ADE-3] [v2.5, v2.6, r9]

Overview   The SCALE command changes the size of existing objects in the current drawing. Objects can be enlarged or reduced relative to a user-specified base point. The selected scale factor changes both the X- and Y-dimensions of the selected object equally.

Procedure   The SCALE command first prompts for the selection of objects. Any acceptable selection method can be used (last, previous, window, crossing, and so on). Objects can be added or removed from the selection set. When you have selected all of the objects to be scaled, press Return a final time in response to the prompt to select objects to indicate that you have finished selecting objects. The SCALE command then prompts for a base point. The base point can be anywhere in the drawing. If the base point is a point within a selected object, that portion of the object remains at the base point as it is scaled.

After you have selected the base point, either by pointing (usually with the help of an object snap) or by entering the coordinates, the SCALE command prompts for the scale factor. The scale factor is entered as a number. For example, to double the size of the selected objects, enter a scale factor of 2. To reduce the objects to one-quarter the current size, enter a scale factor of .25.

```
Command: SCALE
Select objects: (select the desired objects)
Base point: (select a point)
<Scale factor>/Reference:
```

You can enter the scale relative to another object already in the drawing by responding to the last prompt with *R*, indicating reference. The SCALE command then prompts for a reference length and a new length. You can enter the reference length and new lengths as numerical lengths or use the screen cursor, possibly with the help of object snaps, to indicate the reference length.

Examples

In the following examples, both SCALE methods are used.

The SCALE command is activated and the object selected using the window method of entity selection. The base point is selected using the midpoint Object Snap. A scale factor of 2 (double the size of the object) is entered.

In the second part of the example, the object is selected again (by the window method) and again the base point is selected using the midpoint Object Snap. This time, a reference length is specified. The reference length is provided by using the end point Object Snap to select the end points of the object (points G and H). The new length is provided by snapping to the end point (point H) again. The scale factor in this case is the ratio of the reference length (from G to H) and the length from the base point (point F) to the last point selected (point H).

```
Command: SCALE
Select objects: W
First corner: (select point A)
Other corner: (select point B)
Select objects: ( RETURN )
Base point: MIDPOINT
of (select point C)
<Scale factor>/Reference: 2

Command: SCALE
Select objects: W
First corner: (select point D)
Other corner: (select point E)
Select objects: ( RETURN )
```

```
Base point: MIDPOINT
of (select point F)
<Scale factor>/Reference: R
Reference length <1>: ENDPOINT
of (select point G)
Second point: ENDPOINT
of (select point H)
New length: ENDPOINT
of (select point H)
```

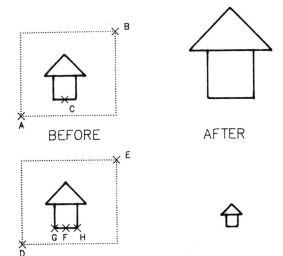

BEFORE   AFTER

Warnings   You can use AutoCAD's Drag feature to visually see the objects as they are scaled. If the object is complex, however, you may want to turn dragging off. Dragging complex objects often results in slow drawing response (see DRAGMODE).

Tips   Surveyors and civil engineers usually draw using AutoCAD's decimal units. Architects almost always use the architectural units. The basic difference between these two settings is that decimal units equate one drawing unit to one foot while architectural units equate one drawing unit to one inch. Thus, a site plan drawn by a civil engineer using decimal units and then edited by an architect using architectural units will be one-twelfth its actual size. In this case, use the SCALE command to rescale the entire drawing using a scale factor of 12. To go from one unit equals one inch to one unit equals one foot (architectural to decimal), use a scale factor of .08333. Most common scaling factors can be found in the chart in Appendix B.

If you make a mistake when scaling an object, you can return the drawing to its condition previous to the command by entering *U* at the next command prompt (see UNDO).

Screen Menu

[all versions]

Overview The screen menu is on the right side of AutoCAD's graphics screen. Unless you purposely configured AutoCAD without a screen menu area, this menu area is present whenever you are editing a drawing. The screen menu area contains either the standard AutoCAD menu (which comes with the program) or another menu that you have loaded. When using the AutoCAD menu, which is provided on the release disks as the file ACAD.MNU (and in its compiled form as ACAD.MNX), you can execute every AutoCAD command by picking the commands within this screen menu area.

Procedure Not every AutoCAD command appears on the screen menu at one time, but every command can be executed directly from the screen menu. You can select commands from the screen menu in several ways. If you are using a pointing device such as a mouse or digitizer, you can simply move the cursor toward the right until the screen menu area is highlighted. Once you have moved into the screen menu area, moving the mouse or digitizer puck will move a highlight bar within the screen menu. Press the pointing device's pick button when the command you want is highlighted.

You can also use the keyboard arrow keys to move the highlight box within the screen menu. To change from the screen cursor to the menu cursor, press the menu cursor key on the keyboard, which on most machines is the Insert key. You should now see the highlight box. Use the up and down arrow keys to move the highlight box to the proper command. To select a screen menu item, press the menu cursor key again. If you decide not to make a selection from the screen menu, you can reactivate the screen cursor by pressing the abort cursor key, which is usually the End key.

A third method is also available. You begin typing on the keyboard. As soon as you type a letter, AutoCAD checks what you have typed against what is currently displayed in the screen menu. As soon as AutoCAD finds a match, or even a partial match, that screen menu item is highlighted. If the screen menu item that AutoCAD has matched is the command that you were typing, you can stop typing the remainder of the command and simply press the menu cursor key.

The screen menu section of a menu file always begins with the line ***SCREEN. Each line that follows may contain a command or a series of commands (macros), which you execute simply by picking the item from the screen menu. The text that appears in the screen menu is either the first eight characters of the line or the first eight characters within square brackets. Text enclosed in square brackets represents a title and is not executed. The actual menu instructions begin after the closing bracket.

Screen menus can be paged, that is, as you select one item from the menu, that selection can cause a new screen menu to appear. This is the method used by the standard AutoCAD menu. For example, when you select the Draw option from the main screen menu, the main screen menu is replaced by a Draw screen menu, which comprises various entity draw-ing commands. This is accomplished by a special menu command embed-ded within the menu macro. The construction $S = N$ (where N is the name of another screen submenu in the current menu file) is used to call other screen submenus. The command $S =$ without a menu name causes the previous screen menu to be restored. Eight submenus can be nested and restored in this fashion.

Each screen submenu is a name prefaced with two asterisks, for example, **DRAW. The name can be up to thirty-one characters long. All of the lines that follow the submenu name until the next menu name (or menu section header) belong to that submenu. Since most graphics screens display only twenty menu items at a time, you should limit each submenu to this number. Otherwise, the additional menu items would not be accessible.

When a submenu is activated, it normally replaces the previous screen menu. However, if the line containing the submenu name is followed by a space and then a number, the submenu begins that number of lines down from the top of the screen menu area. The screen menu lines occu-pying the lines above remain active. For example, if your menu contains the line $S = SAMPLE$ and the Sample submenu title appears **SAMPLE 4, the Sample submenu will begin on the fourth line.

Examples

The screen menu occupies the right side of the graphics screen, to the right of the active drawing area, as shown in Figure 75.

The sample menu that follows is a portion of the standard AutoCAD menu file ACAD.MNU. It illustrates typical screen menu construction. Notice that blank lines appear as blank lines in the actual screen menu (some lines were removed to save space). Each of the menu selections calls other screen submenus.

```
**DR2 3
[POINT:]$S=X $S=POINT ^C^CPOINT
[POLYGON:]$S=X $S=POLYGON ^C^CPOLYGON
[SHAPE:]$S=X $S=SHAPE ^C^CSHAPE
[SKETCH:]$S=X $S=SKETCH ^C^CSKETCH
[SOLID:]$S=X $S=SOLID ^C^CSOLID
[TEXT:]$S=X $S=TEXT ^C^CTEXT
[TRACE:]$S=X $S=TRACE ^C^CTRACE
[3DLINE:]$S=X $S=3DLINE ^C^C3DLINE
[3DFACE:]$S=X $S=3DFACE ^C^C3DFACE

[previous]$S=DR

[__LAST__]$S=DR
```

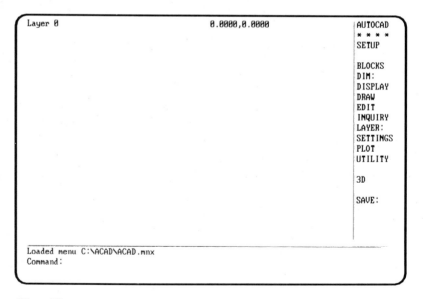

Figure 75

Tips　　　AutoCAD has a standard menu file, ACAD.MNU, that includes a screen menu. As mentioned previously, every command is accessible from this standard menu, although space does not permit all of them to be displayed on the screen at once. To get around this space limitation, AutoCAD places the various commands on different pages. When the menu first loads, you see the root page of the menu. The word AutoCAD is displayed at the top of every page and can always be selected to return the root page to the screen. The group of four asterisks (* * * *) is also displayed on every page. Selecting this item displays an object snap selection menu. Each AutoCAD command is arranged so that it appears within a logical command grouping. For example, if you want to draw a line or a circle, you can find the LINE or CIRCLE command by first selecting the DRAW item on the root page. Likewise, editing commands such as ERASE, COPY, and MIRROR can be found by first selecting the EDIT item. The item on each page labeled _ LAST _ will display the previous menu page. Screen menu items that appear in uppercase and end in a colon (:) activate a command and change menu pages. All capitalized entries without a colon simply change menu pages. Subcommands or command options are always either all lowercase or uppercase and lowercase combined and should be used only in conjuction with another command.

　　　You can include control codes in menus by preceding the character with a caret (^). Thus, ^C is the same as Ctrl-C and ^H is the same as Backspace.

　　　You can spread out long menu macros over more than one line by placing a plus sign as the last character of the line and continuing the macro on the next line.

　　　A semicolon (;) has the same effect as a Return.

　　　A backslash (\) within a menu macro causes the macro to pause for user input. Once the input is provided, the macro continues.

　　　If your version of AutoCAD includes ADE-3, you can include AutoLISP macros within your menu files.

　　　If you intend to customize AutoCAD's ACAD.MNU standard menu, make a copy of it first. The screen menu section of any AutoCAD menu file follows immediately after the line ***SCREEN.

469

SCREENSIZE

[v2.5, v2.6, r9]

Overview The SCREENSIZE system variable contains a point value that represents in pixels the size of the graphics screen. It is expressed as an X- and Y-value. The SCREENSIZE variable is a read-only variable.

Procedure You can use the SETVAR command and (getvar) AutoLISP function to look at its value, but the value cannot be altered.

Examples In the following example, the SETVAR command is used to return the SCREENSIZE value.

```
Command: SETVAR
Variable name or ?: SCREENSIZE
SCREENSIZE = 572.0000,419.0000 (read only)
```

Tips The SCREENSIZE value is a function of your graphics screen and cannot be changed by the program. It is provided as an AutoCAD system variable for use by experienced programmers.

SCRIPT

[all versions]

Overview One of AutoCAD's powerful features is that you can preprogram a series of sequences into a command script. A command script is nothing more than a text file. It may contain a series of AutoCAD commands or a sequence of commands for displaying AutoCAD slide files as part of a self-running demonstration. Script files are started while within a drawing through the use of the SCRIPT command.

Procedure To start the execution of a script file, enter the SCRIPT command. AutoCAD prompts for the name of the script file (which must have been created previously).

```
Command: SCRIPT
Script file: (enter the file name)
```

470

You can include a drive designation and subdirectory name. Do not include a file extension. The file type .SCR is assumed.

Examples   In this example, the SCRIPT command calls a script file named DEMO, which has already been saved to the \SAMPLES subdirectory on drive D.

```
Command: SCRIPT
Script file: D:\SAMPLES\DEMO
```

Warnings   Make sure that any other files called within your script file are in the current subdirectory, are on the computer's path, or have their paths included within the script file. Otherwise, AutoCAD will issue an error message and terminate the script file.

Tips   There are several additional commands you can use with script files (see DELAY, RESUME, and RSCRIPT).

    If a script file calls another script file, the calling script file is terminated and the called file takes over.

    You can also have a script file start automatically when you load AutoCAD. (See AutoCAD.)

SELECT
[v2.5, v2.6, r9]

Overview   The SELECT command creates a selection set of objects that you choose. This selection set can then be used by another command.

Procedure   The SELECT command prompts you to select objects, just like any other editing command. You can select objects using any method (previous, window, crossing, last, and so on). Objects can be added and removed from the selection set as with any other command. When you have finished building the selection set, press Return a final time. At this point, the SELECT command is completed.

```
Command: SELECT
Select objects: (select desired objects)
```

Examples   The SELECT command is useful in preselecting objects for use by a subsequent command. Entities grouped into a selection set by the SELECT command are operated on by a subsequent command when you enter *P* (previous) in response to the command's prompt to select objects.

For example, in a menu, the following command sequence could be used to select a group of objects and then use the CHANGE command to alter one or more of their properties:

```
[CHANGE P]select \change P ;P
```

This macro first prompts you to select objects. After you have selected the objects and pressed Return to indicate completion, the CHANGE command is activated, the previous selection set is entered in response to the prompt to select objects, and the response *P*, for properties, is entered. You then enter the property or properties you want to change. The SELECT command is the only command that, when used within a macro, allows the complete formation of a selection set before the remainder of the macro proceeds. This is how the previous example could build a selection set of many objects.

Tips   Normally the backslash (\) pauses for only one user selection. The SELECT command, followed by the backslash, causes the macro to allow multiple objects to be selected. This is a useful feature and is the reason the SELECT command is included.

Selection Set   [v2.5, v2.6, r9]

Overview   When an AutoCAD command prompts you to select objects, the objects you select are added to a selection set. Think of a selection set as a holding area in memory. Objects are added to or removed from this selection set area of memory until you are satisfied that the proper objects have been selected. As each object is selected, AutoCAD normally highlights it on the screen to indicate that it is now part of the selection set. If you remove an object from the selection set, the object is no longer highlighted. When you have finished selecting objects, pressing Return indicates to AutoCAD that you are ready to proceed with the command.

When you select objects, you are placing them into the current selection set. The previous selection set is also held in memory under many circumstances and can even be added to the current selection set by responding to the prompt to select objects with *P*, for previous (see Previous Selection Set).

Procedure   Every AutoCAD command that first prompts you to select objects creates a selection set prior to completion of the command. When prompted, you can select objects using any available AutoCAD method. You can point at individual objects using the keyboard cursor keys or a pointing device. You can put objects in a window on the screen or select them with a crossing type window. You can select the last object added to the drawing displayed on the current screen view by entering *L* or *Last*. You can add the previous selection set to the current selection set by entering *P* or *Previous*. You can remove objects from the selection set by entering *R* or *Remove* and then selecting from among the highlighted objects. (See Entity Selection for more detail.)

Examples   In the following sequence of commands, entities are added to and removed from the selection set before completion of the MOVE command. Notice that the previous selection set and the last entity are also chosen.

```
Command: MOVE
Select objects: L (select LAST entity)
1 found.
Select objects: W (select objects by WINDOW)
First corner: (select point)
Other corner: (select point) 7 found.
Select objects: R (toggle to REMOVE)
Remove objects: P (select PREVIOUS set)
1 found, 1 removed.
Remove objects: A (toggle to ADD mode)
Select objects: (select object) 1 selected, 1 found.
Select objects: (RETURN) (end selection sequence)
Base point or displacement: (select first point) (complete MOVE command)
Second point of displacement: (select second point)
```

Tips   The previous selection set is not maintained across editing sessions.

'SETVAR

Overview   Many AutoCAD commands control the settings of system variables. System variables control such factors as the mirroring of text, the fillet radius, the size of the pick box, and the way point entities are displayed. Although you can use individual commands to change many of these variables, the SETVAR command lets you change or simply view the current value or setting of every system variable. In addition, some system variables can be changed only through the use of the SETVAR command.

Procedure   The SETVAR command is used to view or change any of AutoCAD's system variables. A complete list of the system variables is in Appendix A. You can use the SETVAR command transparently, that is, while another command is active, by preceding the SETVAR command with an apostrophe ('), as in 'SETVAR. The SETVAR command prompts you to enter the name of the variable that you want to view or change. If you respond with a question mark (?), a listing of all of AutoCAD's system variables is displayed, along with their current settings.

To change the value of a system variable, enter the name of the variable that you want to change. AutoCAD displays the current value for that system variable and prompts you for the new value. To keep the current value, press Return or Ctrl-C. To change the value, enter the new value and press Return.

```
Command: SETVAR
Variable name or ?: (enter variable name)
New value for variable-name <current value>: (enter new value)
```

Some system variables cannot be changed and are provided only for your information. The variables TDCREATE (the time and date at which the drawing was created), DWGNAME (the drawing name), and CLAYER (the current layer) are examples of read-only system variables. If you enter a variable name for a read-only variable, AutoCAD will respond

```
variable-name = current value (read only)
```

For example, entering CLAYER might result in

```
CLAYER = "0" (read only)
```

Examples

Sometimes it is helpful to use the transparent SETVAR command to change a system variable while another command is active. An example of this is when you are picking objects and the pick box is too large or too small.

You may sometimes find that you are using the TRIM command to clean up wall intersections and the pick box is larger than the space between the two lines representing the wall. You can use the transparent SETVAR to change the size of the pick box and then continue with the TRIM command.

```
Command: SETVAR
Variable name or ?: PICKBOX
New value for PICKBOX <5>: 3
```

Responding with a question mark (?) displays all of the system variables and their current settings. Because all of the screen variables will not fit on the screen at once, AutoCAD displays them one screen at a time and prompts you to press Return to display each successive screen, as shown in Figure 76.

```
Command: SETVAR
Variable name or ?: ?
```

```
ACADPREFIX      ""                         (read only)
AFLAGS          0
ANGBASE         0
ANGDIR          0
APERTURE        10
AREA            0.0000                      (read only)
ATTMODE         1
AUNITS          0
AUPREC          0
AXISMODE        0
AXISUNIT        0.0000,0.0000
BLIPMODE        1
CDATE           19871006.110821312         (read only)
CECOLOR         "BYLAYER"                   (read only)
CELTYPE         "BYLAYER"                   (read only)
CHAMFERA        0.0000
CHAMFERB        0.0000
CLAYER          "0"                         (read only)
CMDECHO         1
COORDS          0
DATE            2447075.46413315           (read only)
-- Press RETURN for more --
```

Figure 76

Warnings   Prior to AutoCAD version 2.6, the SETVAR command did not automatically display the previous system variable used in the SETVAR command as the default for subsequent SETVAR commands.

Tips   System variables can also be viewed or changed using AutoLISP subject to the same restriction on read-only variables.

SH

[ADE-3] [v2.5, v2.6, r9]

Overview   You can temporarily leave AutoCAD and run other programs from the DOS prompt. The SH (variation of SHELL) command temporarily suspends AutoCAD so that you can use an internal DOS command such as DIR, COPY, or TYPE.

Procedure   At the AutoCAD command prompt, enter the SH command. You temporarily leave AutoCAD and the DOS command prompt appears. At this point, you can run any internal DOS command. When the DOS command has been executed, you will return to your AutoCAD drawing at the place you left off.

Command: **SH**
DOS command:

Examples   This example shows the result of issuing the DIR command to view the .DWG files in the ACAD subdirectory (see Figure 77).

Command: **SH**
DOS Command: **DIR \ACAD*.DWG**

```
   Volume in drive D is DISK1_VOL1
   Directory of  C:\ACAD

   ACAD      DWG     1385   12-14-86    7:20p
   BORDER    DWG     1024   12-18-85    3:06p
   COLUMBIA  DWG    31901    4-15-86    1:53p
   RECTANG   DWG     1441    2-12-86    6:12p
   SOLAR     DWG     6511    9-02-87    2:32a
            5 File(s)   14452736 bytes free
```

Figure 77

S

Tips   For more detailed information, see SHELL.

SHAPE

Overview   Shapes are one of AutoCAD's basic drawing entities. They are economical in their drawing memory requirements. The use of shapes is simple; their creation is not. Creating shape files is described in detail in Appendix B of the *AutoCAD Reference Manual*. The SHAPE command places shape entities into a drawing after the shape file has been created, compiled, and loaded into the drawing. You can also use the SHAPE command to display a list of the currently loaded shapes.

Procedure   Enter the SHAPE command at the command prompt. The command prompts for the name of the shape you want to place into the drawing. If you enter the name of a shape that is not currently loaded, AutoCAD displays a message that the shape has not been found and repeats the original prompt.

     Once you have entered a valid shape name, you are asked to select the starting point, the height, and the rotation angle for the shape. A height of 1.0 is provided as a default. The current snap grid rotation is used as the rotation angle default. The precise effects of the height and the rotation angle vary for different shapes. You can enter a numerical value for these prompts or use the screen cursor to indicate the height and the rotation angle in relation to the starting point. The last shape entity inserted into the drawing is given as the default shape name.

```
Command: SHAPE
Shape name (or ?) <default>: (enter shape name)
Starting point: (select point)
Height <1.0>: (enter number or select point)
Rotation angle <default>: (enter number or select point)
```

If you respond to the prompt for the shape name with a question mark (?), the SHAPE command lists all the currently loaded shapes (as opposed to using the question mark with the LOAD command, which displays a list of the shape files names, which, in turn, contain the individual shapes).

Examples   In this example, the sample shape file PC.SHX is loaded, and two of the shapes are placed into the drawing. Notice that the PC.SHX file was saved in the \ACAD subdirectory, so it must also be loaded from that subdirectory (subject to the computer's path).

```
Command: LOAD
Name of shape file to load (or ?): \ACAD\PC
```

The SHAPE command is then used to get a listing of the currently loaded shapes.

```
Command: SHAPE
shape name (or ?): ?
Available shapes:

File: \ACAD\PC
   FEEDTHRU          DIP8
   DIP14             DIP16
   DIP18             DIP20
   DIP24             DIP40
```

Finally, the SHAPE command is used to place shapes into the following drawing.

```
Command: SHAPE
Shape name (or ?): FEEDTHRU
Starting point: (select point)
Height <1.0000>: ( RETURN )
Rotation angle <0>: ( RETURN )
Command: ( RETURN )
SHAPE Shape name (or ?) <FEEDTHRU>: DIP14
Starting point: (select point)
Height <1.0000>: ( RETURN )
Rotation angle <0>: ( RETURN )
```

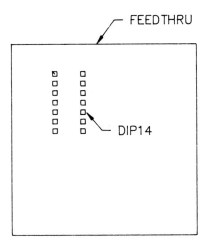

FEEDTHRU

DIP14

Warnings   Before you can use the SHAPE command to insert a given shape entity, the
shape file that contains the definition of that shape must be compiled from
its initial .SHP file type to a .SHX file type. This is done using main menu
task 7. Then you must load the given compiled shape file into the current
drawing using the LOAD command. Only after these initial steps can you
use the SHAPE command to insert the shape entity.

Tips   You can use AutoCAD's Drag feature to visually select the shape's starting
point, height, and rotation angle.
        The creation of shapes is beyond the scope of this book. For more infor-
mation on the actual creation of shapes, read the *AutoCAD Reference Manual*, Appendix B.

SHELL   [ADE-3] [all versions]

Overview   The SHELL command allows you to suspend AutoCAD temporarily and
run other programs from the DOS prompt. When you finish, you can
return to AutoCAD exactly where you left off.

Procedure

The SHELL command suspends AutoCAD while you are in a drawing. The SHELL command is entered at the command prompt and returns the DOS command line.

```
Command: SHELL
DOS command:
```

At this point, you can enter any DOS command or any other program (depending on the program's memory requirements). When the command or program is completed, you return to AutoCAD's command prompt.

If you respond to the DOS command prompt by pressing Return, the message

```
To return to AutoCAD, type "EXIT
```

appears, along with a DOS prompt with an extra >. This lets you know that you have left AutoCAD and are now at the DOS prompt (that is, you have loaded a secondary copy of COMMAND.COM). You can execute DOS commands or run other programs. When you are ready to return to AutoCAD, type *EXIT* at this double DOS prompt, and you will return to your drawing just where you left it.

```
C:>> EXIT
```

If there is not enough free memory for the SHELL command or for the program you wish to run, the error message

```
SHELL error: insufficient memory for command
```

is displayed. If you are just trying to use the DOS DIR or another internal DOS command, try using AutoCAD's SH command instead. The SH command requires less free memory.

Examples

You can use the SHELL command to reach the DOS prompt to format a floppy disk for saving a copy of your drawing, to use DOS's line editor EDLIN to create a script or menu file, or to run another program.

Warnings

Being able to reach the DOS prompt while running AutoCAD is a helpful and powerful feature, but it is also possible crash your system and lose the changes you have made to your drawing. When you have left AutoCAD and are at the DOS prompt, there are certain things you must *not* do:

1. Do not use the DOS command CHKDSK with the /F option.

2. Do not erase AutoCAD temporary files or any file whose extension starts with the $ symbol.

3. Do not load any RAM resident program or command, such as MODE, PRINT, SIDEKICK, or PLUMP, when you have used SHELL to leave Auto-CAD. These programs must be loaded before you start AutoCAD or your system will crash.

4. Do not run a program that changes or resets the serial ports (BASICA is such a program).

If the program you run outside AutoCAD uses the graphics screen area of the computer's memory, when you return to AutoCAD your drawing may appear incorrectly. A REDRAW usually clears up the drawing area. Sometimes the menu area is also affected. In that case, use the MENU command to reload whatever menu you are using. Other stray bits may appear on the screen, but they will not affect your drawing.

Tips   The SHELL command is one of the commands implemented by Auto-CAD's External Commands feature. You can run any number of external programs from AutoCAD's command prompt by first editing the file ACAD.PGP. This file is in the subdirectory where you copied the AutoCAD program when you loaded it onto your computer's hard disk (normally \ACAD). The ACAD.PGP file contains the names of all the non-AutoCAD commands that can be entered at the command line while in AutoCAD. If you list out this file (using the DOS TYPE command, for example), you will see that the SH and SHELL commands occur in this file. Several other internal DOS commands may be listed here as well.

```
DEL,DEL,25000,File to delete: ,0
DIR,DIR,25000,File specification: ,0
EDIT,EDLIN,40000,File to edit: ,0
SH,,25000,*DOS Command: ,0
SHELL,,125000,*DOS Command: ,0
TYPE,TYPE,25000,File to list: ,0
```

The ACAD.PGP file contains the name of the command as well as information that tells AutoCAD how much memory it will need to free for the external command to be run. Each line in the ACAD.PGP file represents one external command. Each line contains five separate entries, or fields, separated by commas. The first field contains the command name. This name must not duplicate an AutoCAD command name. The command name is what you enter at AutoCAD's command prompt to run the external command. For example, it is possible to include the DOS COPY command as an external command, but because AutoCAD has a COPY command, you must give the DOS command a different command name.

The second field is the file command. It is the command that is actually sent to DOS. In the previous example, you could have given the DOS COPY command the command name FCOPY for AutoCAD's benefit. Then, the file command would actually be the DOS COPY command. If the file command field is left blank, the command name is sent to DOS. Also, when this field is blank, the command name must be followed by two commas.

The third field is the memory reserve. This is a number, in bytes, that tells AutoCAD how much RAM to free to load and run the external program. Notice in the previous sample file that the SHELL command frees 125,000 bytes of RAM, while the SH command frees only 25,000 bytes. This is why sometimes there is not enough free memory to execute the SHELL command, yet you can still use an internal DOS command with the SH command, which does not require as much memory to be freed. This may also be why you get an insufficient-memory error when you try to run another program from the DOS prompt that you used the SHELL command to get to. The SHELL command may not have freed enough memory to load and run the external program. Increasing the value for the memory reserve for the SHELL command in the ACAD.PGP file may correct this problem.

The fourth field is the prompt field. If it is included, when you enter the external command name at AutoCAD's command prompt, the prompt field displays. Any response is included with the file command. If the first character in the prompt field is an asterisk (*), the response to the prompt may contain spaces and must be concluded by pressing Return. If the prompt field is omitted, the memory reserve field must be followed by two commas.

The fifth field is the return code. The numbers 0, 1, 2, and 4 are valid return codes. Numbers 1 and 2 are special codes used with DXB files (such as when you are using CAD/camera) and are not of much interest here. A code of 0 causes AutoCAD to remain in the text screen when you are returned to AutoCAD. A return code of 4 causes AutoCAD to return to whichever screen mode was in effect when the external command was activated. In dual-screen configurations, the return codes 0 and 4 have no effect.

SKETCH

[ADE-1] [all versions]

Options

C	Connects: restarts sketch at end point
E	Erases (backs up over) temporary lines
P	Raises/lowers sketching pen
Q	Discards temporary lines, remains in Sketch
R	Records temporary lines, remains in Sketch
X	Records temporary lines, exits from Sketch
.	Draws line to current point

Overview   The SKETCH command allows you to enter freehand drawings as opposed to drawings made up of combinations of other AutoCAD drawing entities. When you use the SKETCH command, only the motion of the mouse or digitizer determines the drawing created on the screen. You cannot enter the cursor location from the keyboard. The SKETCH command creates the drawing as a series of either very short line segments or very short polyline segments. As a result, you can enter a great number of drawing entities in a short period of time. The SKETCH command should be used only when no other drawing command will suffice.

Procedure   When you activate the SKETCH command, the first thing you are prompted for is the record increment. This is the length of the individual segments that make up the freehand drawing. A larger record increment results in the creation of fewer individual segments, but a coarser looking drawing. For example, if you are using architectural units, a record increment of 1.0 will result in your freehand drawing comprising individual segments each 1 inch long. If you use a record increment of 0.1, each segment is 1/10 inch long. You must move the cursor a distance greater than the record increment before a new segment is generated. Obviously, the record increment you use depends on what you are drawing and at what scale it will ultimately be plotted. The default value of the record increment is the last value used and is initially determined by the proto-type drawing.

Command: **SKETCH**
Record increment <current>: **(enter value)**

When you are operating within the SKETCH command, a separate series of subcommands or options apply. AutoCAD lets you know that you are in Sketch mode by displaying the prompt

Sketch.   Pen eXit Quit Record Erase Connect .

The cursor represents the pen. The pen can be in either an up or down position. When the pen is in the up position, no lines appear on the screen. When the pen is in the down position, every movement of the cursor causes an image to be drawn on the screen. When you enter the SKETCH command, the pen is initially in the up position. You place the pen in the down position by pressing the mouse or digitizer's pick button, pressing down on the stylus, or typing *P*. The pen is raised by repeating this process. As you sketch, the lines are shown in a highlighted fashion.

When the pen is in the up position, you can draw a straight line from the end of the last sketched line to the current cursor position by pressing the period (.) on the keyboard or button 1 on a multibutton cursor.

Freehand lines are not added to your drawing until you instruct the SKETCH command to record them. After sketching, record your drawing by pressing *R* or button 2 on a multibutton cursor. The sketched lines change to the current color, and AutoCAD displays the number of lines that were added to the drawing.

To record the freehand lines and to exit from the SKETCH command, press *X*, the Space bar, Return, or cursor button 3. AutoCAD displays the number of lines that were added to the drawing and returns to its command prompt.

If you aren't happy with your freehand sketch, you can discard the temporary sketch before the line segments are recorded and then exit from the SKETCH command. To do this, press *Q*, Ctrl-C, or cursor button 4.

You can erase as much of the temporary sketched line as you wish, starting from its end. To use the ERASE subcommand, press *E* or cursor button 5. If the pen was down, it is now raised. AutoCAD prompts

```
Erase: Select end of delete.
```

You can move the cross-hairs to any point along your sketched line. All of the line from the end to the selected point will be blanked out. To erase this portion of the freehand line, press *P* or the pick button. Only lines that have not yet been recorded can be erased using the ERASE subcommand. If you decide not to erase part of the sketched line, press the pick button or any command key other than *P* (you can press C, E, Q, R, or X). AutoCAD will respond

```
Erase aborted.
```

When the pen is in the up position, you can reconnect the cursor to the end of the last sketched line not yet recorded and continue sketching from its position by pressing C on the keyboard or button 6 on a multibutton cursor. Move the cross-hairs to the end point of the last line. When you are within one unit, the pen automatically lowers. If you attempt to use the CONNECT subcommand when the pen is down, AutoCAD displays the error message

```
Connect command meaningless when pen down.
```

If there is nothing to connect to, AutoCAD displays the error message

```
No last point known.
```

If you decide not to connect to the last sketched line, press any other command key (E, P, Q, R, or X). AutoCAD will respond

```
Connect aborted.
```

Examples   A useful application for the SKETCH command is in conjunction with Auto-CAD's Tablet mode. When Tablet mode is on, you can use the pointing device to digitize a drawing attached to the tablet. If the drawing or object you are digitizing requires it, you can use the SKETCH command to trace the object on the digitizer (see TABLET).

Warnings   Sometimes the temporary sketch lines do not all fit in the computer's memory. When this happens, AutoCAD sounds a continuous beep and displays the message:

```
Please raise the pen!
```

When this happens, raise the pen by pressing *P* or the pick button. Auto-CAD saves part of your drawing to disk and then displays the message

```
Thank you.  Lower pen and continue.
```

At this point, press *P* again (or the pick button) to lower the pen and continue.

The SKETCH command can create hundreds of line entities or polyline segments in a very short time. Use the SKETCH command only when another method of drawing the object will not suffice. Otherwise, your drawing will get too large and drawing regeneration time will become exceedingly lengthy.

Generally you should use the SKETCH command with Ortho and Snap modes turned off. If Snap mode is on, the record increment should not be set smaller than the snap value, since the minimum movement of the cursor is the snap value. If Ortho mode is on, only vertical and horizontal lines can be drawn, so your sketched line will have a stair-stepped appearance.

In AutoCAD release 9, you cannot access the menu bar and pull-down menus from the SKETCH command once the record increment has been set (see Advanced User Interface).

Tips   You can make the SKETCH command generate either individual line entities for each record increment or a continuous polyline composed of one segment for each record increment. The system variable SKPOLY controls which method is used. A SKPOLY value of 0 results in individual line entities; any other value results in polyline sketches. You can view or change the value for SKPOLY by using either the SETVAR command or AutoLISP. Remember that you can run the SETVAR command transparently once the SKETCH command has been activated.

Sketching with very small record increments or on a slow computer can cause inaccurate sketches. To compensate, enter the record increment as a negative number. AutoCAD will use the positive value of the record increment but will sound a single beep whenever you are moving the pen too fast for the computer to keep up.

SKETCHINC   [ADE-1] [all versions]

Overview   The record increment value that you must provide whenever you begin the SKETCH command is stored in the SKETCHINC (Sketch Increment) system variable. This value represents the length of the individual line segments that make up a freehand sketch and is stored as a real number.

Procedure   The SKETCH command is normally the only one that you use to set the sketch record increment, but you can also use the SETVAR command or AutoLISP to set this value by changing the SKETCHINC value.

Examples   In this example, the SETVAR command allows you to reset the sketch record increment.

```
Command: SETVAR
Variable name or ?: SKETCHINC
New value for SKETCHINC <0.1000>: .5
```

Tips   The initial SKETCHINC value is determined from the prototype drawing (default value is 0.1000). The current value is saved with the drawing.

SKPOLY   [ADE-3] [v2.5, v2.6, r9]

Overview   The value of the system variable SKPOLY (Sketch Polylines) determines whether freehand sketches drawn using the SKETCH command are composed of hundreds of individual short line segments or single polyline entities made up of many vectors. If a sketch is made using polylines, each completed sketch line is made up of one polyline that can be edited using the PEDIT command. If the sketch line is made up of line entities, however, each individual line segment has to be edited as an individual entity.

Procedure   If the value of SKPOLY is 0, the SKETCH command creates freehand sketches using lines. If the value is 1, polylines are created. These are the only two values that are permitted.

You can use the SETVAR command or the AutoLISP function (setvar) to change the SKPOLY value.

Examples   The following example uses the SETVAR command to set SKPOLY to 1. All subsequent sketches will be created using polylines rather than lines.

```
Command: SETVAR
Variable name or ?: SKPOLY
New value for SKPOLY <0>: 1
```

In the following example and drawing, the ERASE command has been activated and the sketch selected. Notice that when the sketch has been created using lines (SKPOLY is 0), only the small individual line segment is selected by AutoCAD. But when the sketch is created using polylines (SKPOLY is 1), selecting any point on the sketch causes AutoCAD to select the entire sketch.

```
Command: ERASE
Select object: (select point A)
Select object: ( RETURN )
```

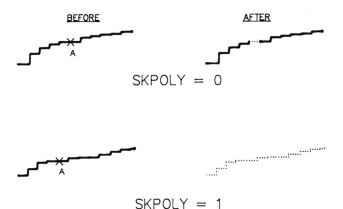

BEFORE   AFTER

SKPOLY = 0

SKPOLY = 1

Tips

Obviously the SKPOLY system variable is an important one to set if you are going to be using the SKETCH command. Setting its value is the only way to determine whether sketching will be done using lines or polylines. Most users recommend that you use polylines when sketching. They are a bit more involved to edit, but you can edit an entire sketched line at once if it is a polyline. A sketched line composed of hundreds of individual line entities is even more difficult to edit.

The initial SKPOLY value is determined from the prototype drawing (default value is 0, use lines). The current value is saved with the drawing.

Slides

[all versions]

Overview   Slides are images of the current AutoCAD graphics screen, captured to a slide file, which can be quickly recalled to the screen later for viewing. Slides are only images. They cannot be edited. You can change a slide only by editing the original drawing and then recapturing the graphics screen. Slides are also the basis of icon menus (available in AutoCAD release 9 only) and are often used to incorporate AutoCAD drawings into documents created with a desktop publishing program.

Procedure   You create slides using the MSLIDE command and view them using the VSLIDE command (see MSLIDE and VSLIDE). Although the standard version of AutoCAD can be used for viewing slide files (with the VSLIDE command), only versions that include ADE-2 can make slides (with the MSLIDE command).

Release 9 Only   In AutoCAD Release 9, you can use slides to create icon menus (see Icon Menus). In addition, you can combine individual slides into a slide library file. The utility program SLIDELIB.COM is provided for this purpose.

Examples   The slide file POINTS.SLD, which illustrates the various PDMODE settings, is provided with the AutoCAD program. You can view this file using the VSLIDE command.

```
Command: VSLIDE
Slide file: POINTS
```

To remove the slide from the screen (thus restoring the AutoCAD drawing that was on the screen), simply issue the REDRAW command.

Release 9 Only   To combine several slide files into a slide library, you must first create the individual slide files using the MSLIDE command. Then from the DOS prompt, issue the SLIDELIB command. The SLIDELIB.COM program reads a list of slide file names from a list file. This list file should be created with any word processor capable of creating ASCII text (for example, EDLIN). Each individual slide file should occur on a separate line. You do not need to include a file extension within the list file (although if you do, it must be .SLD). You can include a device name and path if necessary. The slide file name (but not the drive designation or path) is saved in the library file. The SLIDELIB.COM program makes no provision for updating slide library files. To add or delete a slide file from a slide library file, simply edit the list file and recreate the library. All the original slide files must be available to do this.

To use the SLIDELIB.COM utility to create the slide library SAMPSLD.SLB, use the sample list file named SAMPLE.TXT and the following command structure:

```
slide1
slide2
d:\sample\slide3
slide4.sld
```

```
C>slidelib SAMPSLD <SAMPLE.TXT
```

Warnings　If you anticipate that you will have to edit your slides in the future, be sure to save the original drawings from which they were generated. AutoCAD cannot edit slide files.

The ability to use slide library files and icon menus is available only in AutoCAD release 9. Slide files produced by earlier versions of AutoCAD are readable by release 9, but slides produced by release 9 are not readable by earlier versions of AutoCAD.

Tips　You can use slides as an additional form of Help screen, for example, when using complex menu macros.

A slide incorporates all the colors and linetypes and the screen resolution in effect at the time it was created.

In release 9 only: Although the slide library file that results from using the SLIDELIB.COM utility is approximately the same size as the sum of the slide files included in it, it is a more efficient method of copying and storing the slide files on disk.

SNAP

Options

number	Sets snap alignment resolution
On	Aligns designated points
Off	Does not align designated points
A	Sets aspect (differing X-Y spacings) [ADE-2]
R	Rotates snap grid [ADE-2]
S	Selects style, standard or isometric [ADE-2]

Overview

The SNAP command controls AutoCAD's Snap mode. When the Snap mode is on, the cursor is automatically locked onto an imaginary grid within the drawing. As you move the pointing device, the screen cursor snaps to these invisible grid points. Other features of the SNAP command allow you to change the angle of rotation of this imaginary grid, to change the spacing of the grid points, and to allow different grid spacings along the X- and Y-axes. In addition, the SNAP command can also alter the snap style to enable AutoCAD's Isometric Drawing feature.

Procedure

Enter the SNAP command at the AutoCAD command prompt. The command displays the available options. The current snap spacing is displayed as the default value. Snap mode can be turned on or off by entering the appropriate response to the prompt. If you turn Snap mode on and the current snap spacing is 0, AutoCAD prompts you for a new snap spacing.

```
Command: SNAP
Snap spacing or ON/OFF/Aspect/Rotate/Style <0>: ON
Spacing/Aspect <0>: (enter new value)
```

If Snap mode is currently off and you respond to the SNAP command prompt with a numeric value, Snap mode is turned on and the snap spacing set to the value you entered. If Snap mode is on, the spacing is set to your new value.

You can set the Snap resolution along the X- and Y-axes independently using the Aspect option. Respond to the Snap prompt with *A*. AutoCAD then prompts for the horizontal spacing and the vertical spacing.

```
Command: SNAP
Snap spacing or ON/OFF/Aspect/Rotate/Style <1.5>: A
Horizontal spacing <1.5>: (enter X-axis snap spacing)
Vertical spacing <1.5>: (enter Y-axis snap spacing)
```

The snap grid is normally orthogonal. You can rotate the snap grid using the Rotate option. Enter *R* in response to the Snap prompt. AutoCAD then asks for the base point. This is the reference point about which the snap grid will be rotated. The snap spacing is measured in increments from this base point. The default base point is the 0,0 point. You can enter coordinates for the base point or use the screen cursor to select a point. Next, AutoCAD asks for the rotation angle. This is the angle at which the snap grid will be rotated about the base point. The default value is 0 degrees. Values from -90 to $+90$ degrees are valid. Positive values rotate the snap grid in a counterclockwise direction, negative numbers in a clockwise fashion. A rotation angle of 0 returns the snap grid to an orthogonal orientation.

```
Command: SNAP
Snap spacing or ON/OFF/Aspect/Rotate/Style <1.5>: R
Base point <0,0>: (select base point)
Rotation angle <0>: (enter angle from -90 to +90)
```

You can use AutoCAD's Snap mode as a drawing aid when using the Isometric Drawing feature by responding to the Snap prompt with *S*, for style. The Style option allows two choices, standard and isometric. The isometric Snap mode arranges the snap grid points so that the orientation of the grid matches the orientation of the isometric drawing planes (30, 90, 150, 210, 270, and 330 degrees). AutoCAD then requests the vertical spacing between grid points (see ISOPLANE).

```
Command: SNAP
Snap spacing or ON/OFF/Aspect/Rotate/Style <1.5>: S
Standard/Isometric <current>: I
Vertical spacing <current value>: (enter value)
```

Release 9 Only   If you are using AutoCAD Release 9, you can execute all the features of the SNAP command from within the dialog box displayed when the DDRMODES command is activated. This dialog box allows the

Snap mode to be turned on and off. The snap style can be changed to isometric and the current isoplane chosen. In addition, you can select the X- and Y-snap spacing, the snap rotation angle, and the snap base while within this dialog box. (Available only with AutoCAD release 9; see Advanced User Interface and DDRMODES.)

Examples

In this example, the SNAP command sets a predetermined snap grid. All subsequent lines will always be drawn to these grid points when Snap mode is on. Use the SNAP command to rotate the screen cross-hairs before you use the ARRAY command to generate a rectangular array at an angle other than 90 degrees (see ARRAY). In this example (and in Figure 78), the SNAP command is used to rotate the snap grid (and the screen cross-hairs) to a 45-degree angle. The ARRAY command is then used to create a rectangular array. Notice, however, that since the snap grid is rotated, the resulting array is also rotated to the same 45-degree angle.

```
Command: SNAP
Snap spacing or ON/OFF/Aspect/Rotate/Style <1.0000>: R
Base point <0.0000,0.0000>: ( RETURN )
Rotation angle <0>: 45

Command: ARRAY
Select objects: (select object) 1 selected, 1 found.
Select objects: ( RETURN )
Rectangular or Polar array (R/P): R
Number of rows (---) <1>: 3
Number of columns (|||) <1>: 4
Unit cell or distance between rows (---): 1
Distance between columns (|||): 1
```

Tips

You can turn the Snap mode on and off using the toggle keys Ctrl-B or F9.

You can control the various Snap settings by using either the SETVAR command or AutoLISP to change the associated system variables:

VARIABLE	EXPLANATION
SNAPANG	Value = snap grid rotation angle
SNAPBASE	Coordinate = snap rotation base point
SNAPISOPAIR	Current isometric plane (0 = left, 1 = top, 2 = right)
SNAPMODE	Snap mode (0 = off, 1 = on)
SNAPSTYL	Snap style (0 = standard, 1 = isometric)
SNAPUNIT	Coordinate (X-value = X-spacing, Y-value = Y-spacing)

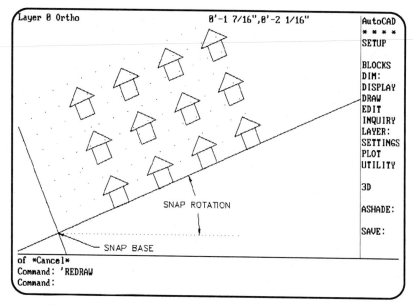

Figure 78

The screen cross-hairs adopt the orientation of the current snap rotation angle. When the cross-hairs are rotated, Ortho mode forces lines to be drawn orthogonally in relation to the cross-hair orientation. The ARRAY command will generate rectangular arrays with this rotated orientation.

The screen cross-hairs follow the current isometric plane when snap is set to isometric style.

Using the SNAP command to change any Snap mode setting also activates the Snap mode if it was previously off.

SNAPANG

[ADE-2] [all versions]

Overview   The snap rotation angle set by the SNAP command is stored as a real number in the SNAPANG (Snap Angle) system variable. The snap grid is normally orthogonal, and SNAPANG would normally have a value of zero. But if you have changed the snap grid rotation angle, the angle you specified will be stored in this variable.

Procedure Normally the SNAP command is used to change the snap grid rotation angle. But you can also use the SETVAR command or AutoLISP to change the snap rotation angle by changing the SNAPANG variable value.

Examples This example uses the SETVAR command to change the snap rotation angle. As soon as the new angle is entered and you press Return, the screen cross-hairs take on the new orientation. But unlike using the SNAP command to change the orientation, the Snap mode is not automatically turned on.

```
Command: SETVAR
Variable name or ?: SNAPANG
New value for SNAPANG <0>: 45
```

Tips The initial SNAPANG value is determined from the prototype drawing (default value is 0). The current value is saved with the drawing.

In AutoCAD release 9, you can change the SNAPANG system variable from within the dialog box displayed by the DDRMODES command (only with AutoCAD release 9; see Advanced User Interface and DDRMODES).

SNAPBASE

[ADE-2] [all versions]

Overview The point specified as the base point when you use the SNAP command to change the snap rotation angle is stored as a point value in the SNAPBASE system variable.

Procedure Using the SNAP command to change the snap rotation angle and to specify a new base point changes the SNAPBASE value. You can also change the value using the SETVAR command or the (setvar) AutoLISP function.

Examples This example uses the SETVAR command to change the snap base point. As soon as the new point is entered and you press Return, the screen cross-hairs move to the new coordinate. Unlike using the SNAP command to change the snap base point, the Snap mode is not automatically turned on.

```
Command: SETVAR
Variable name or ?: SNAPBASE
New value for SNAPBASE <0.0000,0.0000>: 1,1
```

Tips   The initial SNAPBASE value is determined from the prototype drawing (default value is 0.0000,0.0000). The current value is saved with the drawing.

In AutoCAD release 9, you can change the SNAPBASE system variable from within the dialog box displayed by the DDRMODES command (only with AutoCAD release 9; see Advanced User Interface and DDRMODES).

SNAPISOPAIR

[ADE-2] [all versions]

Overview   When you create an isometric drawing using AutoCAD's Isometric Drawing mode, AutoCAD keeps track of which isometric plane it is currently drawing in by storing an integer value in the SNAPISOPAIR system variable. There is always a value in this variable, but it has meaning only when you are drawing in isometric mode.

Procedure   Normally you use the ISOPLANE command or the isoplane toggle (see ISOPLANE) to change between isometric planes. In addition to these two methods, you can use the SETVAR command or the (setvar) AutoLISP function to change the SNAPISOPAIR value. Changing this value in effect toggles you from one isometric plane to the next. A value of 0 represents the left plane, the top plane is represented by a value of 1, and the right plane by a value of 2.

Examples   The SNAPISOPAIR value has no immediate effect on the drawing unless you are currently in Isometric mode. In this example, SNAPSTYLE is first set to Isometric mode and then the isometric plane is changed using the SNAPISOPAIR variable. This same procedure could have been accomplished by using the SNAP command and the isoplane toggle key.

```
Command: SETVAR
Variable name or ?: SNAPSTYL
New value for SNAPSTYL <0>: 1
```

```
Command: SETVAR
Variable name or ? <SNAPSTYL>: SNAPISOPAIR
New value for SNAPISOPAIR <1>: 2
```

Tips   The initial SNAPISOPAIR value is determined from the prototype drawing (default value is 0, left isometric plane). The current value is saved with the drawing.

In AutoCAD release 9, you can change the SNAPISOPAIR system variable from within the dialog box displayed by the DDRMODES command (only with AutoCAD release 9; see Advanced User Interface and DDRMODES).

SNAPMODE   [all versions]

Overview   The SNAPMODE system variable determines whether AutoCAD's Snap feature is on or off. A value of 0 indicates that it is off, and a value of 1 means that it is on.

Procedure   The SNAP command controls the on and off setting of AutoCAD's Snap mode, as does the Snap Toggle. You can also use the SETVAR command or AutoLISP's (setvar) function to control the Snap mode setting by changing the SNAPMODE system variable.

Examples   The SETVAR command is used here to turn the Snap mode on.

```
Command: SETVAR
Variable name or ?: SNAPMODE
New value for SNAPMODE <0>: 1
```

Tips   It is particularly helpful to retrieve and store the SNAPMODE value whenever you write an AutoLISP routine that needs to set a particular Snap mode setting. In this way, you can reset the original Snap mode when your routine is completed.

The initial SNAPMODE value is determined from the prototype drawing (default value is 0, Snap mode off). The current value is saved with the drawing.

In AutoCAD release 9, you can change the SNAPMODE system variable from within the dialog box displayed by the DDRMODES command (see Advanced User Interface and DDRMODES).

SNAPSTYL

[ADE-2] [all versions]

Overview   The SNAPSTYL system variable stores an integer value that determines whether you are currently drawing in Standard mode or Isometric mode.

Procedure   Normally you select the Isometric Drawing mode from within the SNAP command using the Style option to select the Isometric mode (see SNAP). You can also use the SETVAR command and the AutoLISP function (setvar) to change the SNAPSTYL value and thus select the Standard or Isometric Drawing mode. If SNAPSTYL is 0, AutoCAD is in Standard mode. If the value of SNAPSTYL is 1, AutoCAD is in Isometric mode. The numeric value of the variable SNAPISOPAIR then determines which isometric plane you are currently drawing in.

Examples   The SETVAR command is used here to change the Snap style to Isometric mode.

```
Command: SETVAR
Variable name or ?: SNAPSTYL
New value for SNAPSTYL <0>: 1
```

Tips   The initial SNAPSTYL value is determined from the prototype drawing (default value is 0, Standard mode). The current value is saved with the drawing.

In AutoCAD release 9, you can change the SNAPSTYLE system variable from within the dialog box displayed by the DDRMODES command (see Advanced User Interface and DDRMODES).

SNAPUNIT

[all versions]

Overview   The SNAP command, through the Aspect option, permits you to specify different X-axis and Y-axis snap spacings. The values for these two snap spacings are stored in the SNAPUNIT system variable. The SNAPUNIT value is a coordinate point representing the X- and the Y-spacing. If the spacing is the same along both axes, the X- and Y-values will be the same.

Procedure   Normally you use the SNAP command to change the snap spacing. You can also use the AutoLISP (setvar) function or the SETVAR command to change the snap spacing by changing the SNAPUNIT value.

Examples   The following example uses the SETVAR command to change the snap spacing by changing the coordinate pair value of SNAPUNIT. Unlike using the SNAP command to change the snap spacing, the Snap mode is not automatically turned on.

```
Command: SETVAR
Variable name or ?: SNAPUNIT
New value for SNAPUNIT <1.0000,1.0000>: 2,2
```

Tips   The initial snap spacing, and thus the value for SNAPUNIT, is determined from the prototype drawing (default value is 1.0000,1.0000). The current value is held in the SNAPUNIT variable and is saved along with the drawing file.

In AutoCAD release 9, you can change the SNAPUNIT system variable from within the dialog box displayed by the DDRMODES command (see Advanced User Interface and DDRMODES).

SOLID

[all versions]

Overview   The SOLID command draws solid filled polygons, entering them as either triangular or quadrilateral sections. Polygons created by the SOLID command are special AutoCAD entities called solids.

Procedure   The SOLID command prompts for first point, second point, third point, fourth point, and then, if you continue to enter points, alternates back and forth from third point to fourth point to third point, and so on. The points are entered in a triangular fashion. The first and second points establish a starting edge. After that, enter points diagonally across from each other at opposite corners of the object to be drawn. You can continue to enter points, building up a complex solid entity. When done, respond to the fourth and/or third point prompts by pressing Return.

Examples   This example illustrates the diagonal method of entering points in response to the SOLID command's prompts, to build up the solid entity shown in the drawing.

```
Command: SOLID
First point: (select point A)
Second point: (select point B)
Third point: (select point C)
Fourth point: (select point D)
Third point: (select point E)
Fourth point: (select point F)
Third point: (select point G)
Fourth point: ( RETURN )   (completes triangular section)
Third point: ( RETURN )   (ends SOLID command)
```

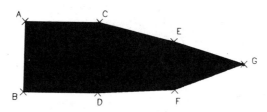

Tips   Solids are drawn as solid-filled objects unless Fill mode is off. When fill mode is off, only the outlines are shown. This results in faster drawing display times. When Fill mode is turned back on, the drawing must be regenerated before the solids once again appear solid-filled (see FILL).

SPLFRAME

[ADE-3] [r9]

Overview   AutoCAD release 9 contains a new PEDIT option that allows the creation of cubic B-spline curves. When you use this option, the original polyline is retained as the frame from which the spline curved polyline is created. This frame is normally not visible. The SPLFRAME (Spline Frame) system variable enables you to control the visibility of this frame.

Procedure   When the SPLFRAME value is 0, the spline frame is invisible. If the value of SPLFRAME is 1, both the spline curve and the spline frame are visible.

   You can change the SPLFRAME system variable by using the SETVAR command or the (getvar) and (setvar) AutoLISP functions.

Examples   In the drawings below, the spline curved polyline is shown with the frame invisible (top) and visible (bottom). The SETVAR command is used to change the SPLFRAME setting.

```
Command: SETVAR
Variable name or ?: SPLFRAME
New value for SPLFRAME <0>: 1
```

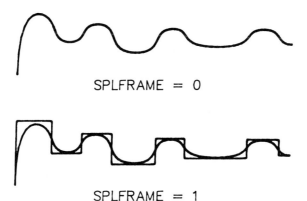

SPLFRAME = 0

SPLFRAME = 1

Warnings   The SPLFRAME system variable is available only in AutoCAD release 9 with ADE-3.

Tips

The spline frame is not displayed after you change the SPLFRAME value until the polyline is regenerated.

The initial SPLFRAME value is determined from the prototype drawing (default value is 0, spline frames not shown). The current value is saved with the drawing.

SPLINESEGS

[ADE-3] [r9]

Overview

AutoCAD release 9 contains a new PEDIT option that allows the creation of cubic B-spline curves. When you use this option, the resulting spline curve is generated with a predetermined number of line segments between each pair of control points to approximate the spline curve. The SPLINESEGS (Spline Segments) system variable contains an integer value that specifies the number of segments to use.

Procedure

You can change the SPLINESEGS value using the SETVAR command or the (getvar) and (setvar) AutoLISP functions. The greater the number, the more line segments used between each spline frame control point to approximate the curve, and the more precise the curve will become.

If the SPLINESEGS variable is set to a negative number, the spline curve will be fit first using the absolute value of the SPLINESEGS variable and then fit using the segments as control points. The result is a smoother representation of the spline curve than when positive values are used.

Examples

The two drawings that follow show the difference in appearance between two different SPLINESEGS values. The SETVAR command is used to change the value from 8 to −8.

```
Command: SETVAR
Variable name or ?: SPLINESEGS
New value for SPLINESEGS <8>: -8
```

As you can see from the two drawings, changing the number of segments should be necessary only when you are doing exacting work using spline curves. For most purposes, the default setting should suffice (see "Warnings").

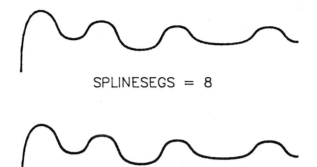

SPLINESEGS = 8

SPLINESEGS = −8

Warnings   The SPLINESEGS variable is available only in AutoCAD release 9 with ADE-3.

    The greater the number of segments used for drawing the spline curve, the longer the curve takes to generate. Higher numbers also result in the spline curve taking up more space in memory and in the drawing file.

    Spline curves generated with a negative SPLINESEGS value take longer to generate than those with positive values, because approximately twice the number of segments are actually used.

Tips   If you change the SPLINESEGS value, you can change a polyline already spline fit by simply resplining it. You do not need to decurve it first.

    The initial SPLINESEGS value is determined from the prototype drawing (default value is 8). The current value is saved with the drawing.

STATUS

[all versions]

Overview   The STATUS command displays a list of statistics about the current drawing. These statistics include the number of entities in the drawing, the drawing limits, the current layer, the current color, and the amount of space left on the disk containing the drawing file.

Procedure   To view statistics about the current drawing, enter the STATUS command at AutoCAD's command prompt. A listing of statistics displays on the text screen.

Examples   Most of the items in Figure 79 are self-explanatory (see the individual commands for a complete description). In this example, limit checking is currently turned off and the drawing extends outside the drawing limits (see LIMITS).

Free RAM and I/O page space relate to the amount of memory within and above DOS's 640K-byte range available as temporary drawing file storage. You can usually ignore these values, although very low values mean that AutoCAD has to access the disk often, resulting in very slow performance. Free disk space is the amount of storage space remaining on the disk containing the drawing file. If you run out of free disk space, AutoCAD cannot continue. You are prompted to save your drawing before AutoCAD terminates.

If you have configured AutoCAD to use a RAM disk for storage of the temporary files created while you edit a drawing, the free disk space within this RAM disk is also displayed (see Configuring AutoCAD). Running out of free disk space in this RAM disk has the same basic effect as running out of regular free disk space.

Tips   Using a RAM disk for placement of temporary files has one drawback. If you run out of RAM in the RAM disk, AutoCAD aborts with an error message that

Command: **STATUS**

```
    150 entities in DRAWING
  Limits are          X:     0.0000    12.0000   (Off)
                      Y:     0.0000     9.0000
  Drawing uses        X:     2.5000    18.0000   ** Over
                      Y:     6.6187     5.7500
  Display shows       X:     2.0000     7.3333
                      Y:     4.0000     8.0000
  Insertion base is   X:     0.0000   Y:    0.0000   Z:    0.0000
  Snap resolution is  X:     1.0000   Y:    1.0000
  Grid spacing is     X:     1.0000   Y:    1.0000

  Current layer:     0
  Current color:     BYLAYER -- 7 (white)
  Current linetype: BYLAYER -- CONTINUOUS
  Current elevation:    0.0000   thickness:    0.0000
  Axis off  Fill on  Grid off  Ortho off  Qtext off  Snap off  Tablet off
  Object snap modes: Intersection, Perpendicular
  Free RAM:  5680 bytes       Free disk: 9676800 bytes
  I/O page space: 124K bytes   Extended I/O page space:1024K bytes

  Command:
```

Figure 79

it is out of RAM. You are given a chance to save your drawing, but AutoCAD then exits to the DOS prompt. To avoid this problem, use the STATUS command from time to time to check the amount of free disk space in the RAM disk.

STATUS LINE

[ADE-1] [all versions]

Overview
The status line is the area across the top of AutoCAD's graphics screen. When the status line is enabled, it reports the current layer, the current Ortho mode, Snap mode, and Tablet mode settings, and the current screen cursor coordinates.

Procedure
If your version of AutoCAD contains ADE-1 (see ADE), the status line area can be enabled when configuring the display device (see Configuring AutoCAD).

Examples
The AutoCAD status line is shown in Figure 80.

Tips
For additional information about the status line, see LAYER, ORTHO, SNAP, and TABLET (and their associated system variables) and COORDS.

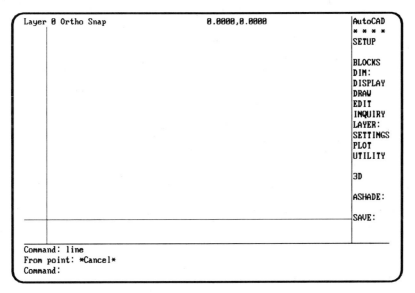

Figure 80

505

STRETCH

[ADE-3] [v2.5, v2.6, r9]

Overview   You may find that you need to move an object in your drawing while preserving its connection to other parts within the drawing. For example, you may need to relocate a door through a wall. The STRETCH command moves a portion of a drawing while maintaining connections to other parts of the drawing. Objects drawn with lines, arcs, traces, solids, and polylines can be stretched.

Procedure   When the STRETCH command is activated, you are prompted to select objects by windowing. This is the only AutoCAD command that requires that you select objects only by including them within a window. In most cases, you will use the Crossing option (all objects within the window or crossing it are selected). Objects that fall entirely within the window are simply moved as if you had used the MOVE command. Objects that cross the window are stretched, that is, the end points that fall within the window are moved while the end points outside the window remain in their original location.

Once you have selected the objects to be stretched, press Return in response to the prompt to select objects, to indicate that you have finished selecting. AutoCAD prompts for a base point and a new point. These two points determine the distance and the direction that the selected objects will be stretched. You can enter actual coordinates or use the screen cursor to indicate the points. For example, if you know you want to stretch the selected objects exactly 2 feet to the right, you would respond to the prompt for the base point by entering the coordinate 0,0 and to the prompt for the new point by entering 2',0.

```
Command: STRETCH
Select objects to stretch by window...
Select objects: (select the objects)
Select objects: ( RETURN )
Base point: (select base point)
New point: (select new point)
```

The STRETCH command has several additional rules. You must use either the crossing or window method to select objects at least once, although you can use other methods to add or remove objects from the selection set. If you use one of the windowing methods more than once, the last window specified will be the window moved by the STRETCH command. Only selected objects are stretched. Objects outside the window are not affected.

The effect of the STRETCH command depends on the type of entity selected. End points of lines that fall within the selection window are moved, but end points outside the window remain where they were. The same is true for arcs, except that the center point and the starting point and end point angles are adjusted so that the distance from the center point to the midpoint of the chord remains constant. The vertices of traces and solids that fall within the window are moved, but those outside the window remain where they are. Polyline segments that cross the window are stretched as if they were line or arc entities. All other AutoCAD entities (points, circles, text, blocks, and so on) are not modified by the STRETCH command. They are moved if included within the selection window.

Examples   In the preceding discussion, you learned about using the STRETCH command to relocate a door in a wall. This is the most obvious use for the command but also illustrates its power. The door can be moved quickly without your having to do any of the clean-up that would otherwise be necessary. Activate the STRETCH command and window the area immediately around the door using the Crossing option. Select the current hinge point as the base point and the new location for the hinge point as the new point. If you know how far you want to move the door, you could simply enter two coordinates, exactly that distance apart, for the two points. For example, if you want to move the door 6 feet, you could enter the coordinate 0,0 as the base point and 6',0 as the new point. You can also use relative coordinates to specify the distance to stretch the object. In this example, you could select the current location as the base point and then move the door over exactly 6 feet by specifying the new point as @6',0. Or you could enter any coordinate as the base point and then specify the new point as @6',0. All of these methods will work and all produce the same result. Choose whichever method makes more sense to you or seems easier to use.

In the following example and drawing, the door and opening are selected by the crossing method and then moved to the right 6 feet by entering 0,0 as the base point and @6',0 as the new point.

```
Command: STRETCH
Select objects to stretch by window...
Select objects: C
First corner: (select point A)
Other corner: (select point B) 8 found.
Select objects: ( RETURN )
Base point: 0,0
New point: @6',0
```

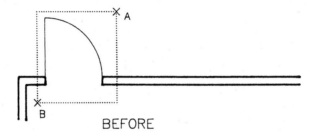

BEFORE

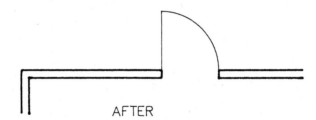

AFTER

Tips

Because the STRETCH command has several rules, it is often helpful to use AutoCAD's Drag feature to see the result of the STRETCH command before you enter the new point.

The Snap and Ortho modes affect the STRETCH command.

If the results of the STRETCH command are not what you had intended, you can return the drawing to its condition prior to the command by entering *U*, for Undo, at the next command prompt (see UNDO).

If you supply a Z-coordinate when entering the base point and the new point, you can stretch a three-dimensional object in three-dimensional space (only with AutoCAD version 2.6 and release 9 when ADE-3 is included).

In AutoCAD release 9 (with the standard menu ACAD.MNU), you can select the STRETCH command from the Edit pull-down menu. The Crossing selection mode is automatically used. The menu item repetition in release 9 causes the command to repeat automatically until you cancel it by pressing Ctrl-C or selecting another command from the menu. (See Advanced User Interface and Entity Selection.)

STYLE

[all versions]

Overview   Before you can place text into a drawing using one of AutoCAD's text commands, you must define a text style to use. The STYLE command creates text styles using one of AutoCAD's supplied font files (or one created by the user or a third party). Each style is given a name by the user. In addition to specifying the font associated with the named style, the STYLE command allows you to determine whether the text will be entered with an obliquing angle, in a vertical orientation, mirrored, or with a different horizontal scale. Text styles can be predefined within the prototype drawing. Initially the prototype drawing has a style called Standard.

Procedure   To define a text style for use in the current drawing, activate the STYLE command. AutoCAD asks for the text style name and displays the current style as the default. You can name the text style anything you like, up to thirty-one characters. After you name the text style, AutoCAD prompts for a font file, again displaying the font used for the current style as the default. A font file is a disk file that contains the description of letters, numbers, and symbols. AutoCAD is supplied with several font files. You can create your own or purchase additional fonts from other sources. Enter the name of the font file you want to use for the text style you are creating. Font files can be used to create more than one style. The style determines the final appearance of the text — the font file contains just the basic definition. The font file name you enter can include a drive designation and a subdirectory. Do not include a file extension. The file type .SHX is assumed. (For examples of all of the fonts supplied with AutoCAD, see Fonts).

Next, AutoCAD asks for the height and again provides a default. You can specify a height if you want, but doing so makes the style you create a fixed-height style; that is, text entered using that style will always be entered at the same height, and you will not be prompted for a text height when you actually enter text using one of AutoCAD's text commands. Normally, you should enter a text height of 0. Specifying a zero height during the creation of the style means that you will be able to control the height of text when you enter it using one of the text commands.

The width factor determines the horizontal scaling of the font (see following examples). A value of 1 uses the width of the font as originally specified within the font file. A factor of .75 causes the text to be compressed 25 percent, and a factor of 1.25 results in the text being stretched 25 percent. Experiment with various width factors to find those most appropriate for your use.

AaBbCc

width = .75

AaBbCc

width = 1.5

The obliquing angle refers to the lean of the individual letters (see following examples). A positive angle will lean the letters toward the right while a negative angle will lean them toward the left. Again, you should experiment to achieve the most pleasing results.

Obliquing angle = 0°

Obliquing angle = +20°

Obliquing angle = −20°

The next two prompts allow you to determine whether text to be entered will be mirrored either about the vertical or horizontal. Normally, to create right reading text, you would respond no to the Backwards? and Upside-down? prompts.

A text style can be defined so that when you enter text it will have a vertical orientation. To do this, respond yes to the Vertical? prompt. If you intend to specify a vertical style, the font-file definition must be capable of a vertical orientation. The vertical parameter appears only if the font file supports vertical orientation. AutoCAD's supplied font files can be oriented vertically. When specifying a vertical style, you should not include an obliquing angle.

The following example shows all of the prompts that you must respond to during the STYLE command.

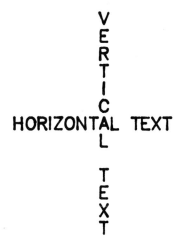

```
Command: STYLE
Text style name (or ?) <current>: (enter name)
Font file <default>: (enter file name)
Height <default>: (enter height) (0 for not fixed)
Width factor <default>: (enter value)
Obliquing angle <default>: (enter value)
Backwards? <Y/N>: (enter Y or N)
Upside-down? <Y/N>: (enter Y or N)
Vertical? <Y/N>: (enter Y or N)
(Name) is now the current text style.
```

Notice that when you finish defining a new text style using the STYLE command, it becomes the current style and will be used the next time you enter one of the text commands. If you type a question mark (?) in response to the prompt for the text style name, AutoCAD displays a list of every style already defined within your drawing, along with its height, width, obliquing angle, and so on.

If you enter the name of an existing style in response to the prompt for the text style name, you can redefine an existing style. Changing a given parameter of a style that has already been used to enter text into the drawing does not affect the existing text, with two important exceptions. If you change the font file or the vertical or horizontal orientation, all text entered with that particular style will be changed the next time your drawing is regenerated. Changing one of these parameters causes an automatic regeneration unless REGENAUTO is off.

Examples In this example, the STYLE command creates a new text style called Dimensions, using the Simplex font and a width factor of 0.92.

```
Command: STYLE
Text style name (or ?) <STANDARD>: DIMENSIONS
New style.

Font file <txt>: SIMPLEX
Height <0.0000>: [ RETURN ]
Width factor <1.00>: .92
Obliquing angle <0>: [ RETURN ]
Backwards? <N> [ RETURN ]
Upside-down? <N> [ RETURN ]
Vertical? <N> [ RETURN ]

DIMENSIONS is now the current text style.
```

In the second example, the Dimensions style is changed to use the TXT font. All previous uses of the Dimensions style are converted to TXT font and take on any new alignment, but the existing width factors are maintained. Only new text entities take on the new 0.75 width factor.

```
Command: STYLE
Text style name (or ?) <DIMENSIONS>: [ RETURN ]
Existing style.

Font file <SIMPLEX>: TXT
Height <0.0000> [ RETURN ]
Width factor <0.92>: .75
Obliquing angle <0>: [ RETURN ]
Backwards? <N> [ RETURN ]
```

```
Upside-down? <N> ( RETURN )
Vertical? <N> ( RETURN )
Regenerating drawing.
```

DIMENSIONS is now the current text style.

Warnings   Remember that changes made to a style definition will not appear in your drawing until the drawing is regenerated.

Only changes made to the horizontal/vertical orientation or the font file cause previously entered text to be changed. Changes to the other parameters affect only subsequent text.

Tips   Although style names can be up to thirty-one characters long, keep them short. Remember, you have to type the name in every time you want to change styles. You can use the RENAME command to change the name of a style.

The same font file can be used by more than one style. This is often the case when you use several different orientations or width factors to change the appearance of the font.

The more complex the font, the longer it takes to display in the drawing. Although the QTEXT mode allows you to turn actual display of text entities into a representation (see QTEXT), sometimes you want to see the text. To save redrawing and regeneration time while you are working on a drawing, define a style using AutoCAD's TXT font. This is the simplest font and displays fairly quickly. Later, when you get ready to plot your drawing, you can use the STYLE command to change the font to the one you want to use.

In AutoCAD release 9, the Options pull-down menu has a fonts selection. This menu item displays an icon menu of all the AutoCAD-supplied text fonts. Selecting one of the icons from this menu automatically executes the STYLE command, selecting the text font within the selected icon box and assigning the font name as the text style name. You are then prompted to select the other style features normally. (See Advanced User Interface and Fonts.)

System Variables   [all versions]

Overview   System variables are internal AutoCAD variables that control individual settings, modes, and parameters within AutoCAD.

Procedure   You can access the system variables while within an AutoCAD drawing through the use of AutoCAD's SETVAR command or from within Auto-LISP. The AutoLISP functions (getvar) and (setvar) are expressly provided for obtaining the current value and changing the value of a system variable, respectively.

   The system variables are all described individually in this book. You can find each variable by looking for it by name. In addition, each command description contains a list of its associated system variables.

   Many system variables are maintained individually with each drawing file. These variables determine such things as whether Ortho mode is on when you begin editing a drawing. Other variables, such as the one that controls the size of the object snap aperture box, are saved in AutoCAD's configuration file. A change to one of these variables, such as changing the size of the pick box, is apparent in every drawing. The third type of system variables is not saved anywhere. They simply provide a temporary holding place for values, such as the most recent area of a selected object, for retrieval and use by other functions. Their values are lost when you quit or end the drawing, and their values change every time you use commands that result in new values.

Examples   An example of the use of each individual system variable is provided under each variable. The SETVAR command is a means to observe and change the system variable values directly from AutoCAD's command prompt. The following example illustrates the use of the SETVAR command to change the size of AutoCAD's pick box.

```
Command: SETVAR
Variable name or ?: PICKBOX
New value for PICKBOX <3>: 5
```

Tips   The initial values of all system variables that are saved with the individual drawing file are determined from the prototype drawings. Other system variables are determined by the the latest AutoCAD configuration or are default values that cannot be altered. (See the individual system variables for more information.)

   A list of all the system variables can be found in Appendix A.

   The dimensioning system variables are not available as system variables (using the SETVAR command) in AutoCAD version 2.5. That version of

AutoCAD allows access only to dimensioning variables while in Dimensioning mode (when AutoCAD displays the dim prompt).

There are ten system variables that are undocumented in the AutoCAD manual. The variables USERI1, USERI2, USERI3, USERI4, and USERI5 can store any user-defined integer value (default value is 0). The variables USERR1, USERR2, USERR3, USERR4, and USERR5 can store any user-defined real number value (default value is 0.0000). All ten of these variables can be entered through the SETVAR command or the AutoLISP functions (getvar) and (setvar). They do not display, however, when you enter a question mark in response to the SETVAR command. These undocumented variables are used by programs like the AutoCAD AEC menu templates for storing parameters used by the menu macros and AutoLISP routines. Users who are not using other AutoCAD add-on programs that make use of these variables can utilize them to store variable information with individual drawing files for use over several editing sessions. These ten variables are saved with each drawing file.

TABLET

[all versions]

Options

On   Turns tablet mode on
Off   Turns tablet mode off
CAL   Calibrates tablet
CFG   Configures tablet menus, pointing area

Overview   If you have a digitizing tablet for use as a pointing device with AutoCAD, the TABLET command allows you to configure the digitizer so that portions of the tablet area can be used for selecting from a tablet menu. A tablet menu can contain AutoCAD commands, predrawn symbols, AutoLISP programs, or any combination of these. You can also use the TABLET command to copy a paper drawing into AutoCAD by first calibrating the tablet so that AutoCAD's drawing scale matches that of your paper drawing. The drawing can then be digitized by first turning AutoCAD's Tablet mode on and then using AutoCAD's drawing commands as you select points within your paper drawing.

Procedure   The TABLET command both configures the digitizer tablet to align up to four distinct tablet menu areas and a screen pointing area and calibrates the digitizing tablet so it can be used to enter a paper drawing. Four options are available within the TABLET command: On, Off, Calibrate, and Configure.

```
Command: TABLET
Option (ON/OFF/CAL/CFG):
```

TABLET ON   When Tablet mode is on, you can digitize a paper drawing into AutoCAD. Before you can turn the Tablet mode on, you must have first used the Calibrate (CAL) option to set two known points on your paper drawing. If you have not yet used the Calibrate option, the on option automatically issues the Calibration prompts.

```
Command: TABLET
Option (ON/OFF/CAL/CFG): ON
```

TABLET OFF   To resume normal screen pointing with the digitizer, use the Off option.

```
Command: TABLET
Option (ON/OFF/CAL/CFG): OFF
```

TABLET CAL (Calibrate)   Before you can use the Tablet mode to digitize a paper drawing, you must calibrate the digitizer. First, fasten your paper drawing onto your digitizer at any angle with drafting tape. Activate the TABLET command and enter the Calibrate option. AutoCAD prompts for two points within your paper drawing and the coordinates those two points represent. Once you enter the coordinates for the two known points, Auto-CAD adjusts for the angle at which you have placed the drawing.

The two points that you choose can be any two different points in your drawing. The coordinates can represent the actual coordinates or simply the relative distance between them.

Once you have calibrated the tablet, you can use any of AutoCAD's commands. As you digitize the paper drawing, AutoCAD displays the coordinates of the drawing. If the entire paper drawing does not fit on your digitizer, you can calibrate the tablet, digitize part of the drawing, and then move the paper drawing. You then need to recalibrate the tablet and digitize the remainder of the drawing. If you are careful when entering the two known points each time, you will have no problem digitizing a drawing, no matter how many times you have to move it to fit all of the drawing on the digitizer.

What you are doing, in effect, is setting your tablet so that it recognizes the scale of your drawing. The cursor movements on the screen correspond to this scale. Normally when you move the cursor, puck, or stylus within the screen pointing area of your digitizer, the relative movement of the cursor within this screen pointing area translates to a relative motion on the screen, independent of the current zoom magnification on the screen. When you use AutoCAD with the Tablet mode on, the motion of the cursor on the digitizer is an absolute movement in relation to real points on the screen.

The following example illustrates the TABLET command sequence for calibrating the tablet to digitize a drawing.

```
Command: TABLET
Option (ON/OFF/CAL/CFG): CAL
Calibrate tablet for use...
Digitize first known point: (digitize point on paper drawing)
Enter coordinates for first point: (enter coordinate)
Digitize second known point: (digitize point on paper drawing)
Enter coordinates for second point: (enter coordinate)
```

TABLET CFG (Configure)   To use the digitizer tablet with either Auto-CAD's supplied tablet menu or a tablet menu you have written yourself or purchased from a third party, you must first use the Configure option of the TABLET command to align the different tablet menu areas. AutoCAD allows up to four different tablet menu areas and a screen pointing area. The Configure option prompts you for the number of tablet menu areas and then prompts you through the configuration routine, asking for the upper left, lower left, and lower right corners for each menu area and the number of columns and rows in each menu area. After configuring each of the tablet menu areas, you are prompted to digitize the lower left and upper right corners of the screen pointing area.

The only restriction is that menu areas and the screen pointing area cannot overlap. As you digitize the corners of the tablet menu areas, the angle formed must be 90 degrees, or you will be prompted to enter the three corners again.

```
Command: TABLET
Option (ON/OFF/CAL/CFG): CFG
Enter number of tablet menus desired (0-4) <default>: (enter number)
Do you want to realign tablet menu area? <N>: Y
Digitize upper left corner of menu area 1: (do so)
Digitize lower left corner of menu area 1: (do so)
```

```
Digitize lower right corner of menu area 1: (do so)
Enter number of columns for menu area 1: (enter proper number)
Enter number of rows for menu area 1: (enter proper number)
```

These prompts are repeated for each of the menus specified. When you have completed this step, AutoCAD prompts

```
Do you want to respecify the screen pointing area? <N>: Y
Digitize lower left corner of screen pointing area: (do so)
Digitize upper right corner of screen pointing area: (do so)
```

Figure 81 shows the tablet menu supplied with AutoCAD. The four menu areas and the screen pointing area are labeled.

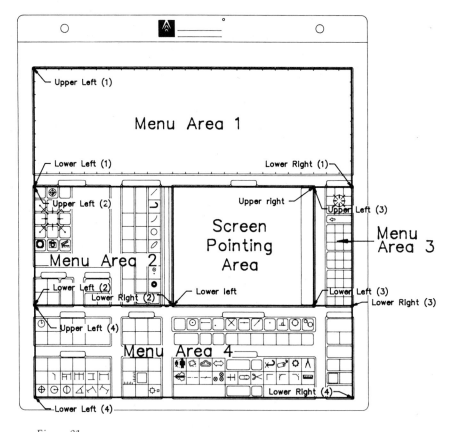

Figure 81

Examples   When using the Calibration option of the TABLET command to match a paper drawing to an AutoCAD drawing, you can enter actual or relative coordinates by typing at the keyboard in response to the prompt to enter the coordinates of the first and second known points. However, if you already have points located on the screen that correspond to points within your paper drawing, you do not need to actually enter these coordinates. Go through the Calibrate option, digitizing the known point on the paper drawing. But when you are prompted for the coordinate, it works equally well (and is much faster) if you simply use one of AutoCAD's object snaps to snap to the object in your drawing that already corresponds to the known point. AutoCAD will read the coordinate from the AutoCAD drawing and match it to the point within the paper drawing on the digitizer.

This is particularly useful when you are digitizing a large paper drawing that does not completely fit on the digitizer. After digitizing the first portion of the drawing, if you select the two known points so they are within both the first part (already digitized) and the second part, instead of entering their coordinates you can object snap to them (since they are already drawn and displayed on the screen).

Warnings   Remember, you cannot turn the Tablet mode on until you have calibrated the digitizer. Also remember that the tablet menu areas and screen pointing areas cannot overlap.

When you calibrate the digitizer, if the coordinates you enter increase in Y-value as you go from bottom to top, then the coordinates must also increase in X-value as you go from left to right.

When the Tablet mode is on, the movement of the digitizer cursor controls the absolute movement within the drawing. It is therefore possible to draw an object that is off the screen. This will happen if the image on the screen does not correspond to the coordinates that are currently mapped on the digitizer. If this happens, pan over or zoom to a lower screen magnification.

Tips   Only the screen pointing area of the digitizer is available for digitizing a paper drawing. If you are using a tablet menu to which you have dedicated a large portion of the digitizer area as menu rather than screen pointing area, you should reconfigure the tablet before using it to digitize a paper drawing. This way, you can use the entire digitizer as screen pointing area for entering your drawing, and you won't have to move the paper drawing and recalibrate as many times.

If you design your own tablet menus, you might want to use the same configuration as the standard AutoCAD tablet menu (the one provided with the program) or match whatever other menu you are using. You can change menus using the MENU command without having to reconfigure the tablet, as long as the menu you change to has the same number of menu areas and the same number and size of rows and columns within each menu area.

If you use more than one tablet menu with differing layouts, rather than reconfigure the tablet every time you change menus, save two different AutoCAD configuration files. Then when you load AutoCAD, you can use a DOS batch file to load the particular AutoCAD configuration pertaining to whichever menu you want to use. This works only if you don't change menus in the middle of editing a drawing. Remember, this is necessary only if the different menus have different physical layouts.

Tablet Menu

[all versions]

Overview   When using a digitizer tablet as your pointing device, you can utilize the tablet as both a pointing device and to input AutoCAD commands. This is done through the use of a tablet menu. A tablet menu is actually one of the parts of any AutoCAD menu file (the other two parts being the screen menu and the button menu). AutoCAD's standard menu contains a tablet menu. When Autodesk receives an AutoCAD registration, they send out a plastic digitizer template for use with the standard menu.

Every AutoCAD menu file can contain up to four tablet menu areas and a screen pointing area. The TABLET command configures AutoCAD so it recognizes the various areas of the digitizer tablet.

Using a tablet menu has several advantages over using a screen menu. The greatest advantage is that you can select AutoCAD commands directly from a well-designed tablet menu instead of having to navigate through AutoCAD's numerous screen-menu pages. A good tablet menu should be logically laid out, grouping similar commands in easy-to-find areas on the template. A tablet menu also allows you to include icons for standard symbols that you regularly include in your drawings.

The one possible disadvantage to using a tablet menu is that you must take your eyes off the screen and refocus on the digitizer. Most users have no trouble developing the eye-hand coordination necessary for this task. After several days using a particular tablet menu, you should be able to find often-used commands with a mere glance at the tablet. The speed increase from using a well-designed tablet menu far outweighs any possible increase in eyestrain.

Procedure　　Before using any tablet menu, you must use the TABLET command to configure AutoCAD to recognize the individual areas on the tablet (see TABLET). Once this has been accomplished, you can select a command from the tablet by simply moving the digitizer's stylus or puck until the pointer is directly over the cell containing the desired command and then press the pick button.

As mentioned previously, any AutoCAD menu file can contain a tablet menu section. AutoCAD's standard menu contains a tablet menu, which is used only if you actually use AutoCAD's supplied digitizer template. If you are not using a digitizer, the screen menu still continues to work normally.

The process of writing a menu is actually a simple process, although you can design very complex menus. The menu file can contain up to four separate tablet menu sections. Each begins with a tablet menu header line ***TABLET*n*, where *n* is a number from 1 to 4. Each line that follows represents one box within the tablet menu, starting in the upper left corner of the individual tablet menu area and proceeding to the right and then down to the next line. Tablet menus can be replaced by tablet submenus in a fashion identical to AutoCAD's other menu types. The construction $T*n* = is used to call a submenu. For example, $T4 = DTEXT1 would call a tablet submenu named DTEXT1 which would replace all or part of tablet menu number 4. If the DTEXT1 submenu header is supplied with a space and a number, the submenu replacement will begin that number of cells from the first cell of tablet menu area 4. The submenu replaces only the number of cells corresponding to the number of lines in the submenu. The following sample from the standard AutoCAD menu illustrates this. The DTEXT1 submenu replaces only cells 33, 34, and 35 of tablet menu area 4.

```
**DTEXT1 33
$S=X $S=DTEXT ^C^CDTEXT
$S=X $S=DTEXT ^C^CDTEXT C
$S=X $S=DTEXT ^C^CDTEXT R
```

Examples

Writing an AutoCAD menu file is not a difficult task. One way to get started is to add some of your standard symbols to AutoCAD's standard menu. You can include your standard symbols in the blank menu area 1, starting in the upper left hand corner and working across one row at a time.

To make each cell of this menu area automatically insert one of your symbols, first save each symbol as a block. Then edit AutoCAD's standard menu file ACAD.MNU. Using a word processor or text editor, find the line in the menu file that reads ***TABLET1. The next 200 lines are labeled [T1-1], [T1-2], [T1-3], and so on. Starting on the line labeled [T1-1], add the following after the right bracket:

`INSERT;symbol-name;\\\\`

where the *symbol-name* is replaced by the name of the symbol that you want to insert. Continue editing this file, including up to 200 of your details and symbols. Remember, each line [T1-1] through [T1-200] represents one cell within tablet menu area 1. Cell numbers match line numbers reading from left to right and from the first line (line A) to the eighth line (line H).

Tips

You can include control codes in menus by preceding the character with a caret (^). Thus, ^C is the same as Ctrl-C and ^H is the same as Backspace.

You can spread out long menu macros over more than one line by placing a plus sign as the last character of the line and continuing the macro on the next line.

A semicolon (;) has the same effect as a Return. A backslash (\) within a menu macro causes the macro to pause for user input. Once the input is provided, the macro continues.

If your version of AutoCAD includes ADE-3, you can include AutoLISP macros within your menu files.

If you intend to customize AutoCAD's ACAD.MNU standard menu, make a copy of it first. The tablet menu sections of a menu file follow immediately after the line ***TABLET*n* where *n* is a number from 1 to 4, representing one of AutoCAD's four tablet menu areas.

TDCREATE
[v2.5, v2.6, r9]

Overview

AutoCAD maintains a record of the time and date that every drawing was created. This information is stored in the system variable TDCREATE

(Time and Date Created) and saved with each drawing file. This variable stores the date and time as a real number with the form < Julian date > . < Fraction > , where the date is stored in days and the fraction represents the number of seconds past midnight. This technique is identical to that described for the DATE variable (see DATE).

Procedure   The TDCREATE variable is a read-only variable. The SETVAR command and (getvar) function can be used to return the value, but the value cannot be altered.

Examples   In this example, the SETVAR command is used to return the date that the current drawing was created.

```
Command: SETVAR
Variable name or ?: TDCREATE
TDCREATE = 2447058.53391775 (read only)
```

Using the sample AutoLISP program illustrated in the example for the DATE variable (see DATE), you will see that this value represents September 19, 1987 at 12:48:50.4936 PM.

Tips   The TDCREATE value is saved with the drawing file.

Unless you are going to utilize the TDCREATE value within an AutoLISP routine, it is much easier to retrieve the drawing's creation date using the TIME command, as shown in Figure 82.

```
Command: TIME
```

```
Current time:          20 Sep 1987 at 15:56:30.160
Drawing created:       19 Sep 1987 at 12:48:50.493
Drawing last updated:  20 Sep 1987 at 15:45:22.910
Time in drawing editor: 0 days 01:05:07.140
Elapsed timer:         0 days 00:30:25.670
Timer on.
Display/ON/OFF/Reset:
```

Figure 82

TDINDWG

[v2.5, v2.6, r9]

Overview   The total length of time that a drawing has been on the screen in Auto-CAD's drawing editor is stored in the TDINDWG (Time and Date in Drawing) system variable. This information is saved with each drawing file. The variable stores this value in a format similar to that used by the DATE variable. The value is a real number representing the elapsed time. The format is < Number of Days >.< Fraction >.

Procedure   The TDINDWG variable is a read-only variable. The SETVAR command and (getvar) function can be used to return the value, but the value cannot be altered. The number before the decimal point, divided by 86400 (number of seconds in a day) and converted to an integer returns the number of days. The number after the decimal point, multiplied by 86400, yields the number of seconds (part of a day).

Examples   In this example, the SETVAR command returns the amount of time the drawing has been on-screen in AutoCAD's drawing editor.

```
Command: SETVAR
Variable name or ?: TDINDWG
TDINDWG = 0.04522157 (read only)
```

This value represents a time in drawing editor of 0 day 01:05:07.140.

Tips   The TDINDWG value is saved with the drawing file.

   Unless you are going to utilize the TDINDWG value within an AutoLISP routine, it is much easier to retrieve the time that the drawing has been in the drawing editor using the TIME command, as shown in Figure 82 (see TDCREATE).

```
Command: TIME
```

TDUPDATE

[v2.5, v2.6, r9]

Overview   The time and date that a drawing was last saved (updated) is stored in the TDUPDATE (Time and Date Updated) system variable. This variable is saved with each individual drawing file. The format used is the same as for the DATE variable. The value is represented as a real number in the format <Julian date>.<Fraction>. The date is in days, and the fraction is in seconds.

Procedure The TDUPDATE variable is a read-only variable. The SETVAR command and (getvar) function can be used to return the value, but the value cannot be altered. To determine the actual date and time of the most recent update, determine the date from the number of days (subtract the known date of 2444240, which is January 1, 1980). Multiply the fraction by 86400 and read the result as the number of seconds past midnight.

Examples In this example, the SETVAR command returns the last time the drawing was updated.

```
Command: SETVAR
Variable name or ?: TDUPDATE
TDUPDATE = 2447059.65651515 (read only)
```

This value represents the time the drawing was last updated as 20 Sep 1987 at 15:45:22.910.

Tips The TDUPDATE value is saved with the drawing file.

Unless you are going to utilize the TDUPDATE value within an AutoLISP routine, it is much easier to retrieve the time that the drawing was last saved (updated) using the TIME command, as shown in Figure 82 (see TDCREATE).

```
Command: TIME
```

TDUSRTIMER

[v2.5, v2.6, r9]

Overview AutoCAD's TIME command allows you to set an elapsed timer. You can also reset this timer to zero. The actual elapsed time displayed by the TIME command is stored in the TDUSRTIMER (Time and Date User Timer) system variable. This variable is saved with the drawing file. The value is stored in a format similar to the DATE variable, except that the number is in the format <Number of days>.<Fraction>. This is the same as the TDINDWG variable.

Procedure   The TDUSRTIMER variable is a read-only variable. The SETVAR command and (getvar) function can be used to return the value, but the value cannot be altered except by using the TIME command.

Examples   In this example, the SETVAR command returns the elasped time.

```
Command: SETVAR
Variable name or ?: TDUSRTIMER
TDUSRTIMER = 0.02113047 (read only)
```

This value represents an elapsed timer value of 0 days 00:30:25.670.

Tips   The TDUSRTIMER value is saved with the drawing file.

Unless you are going to utilize the TDUSRTIMER value within an Auto-LISP routine, it is much easier to retrieve the elapsed time using the TIME command as shown in Figure 82 (see TDCREATE).

```
Command: TIME
```

TEMPPREFIX

[v2.6, r9]

Overview   If you have entered a directory name for placement of temporary files during AutoCAD's configuration, the directory name you specified is stored in the TEMPPREFIX system variable. TEMPPREFIX is a read-only variable.

Procedure   When using AutoCAD's configuration menu to configure the program to match your peripheral devices, you can also configure certain operating parameters. This is accomplished using configuration menu task 8. Selecting this task displays the Operating Parameters submenu. Submenu task 5 allows you to specify the device and/or subdirectory that AutoCAD will use for the placement of its temporary files. (This process is described fully in Configuring AutoCAD.)

If you use this subtask to specify the placement of temporary files, the device and/or subdirectory you provide is saved in the TEMPPREFIX system variable. The value is saved as a string value.

Examples   The TEMPPREFIX value can be returned using the SETVAR command or the AutoLISP function (getvar). Because it is a read-only variable, you cannot change its value from within a drawing. Only by reconfiguring AutoCAD can you change its value. If you have not specified a location for temporary file placement, the variable has a null value (no value shown within quotation marks).

```
Command: SETVAR
Variable name or ?: TEMPPREFIX
TEMPPREFIX = "" (read only)
```

Warnings   TEMPPREFIX is not available in AutoCAD version 2.5.

Tips   The TEMPPREFIX value is not actually saved. It is read from the ACAD.CFG file whenever you load a drawing into AutoCAD's drawing editor.

TEXT   [all versions]

Options:

A   Aligns text between two points, chooses appropriate height
C   Centers text horizontally
F   Fits text between two points, chooses appropriate width
M   Centers text horizontally and vertically
R   Right-justifies text
S   Selects text style

Overview   The TEXT command allows you to annotate your drawing, placing text (text entities) into the drawing, one line at a time. All text entered in the drawing is drawn using the current text style (styles are created using the STYLE command). The TEXT command allows you to select the current text style to be used and to control the way in which the text will be drawn. Text is normally entered as left justified, that is, the left margin is aligned with a preselected point and the right margin is ragged. Text can be entered so that the center point of the base line (bottom of the letters) is centered (Center option) or so that the middle point of the line of text is centered (Middle option) about a preselected point. You can also enter text so that it is right-justified (even right margins, uneven left margins) or so that both the left and right margins are fixed by predetermined points and all text within a given line of text will fit between the points (Fit option). Or text can be aligned between two predetermined points (Aligned option). Further, the TEXT command can be used to determine the angle of orientation and the height of the text as it is entered.

Procedure   The TEXT command prompts for you to select the desired option. Text is always entered using the current text style (see STYLE). If you want to change the current style, you should select that option first. Once you change the style, the TEXT command options are presented again. However, once you select the text insertion point, the TEXT command does not give you another opportunity to alter the current style.

```
Command: TEXT
Start point or Align/Center/Fit/Middle/Right/Style:
```

To select one of the available options, you need enter only the first letter of the option and press Return. When AutoCAD prompts for a height or rotation angle, you can either enter an absolute coordinate or a relative coordinate or use the screen cursor to indicate the height or angle as appropriate. The most recently used height and rotation angle appear as the default values. As always, you can select a default simply by pressing Return.

Examples   **Left-Justified Text**   All AutoCAD text is left-justified unless you specify otherwise. Therefore, when the TEXT command prompts you for the start point, selecting a point causes the text you enter to be placed so that the left end of the baseline of the line of text will occur at that point.

Once you have selected the start point, either by entering an absolute coodinate or a relative coordinate or by pointing on the screen, AutoCAD prompts for the text height, the rotation angle, and finally the text itself. The rotation angle is the orientation of the base line of the line of text in relation to the start point. This is illustrated in the following example and drawing.

```
Command: TEXT
Start point or Align/Center/Fit/Middle/Right/Style: (select point A)
Height <default>: (enter height or press Return)
Rotation angle <default>: (enter angle or press Return)
Text: (enter the text exactly as you want it to read)
```

Aligned Text   Selecting the Align option causes AutoCAD to prompt for the two end points of the line of text. You are then prompted simply to enter the text. The text will be drawn so that its base line starts and ends at the two selected points and fits exactly between them. The distance between the two points determines the height of the text, and the angle from the first point to the second determines the text's rotation angle. The text's height/width ratio remains as set by the current text style. This can be seen in the following example and drawing. Points A and B establish the alignment of the text.

```
Command: TEXT
Start point or Align/Center/Fit/Middle/Right/Style: A
First text line point: (select point A)
Second text line point: (select point B)
Text: (enter the text exactly as you want it to read)
```

Aligned Text

Aligned Text

Centered Text   The Center option causes text to be entered such that the midpoint of the base line of the line of text is placed at the selected center point (as seen in the following drawing). AutoCAD prompts you to select the center point. The command then prompts for the height, the rotation angle, and the text.

```
Command: TEXT
Start point or Align/Center/Fit/Middle/Right/Style: C
Center point: (select point A)
Height <default>: (enter height or press Return)
Rotation angle <default>: (enter angle or press Return)
Text: (enter the text exactly as you want it to read)
```

text height

CENTERED TEXT

A

CENTERED TEXT
rotation angle

A

Fit Text   Fitting text between two points with the Fit option is similar to the Align option, except that you determine the height of the text. The command prompts for the two points, the height, and the text. The two points simply determine the end points of the text base line and the rotation angle. Letters are stretched or compressed to fit between these two points as seen in the following example and drawing.

```
Command: TEXT
Start point or Align/Center/Fit/Middle/Right/Style: F
First text line point: (select point A)
Second text line point: (select point B)
Height <default>: (enter height or press Return)
Text: (enter the text exactly as you want it to read)
```

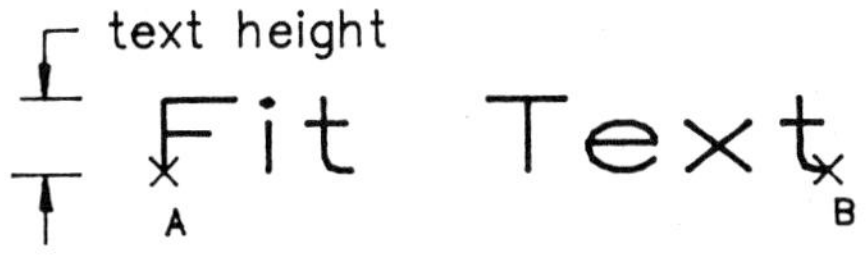

text height

Fit Text

A   B

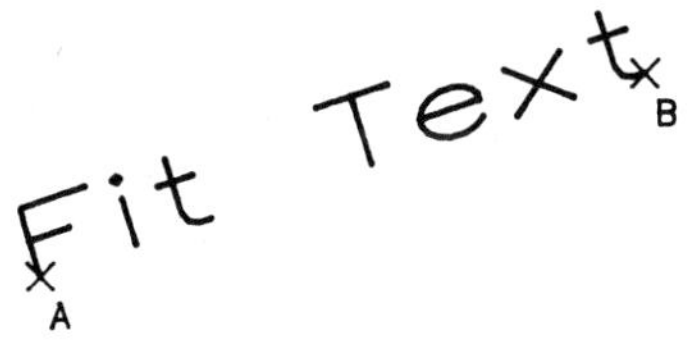

Fit Text

B

A

Middle Text   The Middle option is similar to the Center option, except that instead of placing the midpoint of the baseline at the selected point, the text is centered both horizontally and vertically at the selected point (see the following drawing). AutoCAD prompts for the middle point, the height, the rotation angle, and the text.

```
Command: TEXT
Start point or Align/Center/Fit/Middle/Right/Style: M
Middle point: (select point A)
Height <default>: (enter height or press Return)
Rotation angle <default>: (enter angle or press Return)
Text: (enter the text exactly as you want it to read)
```

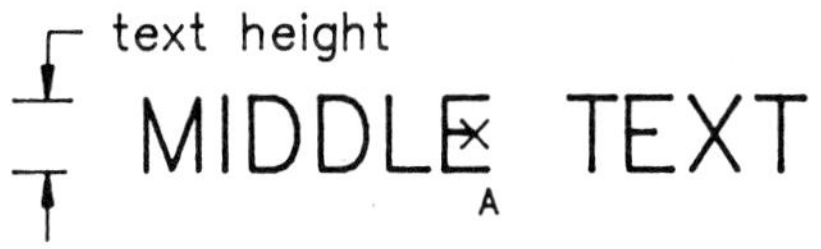

text height

MIDDLE TEXT

A

MIDDLE TEXT   rotation angle

A

Right-Justified Text   Right-justified text is similar to left-justified text except that the point you select defines the right end of the line of text and the text then extends toward the left (see the following drawing). The length of the line of text depends solely on the number of text characters entered. AutoCAD prompts for the end point, the height, the rotation angle, and the text.

```
Command: TEXT
Start point or Align/Center/Fit/Middle/Right/Style: R
End point: (select point A)
Height <default>: (enter height or press Return)
Rotation angle <default>: (enter angle or press Return)
Text: (enter the text exactly as you want it to read)
```

text height

Right Justified

A

rotation angle

Right Justified

A

Changing the Current Text Style   If you want to use a different text style, you must use the Style option before selecting the text insertion point. Once you select the Style option, AutoCAD prompts for the name of the style you want to use. The style must already exist in the drawing. You must have first named the style and selected the font file associated with the named style using the STYLE command (see STYLE). If you forget the names of the styles that have been named within the current drawing, enter a question mark (?). AutoCAD will display a list of the named styles along with their text font, width factors, and so on. You can accept the current style (the default) by pressing Return. You can change styles (make a different style the current style) by entering the valid style name. AutoCAD then repeats the TEXT command prompts.

```
Command: TEXT
Start point or Align/Center/Fit/Middle/Right/Style: S
Style name (or ?) <current style>: (enter new Style name, ?, or press Return)
Start point or Align/Center/Fit/Middle/Right/Style:
```

Entering Multiple Lines of Text   The TEXT command enters text one line at a time. The line of text does not appear on the screen until you press Return after typing in the actual text. There is a method you can use with the TEXT command to enter multiple lines of text.

If you want to enter another line of text directly below the previous line with the same rotation angle and same justification, press Return when the TEXT command asks for you to select an option. The command immediately prompts you to enter text. The text is placed below the previous line. The distance separating the lines of text entered in this fashion is determined by the font definition. Additional lines of text following an aligned text entry are left-justified using the text height established by the previous line. Text entered following a fit entry is fully justified (left and right ends of each succeeding line are in alignment). The following sequence illustrates this method of entering multiple lines of text.

```
Command: TEXT
Start point or Align/Center/Fit/Middle/Right/Style: (select point A)
Height <0.2000>: (RETURN)
Rotation angle <0>:  (RETURN)
Text: This is the first line.
Command: (RETURN)
TEXT Start point or Align/Center/Fit/Middle/Right/Style: (RETURN)
Text: This is the second line.
```

The following drawing shows the text entered in the previous example. Notice the difference when text is left justified, centered, fit, and right justified.

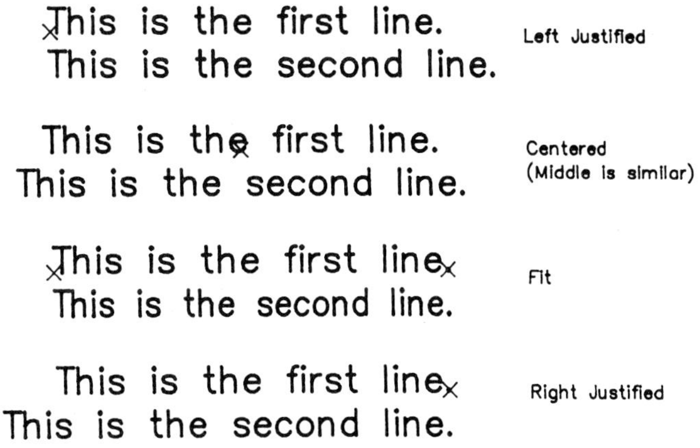

This is the first line.   Left Justified
This is the second line.

This is the first line.   Centered
This is the second line.   (Middle is similar)

This is the first line   Fit
This is the second line.

This is the first line   Right Justified
This is the second line.

Tips   The point you select in response to the start point, center point, or similar TEXT command prompt becomes the text insertion point for the line of text that you enter. You can use AutoCAD's object snap to lock onto that point using the INSERT Object Snap mode.

When you enter a line of text, the Space bar puts spaces between letters. You cannot use it instead of the Return key when AutoCAD expects text. To finish entering a line of text, you must press Return.

TEXTEVAL   [ADE-3] [v2.6, r9]

Overview   The TEXTEVAL system variable determines whether responses to prompts to provide a text string are evaluated literally or as AutoLISP expressions.

Procedure   If the TEXTEVAL value is 0, all responses to AutoCAD prompts requiring a text string are taken literally. That is, when AutoCAD prompts for text, anything that is typed or returned from menus or AutoLISP is considered as a text string. When the TEXTEVAL value is 1, if the first symbol entered in response to a prompt for text is an open parenthesis or an exclamation point, the response to the prompt is evaluated as an AutoLISP expression.

You can change the TEXTEVAL value using the SETVAR command or AutoLISP's (setvar) function.

Examples   In some situations, you may assign a text string to an AutoLISP variable and later want to provide that variable as a response to a command requesting a text string. For this example, the text string FULL SCALE has previously been assigned to the AutoLISP variable USCALE.

In the following sequence, the block TAGSCALE (a previously saved block that contains an attribute) is inserted using the INSERT command. When AutoCAD prompted for the attribute values, !USCALE was entered. In the first part of the example, the TEXTEVAL value is 0, so AutoCAD interprets the entry literally as a text string and causes !USCALE to be written on the drawing. When TEXTEVAL is changed to a value of 1 using the SETVAR command and the INSERT command is repeated, !USCALE is interpreted as an AutoLISP variable and the text string FULL SCALE is written. The results of both interpretations are shown in the drawing.

```
Command: INSERT
Block name (or ?): TAGSCALE
 Insertion point: (select point A)
 X scale factor <1> / Corner / XYZ: [ RETURN ]
 Y scale factor (default=X): [ RETURN ]
 Rotation angle <0>: [ RETURN ]

Enter attribute values
Enter scale <SCALE>: !USCALE

Command: SETVAR
Variable name or ?: TEXTEVAL
New value for TEXTEVAL <0>: 1

Command: INSERT
Block name (or ?) <TAGSCALE>: [ RETURN ]
 Insertion point: (select point A)
 X scale factor <1> / Corner / XYZ: [ RETURN ]
 Y scale factor (default=X): [ RETURN ]
 Rotation angle <0>: [ RETURN ]

Enter attribute values
Enter scale <SCALE>: !USCALE
```

DETAIL
!USCALE
                              TEXTEVAL = 0

DETAIL
FULL SCALE
                              TEXTEVAL = 1

Warnings   The TEXTEVAL variable does not affect the DTEXT command. DTEXT always takes all input literally.

TEXTEVAL is not available in AutoCAD version 2.5.

Tips   The TEXTEVAL value is not saved when you save the drawing. The value always returns to 0 when you return to AutoCAD's main menu.

'TEXTSCR

Overview   The TEXTSCR (Text Screen) command causes AutoCAD to flip from the graphics screen to the text screen when you are running AutoCAD on a single-screen system. If you are using a dual-screen system, the TEXTSCR command has no effect. This command is included in AutoCAD's command set for the explicit purpose of incorporating it into menus and script files.

Procedure   While the TEXTSCR command is usually used within menus or script files, you can also use it to flip the display to the text screen directly from AutoCAD's command prompt.

In the following example, if the graphics screen was previously displayed, AutoCAD would now be flipped to the text screen.

```
Command: TEXTSCR
```

Examples   There are various reasons to use the TEXTSCR command in a menu or script file. Usually you want to force AutoCAD into the Text Screen mode so the user can read more than just the three text lines within the command area below the graphics screen.

Tips   Just as the Flip Screen key (usually F1) can flip between the graphics screen and the text screen without affecting an active command, the TEXTSCR command can be used transparently while another AutoCAD command is active. To make the TEXTSCR command transparent, precede it with an apostrophe, as in 'TEXTSCR. It is helpful to build 'TEXTSCR into a tablet menu for use as another toggle besides the F1 key.

TEXTSIZE

Overview   The TEXTSIZE system variable stores the current text height value. This value is normally set using the height setting within the TEXT or DTEXT command.

Procedure   You can read or change the current text height using the SETVAR command or the (getvar) and (setvar) AutoLISP functions, respectively. This variable is useful for obtaining or changing the current text height from within an AutoLISP routine.

Examples   In the following example, the SETVAR command is used to return the current TEXTSIZE value.

```
Command: SETVAR
Variable name or ?: TEXTSIZE
New value for TEXTSIZE <0.2000>:
```

Tips   The initial TEXTSIZE value is determined from the prototype drawing (default value is 0.2000). The current value is saved with the drawing.

TEXTSTYLE   [all versions]

Overview   The current text style, set by the STYLE command, is stored in the TEXTSTYLE system variable.

Procedure   The TEXTSTYLE system variable stores a string value that represents the current text style. It is a read-only value. You can observe it by using the SETVAR command or the (setvar) AutoLISP function, but you can alter it only by using the STYLE, DTEXT, or TEXT command.

Examples   In the following example, the SETVAR command is used to return the current TEXTSTYLE.

```
Command: SETVAR
Variable name or ?: TEXTSTYLE
TEXTSTYLE = "STANDARD" (read only)
```

Tips   The initial TEXTSTYLE value is determined from the prototype drawing (default value is STANDARD). The current value is saved with the drawing.

THICKNESS

[ADE-3] [all versions]

Overview   The thickness applied to objects as they are drawn is normally set with the ELEV (Elevation) command. This command prompts for a thickness value. The current value or the new value that you supply is stored in the Auto-CAD system variable THICKNESS.

Procedure   The SETVAR command or the (setvar) AutoLISP function can be used to change the current THICKNESS value. This is particularly useful for incorporation in AutoLISP routines that automatically draw three-dimensional objects.

Examples   In the following example, the SETVAR command is used to return the current THICKNESS.

```
Command: SETVAR
Variable name or ?: THICKNESS
New value for THICKNESS <0.0000>:
```

Tips   Changing the THICKNESS value does not affect entities previously drawn. Only new entities inherit the new thickness. To alter the thickness of previously drawn entities, use the CHANGE command.

The initial THICKNESS value is determined from the prototype drawing (default value is 0.0000). The current value is saved with the drawing.

In AutoCAD release 9, you can change the current THICKNESS from within the Entity Creation Modes dialog box, which is displayed when the DDEMODES command is activated (available only with AutoCAD release 9; see Advanced User Interface and DDEMODES).

Three-D Face.   *See* 3DFACE

Three-D Line.   *See* 3DLINE

TIME

[v2.5, v2.6, r9]

Options

D   Displays current times
On   Starts user elapsed timer
Off   Stops user elapsed timer
R   Resets user elapsed timer

Overview   Whenever you boot your computer (turn it on), you are required to enter the current date and time (or you had to set the date and time when you first set up your computer and an internal battery maintains the setting). The computer's operating system marks the date and time of creation on each file you save. AutoCAD also uses the date and time to mark drawing files and to record the amount of time you have spent in the drawing editor on a particular drawing. AutoCAD's TIME command shows you not only the current time and date, but also the date and time the current drawing was created, the last time the drawing was saved, and the length of time you have spent editing the current drawing since its creation. The TIME command also allows you to set a second timer that runs while you are in the drawing editor. You can start, stop, and reset this secondary timer whenever you like.

Procedure   When you activate the TIME command, it immediately displays the current time and other timer information related to the current drawing. All times are displayed using a 24-hour clock. After this information is displayed, you are prompted to select the desired option. If you want only to check the information, pressing Return returns you to the command prompt.

The available options are Display, On, Off, and Reset. Display simply repeats the display of the timer information and the options again. The current time is updated. Selecting On turns the secondary elapsed timer on if it was previously off. Whenever you enter the drawing editor, this timer is initially on. Off turns the elapsed timer off. Reset clears the elapsed timer, setting it back to zero.

Examples   Figure 83 illustrates the display presented by the TIME command.

Command: **TIME**

Tips   The values displayed by the TIME command are saved as AutoCAD system
variables. They can be viewed using the SETVAR command or AutoLISP.
The variables are available only for viewing (read-only); you cannot
change them. The TIME command can reset the elapsed timer, but SETVAR
and AutoLISP cannot. The following system variables are used for the
timer functions:

VARIABLE	FUNCTION
DATE	Current date and time
TDCREATE	Time and date of drawing creation
TDINDWG	Total editing time
TDUPDATE	Time and date of last update/save
TDUSRTIMER	User elapsed time

One feature you may want to implement through AutoLISP is to build an
automatic SAVE routine into a menu macro. Such a routine would check
the elapsed time against the last time the drawing was saved. If you
haven't saved the drawing for a fixed amount of time, the routine would
execute the SAVE command automatically before continuing with the remainder of the command. If the time since the last update is less than the
predetermined period, the command would execute normally.

The following is such a routine. It uses the DATE and TDUPDATE
system variables to determine the amount of time that has elapsed
since the drawing was last saved. If more than 15 minutes have passed,
it prompts Save drawing? (Y/N). If you answer yes, the drawing is saved

```
Current time:            20 Sep 1987 at 15:56:30.160
Drawing created:         19 Sep 1987 at 12:48:50.493
Drawing last updated:    20 Sep 1987 at 15:45:22.910
Time in drawing editor:  0 days 01:05:07.140
Elapsed timer:           0 days 00:30:25.670
Timer on.
Display/ON/OFF/Reset:
```

Figure 83

540

using the name originally used to load it. You can change the length of time between saves by changing the value 0.0104 on the second line so that $n = time\ in\ minutes/1440$.

```
(defun autosave ()
  (if (> (getvar "date") (+ (getvar "tdupdate") 0.0104))
    (setq s (getstring "Save drawing? (Y/N) : ")))
  (setq s (cond ((= s "Y") 1) ((= s "y") 1)))
  (if (= s 1) (command "save" ""))
)
```

Include the routine in the ACAD.LSP file (which loads automatically whenever you enter AutoCAD's drawing editor) or load the routine by entering *(load "autosave")* at AutoCAD's command prompt. Once the routine is loaded, the best way to make use of this routine is to include it in a menu macro. For example, change each occurrence of the REDRAW command in your menu to the following:

```
(autosave)'REDRAW
```

Another way to use this routine with AutoCAD release 9 is to use the UNDEFINE command and substitute the following AutoLISP routine (named REDRAW.LSP):

```
(defun C:REDRAW ()
  (autosave)
  (command ".REDRAW"))
```

After loading the routine, whenever you type the REDRAW command or select it from a menu, the routine will first check the time since the last time the drawing was saved and only then execute the REDRAW command.

TRACE

[all versions]

Overview   The TRACE command draws trace entities, one of AutoCAD's basic drawing entities. Traces are similar to lines except they also have a specified width.

Procedure   To draw a trace entity (a wide line), activate the TRACE command. AutoCAD prompts for the trace width, giving the most recent setting as the default. To specify the width of the trace, you can enter a value or use the screen cursor to indicate a width by selecting two points on the screen (the width being the distance between them). After you either accept the default or enter a new value, the TRACE command prompts you for the first point (the "from point") and successive points ("to points") in a fashion similar to the LINE command. The points selected are always on the centerline of the trace. The width of a trace cannot be varied from one end to the other. You can specify end points either by entering absolute or relative coordinates or by screen pointing.

As the trace is drawn, a segment is not displayed on the screen until the next segment is entered. This is so that the TRACE command correctly bevels the trace at the intersection of each segment. To complete the TRACE command, press Return when prompted for a "to point." The ends of the trace are always cut square.

Examples

The following example and drawing illustrate the use of the TRACE command.

```
Command: TRACE
Trace width <0.0500>: ( RETURN )
From point: (select point A)
To point: (select point B)
To point: (select point C)
To point: (continue selecting points)
To point: (press Return when finished)
```

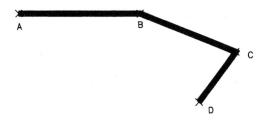

Warnings

There is no Undo feature for use with the TRACE command. The UNDO command will erase all trace segments added in the last TRACE command. You can, however, erase individual segments.

Tips The current trace width is stored in the system variable TRACEWID and can be viewed or changed using the SETVAR command or AutoLISP.

The initial trace width setting is determined by the prototype drawing.

You can use AutoCAD's editing commands (BREAK, TRIM, EXTEND, MIRROR, COPY, ERASE, and so on) to operate on trace entities.

TRACEWID
[all versions]

Overview The current width applied to trace entities (the last width value specified with the TRACE command) is stored in the TRACEWID system variable.

Procedure The TRACEWID value is normally set using the TRACE command itself. You can also change the TRACEWID value using the SETVAR command or (setvar) AutoLISP function.

Examples In the following example, the SETVAR command is used to return the current TRACEWID value. The method used by AutoCAD to measure this width is shown in the following drawing.

```
Command: SETVAR
Variable name or ?: TRACEWID
New value for TRACEWID <0.0500>:
```

Tips The initial TRACEWID value is determined from the prototype drawing (default value is 0.0500). The current value is saved with the drawing.

Transparent Commands
[v2.5, v2.6, r9]

Overview Transparent commands are commands that can be executed while you are actually in the middle of another command. Their execution does not interfere with the original active command. In other words, their action is "transparent" to the original command. If the transparent command displays any options or issues any prompts, the options or prompts are preceded by the symbol > >. This is to remind you that the command is being executed transparently. When a transparent command is completed, AutoCAD indicates that it is returning to the original command by displaying the message

Resuming *XXXX* command.

where *XXXX* is the name of the original command.

AutoCAD has nine commands that can be executed transparently. These commands are

GRAPHSCR
HELP or ?
PAN
REDRAW
RESUME
SETVAR
TEXTSCR
VIEW
ZOOM

Release 9 Only   If you are using AutoCAD release 9, the Advanced User Interface features can be used transparently. Specifically, the menu bar can be accessed and pull-down menus selected while another command is active. If the command selected from the pull-down menu is not a transparent command, however, the command cannot be activated without canceling the current active command.

In addition to being able to access the pull-down menus transparently, the dialog box commands DDEMODES, DDLMODES, and DDRMODES can be used transparently. (These features are available only with AutoCAD release 9; see Advanced User Interface.)

Procedure   Any of AutoCAD's transparent commands can be used normally by simply entering the command at AutoCAD's command prompt. Each command

544

is executed transparently when you precede the command name with an apostrophe, for example, 'ZOOM.

Certain restrictions apply to the transparent usage of the PAN, VIEW, and ZOOM commands (see the individual command descriptions for particulars). These commands, when used transparently, allow you to move around within a drawing while you are actively using another command. For example, you could begin the LINE command, zoom in to attach the line to a portion of a small detail, zoom back out, zoom in to another detail in another area of the drawing, and finally connect the other end of your line while within the second zoomed view.

The transparent use of the GRAPHSCR and TEXTSCR commands permits you an additional method to toggle from text screen to graphics screen besides using the Flip Screen key (F1). This can be useful when embedded in menu selections. The transparent RESUME command allows you to continue a command script that has been interrupted without interfering with the active command. The HELP command, when used transparently, displays a help screen for the current active command.

The SETVAR command can be used transparently to your advantage in many instances. Use this command to change the size of the aperture or pick box even while you are in the middle of a command.

Examples   In the following example, the SETVAR command is used transparently to alter the size of the pick box while the MOVE command is active.

```
Command: MOVE
Select objects: (select an object) 1 selected, 1 found.
Select objects: 'SETVAR
>>Variable name or ?: PICKBOX
>>New value for PICKBOX <5>: 3
Resuming MOVE command.
Select objects: (select an object) 1 selected, 1 found.
Select objects: ( RETURN )
Base point or displacement: (select first point)
Second point of displacement: (select second point)
```

Warnings   You cannot activate a transparent command while another transparent command is active. Also, transparent commands cannot be activated when AutoCAD is prompting for text. In that case, the command would simply be read as a text string. Finally, a transparent command cannot be activated while a normal version of the same command is active.

If you are using AutoCAD release 9, you cannot use transparent commands or the pull-down menu under the following circumstances:

- With the DTEXT command once the command is prompting for text
- With the SKETCH command once the record increment has been set
- With the VPOINT command while the compass and axis tripod are displayed
- During the ZOOM—Dynamic command

Tips   At the AutoCAD command prompt, entering a transparent command with the apostrophe in front of it has the same effect as entering the command without the apostrophe. In other words, 'ZOOM has the same effect as ZOOM when entered at the AutoCAD command prompt. For this reason, when you write a menu, always include the apostrophe before any commands that can be executed transparently. In this way, when you select them from the menu they can be either normal (if being entered in response to AutoCAD's command prompt) or transparent (if entered while within another command).

On a single-screen system, if a transparent command causes the display to flip from the graphics screen to the text screen (such as the 'HELP command), use the Flip Screen toggle key to flip it back to the graphics screen.

TRIM   [ADE-3] [v2.5, v2.6, r9]

Overview   The TRIM command cuts off a portion of an object so that it ends precisely at a predetermined boundary or cutting edge. Both the object to be trimmed and the boundary edge must already exist within the drawing and the object to be trimmed must physically cross the cutting edges.

Procedure   The TRIM command prompts you first to select the cutting edge or edges, and on the next line to select objects. You can use any method to select the cutting edges (window, last, screen picking, and so on). You can pick as many cutting edges as you want. The objects you pick as cutting edges can themselves be trimmed to other cutting edges within the same TRIM command. Lines, arcs, circles, and polylines can serve as cutting edges. Other objects, such as blocks, are not allowed. The selected cutting-edge objects are highlighted so you can remember where they are.

When you have finished selecting cutting edges, adding and removing objects from the selection set as necessary, press Return. The TRIM command now prompts you to select objects to trim. Select these objects by pointing to them using AutoCAD's pick box. You cannot use any other method to select the objects to be trimmed. Point to the part of the object that you want to be deleted. The portion of the object picked is trimmed back to its intersection with the first cutting edge it encounters. If the object picked is between two cutting edges, and you pick the object on the portion lying between the two cutting edges, only the portion between those cutting edges is deleted. If the object picked to be trimmed is also a cutting edge, the deleted portion disappears from the screen and that cutting edge is no longer highlighted. The nondeleted portion of the object can still serve as a cutting edge, however.

The prompt to select an object to trim continues to be displayed, allowing you to trim several objects within one TRIM command. When you have finished trimming objects, press Return to end the command.

```
Command: TRIM
Select cutting edge(s)...
Select objects: (select cutting edge objects)
Select objects: (press Return when done selecting cutting edge objects)
Select object to trim: (select object to be trimmed on the portion to be trimmed)
Select object to trim: (continue selecting objects to be trimmed)
Select object to trim: (press Return when done trimming objects)
```

In the following drawing, notice how the selection of the cutting edges and the position chosen to select the object to be trimmed determine the portions of the object that is erased.

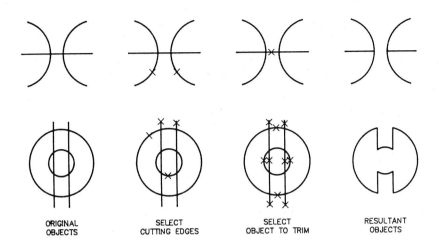

| ORIGINAL OBJECTS | SELECT CUTTING EDGES | SELECT OBJECT TO TRIM | RESULTANT OBJECTS |

Examples   The TRIM command is particularly handy in editing a reflected ceiling grid. For example, you can draw a 2′ × 2′ ceiling grid and then go back and insert 2′ × 4′ fluorescent light fixtures. The TRIM command makes short work of breaking out the sections of the grid that need to be removed where the lights occur (as shown in the first drawing that follows).

Another popular use for the TRIM command is to clean up wall intersections. The TRIM command can quickly clean up T and X wall intersections (as shown in the second drawing that follows). You can also use TRIM to break out portions of double-line walls to insert doors and windows.

```
Command: TRIM
Select cutting edge(s)...
Select objects: (select point A) 1 selected, 1 found.
Select objects: (select point B) 1 selected, 1 found.
Select objects: (select point C) 1 selected, 1 found.
Select objects: (select point D) 1 selected, 1 found.
Select objects: ( RETURN )
Select object to trim: (select point A)
Select object to trim: (select point B)
Select object to trim: (select point C)
Select object to trim: (select point D)
Select object to trim: ( RETURN )
```

Notice in the following drawings that you can select the cutting edges by the window or crossing window method.

548

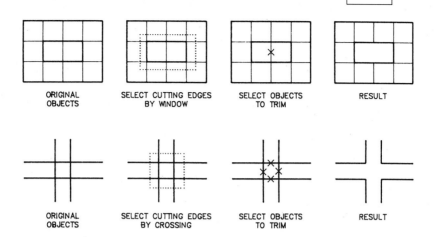

ORIGINAL OBJECTS   SELECT CUTTING EDGES BY WINDOW   SELECT OBJECTS TO TRIM   RESULT

ORIGINAL OBJECTS   SELECT CUTTING EDGES BY CROSSING   SELECT OBJECTS TO TRIM   RESULT

Warnings

If the object selected as a cutting edge cannot be used as a cutting edge, AutoCAD issues the error message

```
No edges selected.
```

and prompts you again to select objects for cutting edges.

If you select an object to trim that cannot be trimmed (a block, text, or a trace entity, for example), AutoCAD issues the error message

```
Cannot TRIM this entity.
```

and prompts you again to select an object to trim.

If none of the current cutting edges intersects with an object selected to be trimmed, AutoCAD issues the error message

```
Entity does not intersect an edge.
```

and prompts you again to select an object to trim.

If you select a circle to trim that intersects a cutting edge only once, AutoCAD issues the error message

```
Circles must intersect twice.
```

and prompts you again to select an object to trim.

If you trim a polyline, the polyline width remains the same at the cutting line, but the cut edge is squared off regardless of the angle of the cutting line to the polyline.

If you trim a polyline that has been curve-fit, the curve information converts to actual curved polyline segments. A polyline thus cut cannot be decurved.

Tips If you make a mistake during the TRIM command, you can restore the drawing to its condition prior to the TRIM command by using the UNDO command. All objects trimmed by the TRIM command will be restored. You cannot restore individual entities deleted within a TRIM command that operated on multiple entities.

In AutoCAD release 9 (with the standard menu ACAD.MNU), you can select the TRIM command from the Edit pull-down menu. In this situation, the Auto selection mode is used to select the first cutting edge(s). The menu item repetition of release 9 causes the command to repeat automatically until you cancel it by pressing Ctrl-C or by selecting another command from the menu. (See Advanced User Interface and Entity Selection.)

U [v2.5, 2.6, r9]

Overview The U command is a single implementation of the UNDO command. The U command reverses the effects of the most recent command.

Procedure To use the U command, enter U at the command prompt. The effects of the most recent command will be reversed. You can continue to enter the U command to step back through the commands one at a time, undoing the effects of each command as you go. As each command is undone, the command name is displayed.

Examples If the last command issued was the COPY command, entering the U command at the next command prompt causes all the objects copied during the last command to disappear, as shown in the following drawing. Auto-CAD then displays the COPY command on the next command line.

```
Command: U
COPY
```

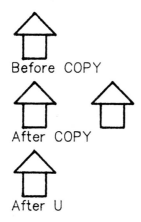

Before COPY

After COPY

After U

Tips   The U command is exactly the same as UNDO 1.

Undoing a drawing command is similar to erasing the last object, except that the OOPS command does not restore it.

The REDO command restores the effects of the last U command (see REDO).

Issuing the U command after a BLOCK command restores the individual entities of the block to the drawing (like the OOPS command) but also deletes the definition of the block.

The U command cannot undo a PLOT command.

In AutoCAD release 9 (with the standard menu ACAD.MNU), you can select the U command from the Tools pull-down menu (only with Auto-CAD release 9; see Advanced User Interface).

UNDEFINE
[ADE-3] [r9]

Overview   The UNDEFINE command allows you to delete an AutoCAD command. When undefined, a command cannot be entered (in a normal fashion) in response to AutoCAD's command prompt. This enables you to substitute AutoLISP functions for regular AutoCAD commands.

Procedure   To undefine a command, activate the UNDEFINE command. The command prompts for the name of the AutoCAD command to be undefined. Entering the command's name completes the UNDEFINE command.

Examples   The following AutoLISP routine is a different method for trimming drawing entities. If you wanted to replace the regular AutoCAD TRIM command with the C:TRIM function that follows, you could undefine the TRIM command.

```
Command: UNDEFINE
Command name: TRIM
```

Now, when you enter *TRIM* at the command prompt, the C:TRIM function is used instead of AutoCAD's TRIM command.

```
(defun C:TRIM (/ SL1 SL2 C1 C2 IN EN PT1 PT2 INT)
  (setvar "cmdecho" 0)
    (prompt "Select cutting edge(s)...")
    (setq SL1 (ssget))
    (prompt "\nSelect objects to trim by selecting two points ")
    (prompt "\ndefining a line through the objects to be trimmed: ")
    (setq C1 (getpoint "\nFirst point: "))
    (setq C2 (getpoint C1 "\nSecond point: "))
    (setq SL2 (ssget "C" C1 C2))
    (setq IN 0)
    (repeat (sslength SL2)
      (setq EN (entget (ssname SL2 IN)))
      (setq PT1 (cdr (assoc 10 EN)))
      (setq PT2 (cdr (assoc 11 EN)))
      (setq INT (inters PT1 PT2 C1 C2))
      (command ".TRIM" SL1 "" INT "")
      (setq IN (1+ IN))
  ))
```

Warnings   The UNDEFINE command is available only with AutoCAD release 9 with ADE-3.

   You can undefine only those commands that can be entered in response to the command prompt. You cannot undefine subcommands. For example, you cannot undefine the LAYER — Set option or any of the dimensioning subcommands. Attempting to do so results in the error message that the command is unknown.

Tips   An undefined command can still be used if it is prefaced by a period, as in .TRIM. In this case, the actual TRIM command would be used.

   If you expect to use command redefinition, you can protect the use of AutoCAD commands called from menus by prefacing each one with a period. If you enter a command prefaced with a period and the command has not been undefined, the period is simply ignored.

Entering a command that previously has been undefined results in the error message

```
Unknown command. Type ? for list of commands.
Enter "Ins" key to select menu item.
```

Undefined commands are still displayed in AutoCAD's help screens, however.

Commands remain undefined only during the current editing session. When you leave the drawing and later reload it or load another drawing, all AutoCAD commands are once again active.

UNDO   [v2.5, v2.6, r9]

Options

number	Undoes the *number* of most recent commands
A	Auto; controls treatment of menu items as undo groups
B	Back; undoes everything to previous Undo mark
C	Control; enables or disables the Undo feature
E	End; terminates an undo group
G	Group; begins a sequence to be treated as one command
M	Mark; places a marker in the Undo file (for back)

Overview The UNDO command reverses the effects of several commands in one step. It also allows you to turn the Undo feature on and off, mark a spot within a drawing to return to using the UNDO command, and reverse the effects of all AutoCAD commands back to a previously marked spot in the drawing.

Procedure The UNDO command has seven options. The default option is to enter a number. This number represents the number of command steps to be undone. This has the same effect as entering the U command multiple times, but it has the added advantage of avoiding additional redrawings

553

or regeneration as you step back through the commands. You can also reverse all of the UNDO with a REDO command.

```
Command: UNDO
Auto/Back/Control/End/Group/Mark/<Number>: (enter number of steps to undo)
```

Auto The Auto option sets the UNDO command to treat menu macros as single commands no matter how complex or how many AutoCAD commands are embedded in the macros. The Auto option prompts for on or off. When on, an undo group marking is placed at the start of each menu macro so that a single UNDO will reverse the entire macro. When the Auto option is set to off, each individual step within a menu macro must be undone individually.

```
Command: UNDO
Auto/Back/Control/End/Group/Mark/<Number>: A
ON/OFF: (enter ON or OFF)
```

Back The Back option takes the drawing back to the state it was in the last time the UNDO — Mark option was used to mark the drawing. You can use the Back option multiple times, stepping back to each preceding mark. If there is no preceding mark, AutoCAD prompts

```
This will undo everything.  OK? <Y>: (enter Y or N or press Return)
```

Entering or accepting the yes response undoes everything you have done to the drawing since entering the drawing editor.

Control The Control option allows you to limit or disable AutoCAD's Undo feature. The Control option presents three suboptions:

```
Command: UNDO
Auto/Back/Control/End/Group/Mark/<Number>: C
All/None/One <All>:
```

When set to All, AutoCAD's full Undo feature is enabled. When enabled, the Undo feature maintains a temporary file that contains all the commands and all the UNDO command's group, mark, and auto markings that have been used since you entered the drawing editor. This temporary file can become quite large. If you are running low on disk space, having the full Undo feature enabled may cause you to get a disk-full error while editing the drawing. You can eliminate this problem by freeing up some additional disk space (by removing some unneeded files) or by disabling part or all of the Undo feature.

You can turn the Undo feature off entirely by responding *none* to the Control suboption. The UNDO command's temporary file will be removed and the disk space recovered. The U and UNDO commands will not be available for use until you use the UNDO — Control option to turn the Undo feature back on again.

You can also set the Undo feature to record only the most recent command. This is accomplished by setting the Control option to one. Only the most recent command can be undone, but the Undo feature's temporary file takes up very little disk space.

When Control is set to none or to one, the UNDO command's prompt is reduced to

```
Command: UNDO
Control/<1>:
```

End   The End option is used in conjunction with the Group option to mark the end of a series of AutoCAD commands that can later be undone in one step. You must first mark the start of the grouping with the UNDO — Group option.

```
Command: UNDO
Auto/Back/Control/End/Group/Mark/<Number>: E
```

Group   The Group option is used to mark the beginning of a grouping of AutoCAD commands that will be treated as a single unit by the UNDO command. The end of this grouping must be marked with the End option. Later, this entire grouping can be undone in one step (similar to the Auto option's effect on menu macros).

```
Command: UNDO
Auto/Back/Control/End/Group/Mark/<Number>: G
```

Mark   The Mark option places a special marker in the Undo feature's temporary file for later use with the UNDO — Back option. Use the Mark option to mark a spot in the drawing prior to making experimental changes to your drawing. Then, if the changes don't meet with your satisfaction, you can restore the drawing to its condition prior to those changes by using the UNDO — Back option.

```
Command: UNDO
Auto/Back/Control/End/Group/Mark/<Number>: M
```

In the example and drawings that follow, a circle and a line are initially drawn. The UNDO Mark option is used to mark a spot in the Undo file at this stage. The COPY command is then used to add two copies of the circle. Since the positions of the added circles were not what you intended, they are removed in one step through the use of the UNDO Back option.

Examples

```
Command: CIRCLE
3P/2P/TTR/<Center point>: (select point A)
Diameter/<Radius>: (select point B)
Command: LINE
From point: (select point C)
To point: (select point E)
To point: (RETURN)
Command: UNDO
Auto/Back/Control/End/Group/Mark/<number>: M
Command: COPY
Select objects: (select point B) 1 selected, 1 found.
Select objects: (RETURN)
<Base point or displacement>/Multiple: M
Base point: (select point C)
Second point of displacement: (select point D)
Second point of displacement: (select point E)
Second point of displacement: (RETURN)
Command: UNDO
Auto/Back/Control/End/Group/Mark/<number>: B
COPY
Mark encountered
```

Before
UNDO "Mark"

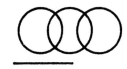

After
COPY

After
UNDO "Back"

Warnings

The Undo feature's temporary file is cleared after a PLOT or PRPLOT command.

Undoing a block has an effect similar to using the OOPS command after the BLOCK command, except that the block definition is also undone.

The UNDO command undoes an entire command. Therefore, everything created by the LINE command or a dimensioning command, for example, is undone at once. You cannot use the UNDO command selectively to undo parts of objects created at one time.

The Group/End options are provided for use within menu macros so that the macro will be treated as a single selection. However, if the Auto option is turned on, it is not necessary to use the Group/End options. In fact, using both Auto and Group/End may cause the menu macro to be treated as individual commands rather than as a single selection.

Tips   When the UNDO command is used to undo a group, the entire group can be restored after the undo with the REDO command.

The UNDO command can save time when you are undoing multiple commands. While issuing the U command multiple times has the same effect as entering the UNDO command and specifying the number of commands to be undone, the U command can cause intermediate redrawings and regenerations to be executed if they were part of the commands being undone. The UNDO command causes only one redrawing or regeneration to be executed at the end of the UNDO command.

Some commands have no effect that can be undone. These commands are AREA, DBLIST, DELAY, DIST, DXFOUT, END, FILES, HELP, ID, IGESOUT, LIST, MSLIDE, QUIT, RESUME, RSCRIPT, SAVE, SCRIPT, STATUS, and WBLOCK.

UNITS
[ADE-1] [all versions]

Overview   AutoCAD is an "open-architecture" CAD package. That means you can custom-tailor AutoCAD to serve the purposes of many different drafting disciplines. Various disciplines use different notations for measuring distances and angles. The UNITS command allows you to select one of five different methods of displaying distances, one of five different methods of displaying angles, the number of decimal places distances and angles are displayed to, the direction to be used as the zero angle, and whether the program considers positive angle directions as clockwise or counterclockwise.

Procedure   The UNITS command takes you through a series of prompts to select the method that AutoCAD will use to display distances and angles.

Command: **UNITS**

Units of Measure   The first thing the UNITS command asks for is your desired units of measure. The menu of selections shown in Figure 84 is presented.

   The format selected determines the way in which AutoCAD displays the distance. For example, a distance of 22.5 units would be displayed differently by each format:

FORMAT	EXAMPLE
Scientific	2.25E + 01
Decimal	22.50
Engineering	1'-10.50"
Architectural	1'-10½"
Fractional	22½

```
Systems of units:      (Examples)

       1.  Scientific     1.55E+01
       2.  Decimal        15.50
       3.  Engineering    1'-3.50"
       4.  Architectural  1'-3 1/2"
       5.  Fractional     15 1/2

Enter choice, 1 to 5 <2>:
```

Figure 84

Notice that engineering and architectural units assume that one drawing unit equals one inch, whereas scientific, decimal, and fractional simply equate one drawing unit to one displayed unit. In those cases, it is up to you to make your own assumption as to what a drawing unit represents. For example, surveyors usually use decimal units to draw site plans and equate one drawing unit to one foot rather than one inch.

You should also note that if you select architectural units, 1'10½ and 1'10.5 are both valid entries. In engineering units, however, only the entry 1'10.5 units would be accepted.

Once you have selected the type of units you want to use, AutoCAD prompts for the precision. If you select format 1, 2, or 3, AutoCAD prompts

```
Number of digits to right of decimal point (0 to 8) <default>:
```

If you select format 4 or 5, AutoCAD prompts

```
Denominator of smallest fraction to display
(1, 2, 4, 8, 16 ,23, or 64) <default>:
```

Enter the value that you want to use and press Return or simply press Return to accept the current default. In either case, you can always enter a number of greater precision (more decimal places to the right) or with a smaller denominator and AutoCAD will preserve that greater accuracy. The number that you select here controls only the numerical precision that is displayed.

Angle of Measure   Next, the UNITS command prompts you to select the method for the display of angles. Again, a menu is displayed (Figure 85).

Select the method you want to use to display angles by entering the menu number associated with it. Decimal degrees are simply displayed as a number with a decimal point. Degrees, minutes, and seconds display with a lowercase *d* to indicate degrees, as in 45d32'26". Grads appends a lowercase *g* as a suffix, as in 32.435g. In a similar fashion, radians appends a lowercase *r*, as in 0.725r. Surveyor's units display a bearing angle displayed as north or south so many degrees, minutes, and seconds, east or west. The angle is always less than 90 degrees and is arrived at by starting at either the north or south compass point and measuring the listed angle from that point toward either the east or west compass point. Thus, an angle of 30 degrees would be represented as N 60d00'00" E.

When you are using decimal degrees, grads, or radians, only that method of angle measure is acceptable. When you are using degrees, minutes, and seconds, angles can be entered in decimal degrees also. With

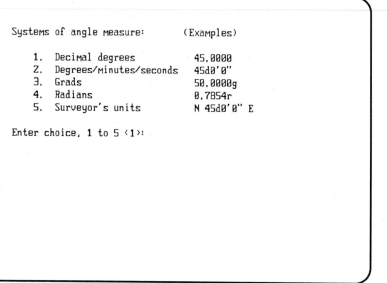

```
    Systems of angle measure:        (Examples)

        1.  Decimal degrees          45.0000
        2.  Degrees/minutes/seconds  45d0'0"
        3.  Grads                    50.0000g
        4.  Radians                  0.7854r
        5.  Surveyor's units         N 45d0'0" E

    Enter choice, 1 to 5 <1>:
```

Figure 85

surveyor's units, all three methods of entering angles (decimal, degrees/minutes/seconds, and surveyor's units) are acceptable.

After you have selected the method used to display angles, the UNITS command prompts for the numerical precision with which angles will be displayed. AutoCAD prompts:

`Number of fractional places for display of angles (0 to 8) <default>:`

If you select grads, radians, or decimal degrees, the number of places simply represents the number of places to the right of the decimal. If you are using either degrees/minutes/seconds or surveyor's units, the number of fractional places has another meaning. Selecting 0 causes only the degrees to be displayed. Selecting 1 or 2 causes only degrees and minutes to display. If you select 3 or 4, degrees, minutes, and seconds are displayed. Selecting any other number from 5 through 8 determines how many places to the right of the decimal to display for seconds of an arc.

Alternate Angle Representations   Normally, AutoCAD uses three o'clock as the zero angle and measures angles in a counterclockwise direction (all examples in this book assume this default). You can, however, change the angle that AutoCAD uses as the zero angle.

The next thing the UNITS command prompts for is the direction for the zero angle. Again, AutoCAD displays a menu, but this time only to remind you of some typical settings. You are prompted to enter an angle. You can enter any number, or you can flip to the graphics screen and use the screen cursor to indicate the desired angle. The angles display in the current mode just selected for display of angles (see Figure 86).

After you have selected the angle for the zero direction, the UNITS command goes on to ask whether you want positive angles to be measured in a clockwise or counterclockwise direction. AutoCAD's default setting is to measure in a counterclockwise direction (all the examples in this book assume this as the default).

```
Do you want angles measured clockwise? <current>:
```

Examples   The following example illustrates the entire sequence used to activate the UNITS command, select architectural units, accept the default denominator of 16, select surveyor's units of angle measure, accept the default of 4 fractional places for angle display, accept the default east angle, and accept the default condition of angles measured counterclockwise.

```
Command: UNITS

Systems of units:      (Examples)

   1.  Scientific     1.55E+01
   2.  Decimal        15.50
   3.  Engineering    1'-3.50"
   4.  Architectural  1'-3 1/2"
   5.  Fractional     15 1/2
```

```
Direction for angle 0:
    East    3 o'clock  =  0
    North  12 o'clock  =  90
    West    9 o'clock  =  180
    South   6 o'clock  =  270
Enter direction for angle 0 <0>:
```

Figure 86

561

```
Enter choice, 1 to 5 <2>: 4
Denominator of smallest fraction to display
(1, 2, 4, 8, 16, 32, or 64) <16>: ( RETURN )

Systems of angle measure:        (Examples)

    1.  Decimal degrees          45.0000
    2.  Degrees/minutes/seconds  45d0'0"
    3.  Grads                     50.0000g
    4.  Radians                   0.7854r
    5.  Surveyor's units         N 45d0'0" E

Enter choice, 1 to 5 <1>: 5
Number of fractional places for display of angles (0 to 8) <4>: ( RETURN )

Direction for angle E:
    East    3 o'clock  =  E
    North  12 o'clock  =  N
    West    9 o'clock  =  W
    South   6 o'clock  =  S
Enter direction for angle E <E>: ( RETURN )

Do you want angles measured clockwise? <N> ( RETURN )
```

Warnings

When you are using surveyor's units, angles greater than 90 degrees are not accepted. You must include the N or S and the E or W.

Prior to AutoCAD version 2.6, the fractional linear dimensioning mode was not available.

Tips

Depending on the system of measure you select, drawing units and angles can always be represented in the default units of measure without your having to enter those units specifically. For example, if you are using architectural units, you must specifically enter the number of feet, but you do not need to include the inch symbol. Therefore, 1'10 is just as valid as 1'10" and saves typing an additional character.

The same is true of angles. The method of measuring angles is understood. You do not need to include the *g*, *r*, or *d* when entering grads, radians, or decimal degrees. If you are using degrees/minutes/seconds or surveyor's units and no minutes or seconds are included, you do not need to include the "d" to indicate degrees. Degrees are understood. Similarly, if seconds are not included, only the *d* and the minute symbol are necessary. Further, decimal degrees can be entered in lieu of minutes and seconds.

When you are entering fractions in either architectural or fractional units, the fraction must be separated from the rest of the distance by any character other than a number, double-quote, slash, period, or comma. A hyphen is generally used.

Although AutoCAD does not display denominators greater than 64, any denominator that is a power of 2 can be used, up to 1,024. AutoCAD maintains the accuracy, although these numbers will not display as fractions.

You can force AutoCAD to accept angles measured in degrees by using 3 o'clock as the zero angle and with positive angles measured counterclockwise regardless of the method of angle measure currently in use and the other current settings. Just precede the angle by a double-angle symbol (< <) rather than a single symbol.

You do not need to go through the entire series of UNITS command prompts just to change one setting. You can proceed through the command until you have changed the specific setting and then press Ctrl-C to end the command. For example, to simply change the system of units from decimal to architectural, activate the UNITS command and answer the first prompt. When the command prompts for the number of places after the decimal point, press Ctrl-C.

The various settings controlled by the UNITS command are stored in system variables. These variables can also be controlled using the SET-VAR command and AutoLISP. The values for these system variables are saved with the individual drawing. The following system variables are used:

VARIABLE	EXPLANATION
ANGBASE	Angle for the 0 direction
ANGDIR	Direction for positive angles (1 = clockwise, 0 = counterclockwise)
AUNITS	Angular units mode (0 = decimal degrees, 1 = degrees/minutes/seconds, 2 = grads, 3 = radians, 4 = surveyor's units)
AUPREC	Angular units decimal places
LUNITS	Linear units mode (1 = scientific, 2 = decimal, 3 = engineering, 4 = architectural, 5 = fractional)
LUPREC	Linear units decimal places or denominator

The initial UNITS settings are determined from the prototype drawing.

'VIEW

[ADE-2] [all versions]

Options

D   Deletes named view
R   Restores named view to screen
S   Saves current display as named view
W   Saves specified window as named view
?   Lists named views

Overview

As you work on a drawing, you will often find that you need to switch from one portion of a drawing to another and then return to the other portions frequently. You could use the PAN and ZOOM commands to accomplish this. But, a far easier way is to use the VIEW command.

The VIEW command allows you to save either the current screen display or a window of the current display as a view. You provide the name for each view that you save. The VIEW command also enables you to return to any previously named view, to obtain a list of all the named views in the current drawing, and to delete any previously named views.

Procedure

The VIEW command has five options: ?, Delete, Restore, Save, and Window. There is no default option. To select an option you need to enter only the first letter of the option and press Return.

The first method of using the VIEW command is to save the current screen display as a named view. Respond to the VIEW prompt by entering *S*. The VIEW command asks for the name that you want to call the current screen display. A valid view name can be up to thirty-one alphanumeric characters.

```
Command: VIEW
?/Delete/Restore/Save/Window: S
View name: (enter name)
```

To return to a named view (that is, to display a previously saved view on the screen), use the Restore option. At the VIEW command prompt, enter *R*. AutoCAD prompts for the name of the view to restore. You must enter the name exactly.

```
Command: VIEW
?/Delete/Restore/Save/Window: R
View name: (enter name)
```

To remove a named view from the current drawing, respond with *D*. The VIEW command prompts you for the name of the view that you want to delete. You must enter the name exactly.

```
Command: VIEW
?/Delete/Restore/Save/Window: D
View name: (enter name)
```

You can also save a portion of the current screen as a view by windowing that portion of the screen rather than first zooming in. To do this, use the Window option. The VIEW command prompts for a view name and then prompts you to select the first and second corners of the window.

```
Command: VIEW
?/Delete/Restore/Save/Window: W
View name: (enter name)
First corner: (select point)
Other corner: (select point)
```

To display a list of all the named views in the current drawing, enter a question mark (?). AutoCAD flips to the text screen and displays a list of every named view in the current drawing, along with their centerpoint coordinate and their magnification.

You can use the VIEW command transparently (while another command is active) by preceding the command with an apostrophe, as in 'VIEW. This is possible only as long as the command does not require the drawing to be regenerated.

Examples   A successful CAD operator makes good use of the VIEW command. Use this command to save every view of your drawing that you need to return to frequently. For example, on an architectural floor plan, saving the zoomed view of each room is often a good practice. Naming the views using either the room name or room number makes it easier to remember the names of the views. At a minimum, save the ZOOM—Extents view and the view that you will ultimately plot. This saves regeneration time.

In the following example and floor plan drawing, the VIEW command is used to save the view named ROOM-1 (a view of the living room of the house). The window method is used.

565

```
Command: VIEW
?/Delete/Restore/Save/Window: W
View name to save: ROOM-1
First corner: (select point A)
Other corner: (select point B)
```

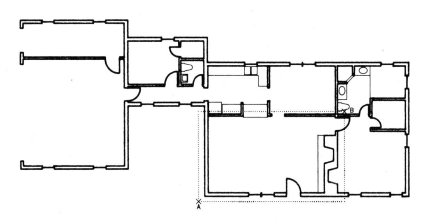

Warnings   You cannot use the VIEW command transparently while a VPOINT, ZOOM, PAN, or VIEW command is active. It also cannot be used transparently if the operation would require a regeneration of the drawing. Fast Zoom mode, as set by the VIEWRES command, must be on for transparent operation.

The transparent VIEW command is not available with AutoCAD version 2.5.

Tips   Although the names you use to save views can be up to thirty-one characters, remember that you will need to enter these names again when you restore the views. So keep the names descriptive and short.

You can change the name of named views with the RENAME command.

When plotting a drawing, you can plot a named view.

When you restore a windowed view, the screen may display more than what was actually included within the window. But if you plot the named windowed view, only what was actually included within the window is plotted.

566

When first entering the drawing editor from AutoCAD's main menu, using task 2 (Edit an Existing Drawing), you can instruct AutoCAD to load a previously saved named view (see ACAD).

If your current display is using AutoCAD's three-dimensional mode, the VIEW command will save and later restore the current three-dimensional viewpoint.

The current view center point and size are stored in the system variables VIEWCTR and VIEWSIZE, respectively. They can be viewed using the SET-VAR command or AutoLISP but cannot be changed (they are read-only variables). These variables are saved with the drawing. The initial values are determined by the prototype drawing.

VIEWCTR

Overview   The coordinate of the center of the current view displayed on AutoCAD's graphics screen is stored in the system variable VIEWCTR. This variable is a read-only value.

Procedure   The SETVAR command or (getvar) function can be used to return the VIEWCTR value. Only actually changing the current view, however, can change the value. Occasionally, you may want to include a function that uses the VIEWCTR value in an AutoLISP routine.

Examples   In the following example, the SETVAR command is used to return the VIEWCTR value.

```
Command: SETVAR
Variable name or ?: VIEWCTR
VIEWCTR = 6.5998,4.8344 (read only)
```

Tips   The initial VIEWCTR value is determined by the initial displayed view when a drawing is loaded. This, in turn, is determined by the prototype drawing. The current value is saved with the drawing file.

VIEWRES

[v2.5, v2.6, r9]

Overview The VIEWRES command controls whether AutoCAD maintains a "virtual" screen. When a virtual screen is maintained, AutoCAD attempts to perform all PAN, VIEW, and ZOOM commands at redraw speed rather than regenerating the drawing. The VIEWRES command allows you to further control the degree of accuracy in displaying arcs and circles on the graphics screen.

Procedure The VIEWRES command first asks whether you want fast zooms. Answering yes enables the virtual screen. Answering no disables the virtual screen and causes the PAN, VIEW — Restore, and ZOOM commands to force the drawing to be regenerated every time. Disabling fast zoom also disables the ZOOM — Dynamic feature.

If fast zoom is enabled (you answered yes), the VIEWRES command prompts further for the circle zoom percent. This value, any number from 1 to 20,000, controls the degree of accuracy with which arcs will be displayed on the graphics screen. The circle zoom percent determines the number of straight line segments that are used to display circles. The lower the number, the faster drawings will be redrawn but at the expense of fewer segments being used to display circles and arcs (resulting in a coarser appearance). The higher the number, the more segments used to display curves but at the expense of the drawing taking longer to be redrawn.

```
Command: VIEWRES
Do you want fast zooms? <Y>:
Enter circle zoom percent (1-20000) <100>:
```

If fast zoom is enabled, AutoCAD tries to complete all PAN, VIEW — Restore, and ZOOM commands without regenerating the drawing. The drawing is redrawn from the image stored in the virtual screen. If you zoom in too close, you may exceed the degree to which AutoCAD can display curves based on the current circle zoom percent setting, and AutoCAD is forced to regenerate the drawing. This happens if the current virtual screen doesn't contain enough segments to display circles with the degree of accuracy you have requested.

At other times, the PAN, VIEW, or ZOOM command may simply cause AutoCAD to display a portion of the drawing that is not part of the current virtual screen. In this case, the drawing also needs to be regenerated.

Examples   The default value for circle zoom percent is 100. This setting tells AutoCAD to use its internal algorithm without change. To display circles that would have enough precision to be displayed without regenerating at a zoom magnification of 5X, set the circle zoom percent to 500. For most applications, a setting of 500 is acceptable. The 100 default setting is usually a bit too coarse.

If you set REGENAUTO to off, whenever you exceed the circle zoom percent (thus requiring AutoCAD to perform a regeneration), you will be prompted:

`About to regen, proceed?` `<Y>:`

A no response cancels the PAN, VIEW— Restore, and ZOOM commands.

Tips   Regardless of the way curves are displayed on the graphics screen, they are always plotted as accurate curves.

Regardless of the circle zoom percent setting, circles are always displayed with at least eight sides.

VIEWSIZE

[all versions]

Overview   The current view height, in drawing units, is stored in the VIEWSIZE system variable. The value controls the size of the view of the drawing when a drawing is loaded into AutoCAD's drawing editor. The current value is saved with the drawing file. Whenever you reload an existing drawing, VIEWSIZE, along with the VIEWCTR variable, enables AutoCAD to remember what the last view looked like.

Procedure   The VIEWSIZE variable is a read-only variable. The SETVAR command and (getvar) AutoLISP functions can be used to read the value, but only actually changing the current view size using one of AutoCAD's display commands (such as ZOOM or VIEW) changes the VIEWSIZE value. However, the VIEWSIZE value can be used in AutoLISP routines that need to know the current height of the screen.

Examples   In the following example, the SETVAR command is used to return the current VIEWSIZE value.

```
Command: SETVAR
Variable name or ?: VIEWSIZE
VIEWSIZE = 9.6689 (read only)
```

Tips   The initial VIEWSIZE value is determined from the prototype drawing (default value is 9.0000). The current value is saved with the drawing file.

Virtual Screen
[v2.5, v2.6, r9]

Overview   You can instruct AutoCAD to maintain a virtual screen in addition to the displayed screen. The virtual screen is a pixel map of the current regenerated view, which is stored in memory. When the virtual screen is being stored, AutoCAD executes most PAN, VIEW, and ZOOM commands at redraw speed rather than requiring the drawing to be regenerated. A regeneration is required only if the command causes AutoCAD to display a portion of the drawing that is not part of the current virtual screen. A regeneration would also be required if the zoom magnification resulted in arcs and circles being drawn with too few vectors.

Procedure   You can enable or disable AutoCAD's virtual screen through the use of the VIEWRES command (see VIEWRES). Answering yes to the VIEWRES command prompt if you want fast zooms turns the virtual screen on. When the virtual screen is enabled, AutoCAD scans the entire virtual screen when searching for selected objects.

Examples   When the virtual screen is enabled, AutoCAD performs most PAN and ZOOM commands without having to regenerate the drawing. When the virtual screen is disabled, however, every PAN and ZOOM command requires a regeneration. This is illustrated in the following command sequence. In the first part of the example, the virtual screen is enabled and the ZOOM command does not cause the drawing to be regenerated. Then, the virtual screen is disabled (by turning the fast zoom feature off). This forces a drawing regeneration. The ZOOM command is then repeated, using the same window as before (see Figure 87). But now, with the virtual screen disabled, a drawing regeneration is required.

```
Command: VIEWRES
Do you want fast zooms? <Y> ( RETURN )
Enter circle zoom percent (1-20000) <100>: ( RETURN )
Command: ZOOM
All/Center/Dynamic/Extents/Left/Previous/Window/<Scale(X)>: W
First corner: (select point A)
Other corner: (select point B)

Command: VIEWRES
Do you want fast zooms? <Y> N
Enter circle zoom percent (1-20000) <100>: ( RETURN )
Regenerating drawing.
Command: ZOOM
All/Center/Dynamic/Extents/Left/Previous/Window/<Scale(X)>: W
First corner: (select point A)
Other corner: (select point B)
Regenerating drawing.
```

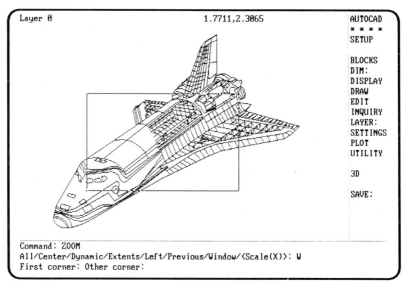

Figure 87

VPOINT

[ADE-3] [all versions]

Options

(RETURN)	Selects view point via compass and axis tripod
X, Y, Z	Specifies view point

Overview The VPOINT command determines the position from which you view drawings. The VPOINT command changes the position of view of any drawing but is included mainly for use with drawings that incorporate three-dimensional information. The view point location, once set, is maintained until you change it by using the VPOINT command again or by using the VIEW or ZOOM command to change the current view.

Procedure To change the current view point, activate the VPOINT command. Auto-CAD prompts for a new view point, displaying the current X-, Y-, and Z-coordinates of the current view point as the default. You can change the view point by entering new X-, Y-, and Z-coordinates. The coordinate that you enter tells AutoCAD the orientation of the new view point in relation to the 0,0,0 point. It does not represent a distance from the origin point and the drawing will always be oriented looking back toward the origin point.

```
Command: VPOINT
Enter view point <current X, Y, Z view point>: (enter X,Y,Z)
```

If you want, you can use AutoCAD's three-dimensional compass and tripod to graphically select the view point. To do so, press Return in response to the prompt to Enter the view point. The compass and tripod are displayed on the graphics screen, as shown in Figure 88.

The compass has a small cursor within it that you can move by moving the pointing device. The center of the compass represents the North Pole (0,0,1). The inner circle is the equator ($n,n,0$). The outer circle is the South Pole (0,0,−1). Thus, any point between the center and the inner circle means that you are looking down on the three-dimensional object from above. The orientation (east, north, west, or south) is determined by the angle from the center to the compass cursor. If you are exactly on the inner circle, you are viewing your three-dimensional object from ground level. Between the

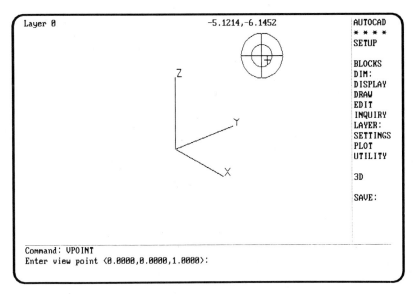

Figure 88

inner and outer circles you are looking up at the object from the underside (as if the ground plane were transparent). Placing the cursor on the outer circle results in looking straight up at it from directly below.

As you move the compass cursor, the tripod also changes its orientation to further assist you in selecting the desired view. If, after pressing Return to obtain the compass and tripod, you decide to enter the view point coordinates explicitly, you may still do so. If you decide to accept the default view point coordinates, simply press Return again. To make a selection using the compass and tripod, place the compass cursor in the desired position and/or orient the tripod as you wish and then press the pick button.

Release 9 Only If you are using AutoCAD release 9, the VPOINT command presents a Rotate option in addition to the previous method of entering view point coordinates. Entering VPOINT coordinates is the default method, and the current view point coordinates are displayed as the default. Pressing Return still causes the tripod axis and compass to be displayed.

```
Command: VPOINT
Rotate/<View point> <current view point>:
```

If you enter *R* to indicate that you want to use the Rotate option, the VPOINT command prompts for an angle on the X-Y plane in relation to the X axis, displaying the current viewpoint's angle as a default. After responding to this prompt, the command prompts for an angle from the X-Y plane, again displaying the current viewpoint's angle in this direction as the default. The angles can be provided by entering a number value or by selecting points on the screen. The following drawing shows how these angles can be visualized.

```
Command: VPOINT
Rotate/<View point> <current view point>:
Enter angle in X-Y plane from X axis <current angle>:
Enter angle from X-Y plane <current angle>:
```

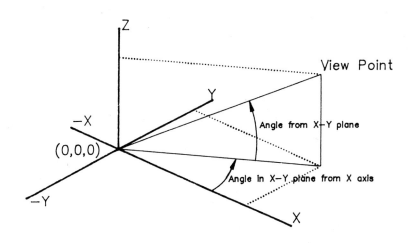

Examples   The following example shows the AutoCAD sample drawing ROBOT as seen in a two-dimensional and then a three-dimensional view (Figure 89).

```
Command: VPOINT
Enter view point <2.7996,-4.0835,1.0489>: 0,0,1

Regenerating drawing.

Command: VPOINT
Enter view point <0.0000,0.0000,1.0000>: 2.7996,-4.0835,1.0489
Regenerating drawing.
```

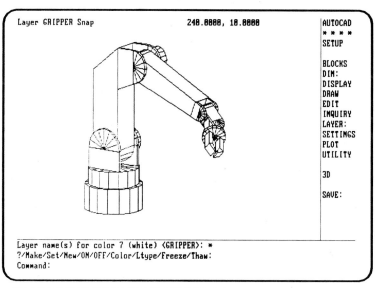

```
Layer GRIPPER Snap            248.8888, 18.8888        AUTOCAD
                                                       * * * *
                                                       SETUP

                                                       BLOCKS
                                                       DIM:
                                                       DISPLAY
                                                       DRAW
                                                       EDIT
                                                       INQUIRY
                                                       LAYER:
                                                       SETTINGS
                                                       PLOT
                                                       UTILITY

                                                       3D

                                                       SAVE:


Layer name(s) for color 7 (white) <GRIPPER>: *
?/Make/Set/New/ON/OFF/Color/Ltype/Freeze/Thaw:
Command:
```

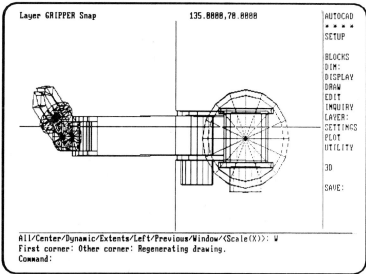

```
Layer GRIPPER Snap            135.8888,78.8888         AUTOCAD
                                                       * * * *
                                                       SETUP

                                                       BLOCKS
                                                       DIM:
                                                       DISPLAY
                                                       DRAW
                                                       EDIT
                                                       INQUIRY
                                                       LAYER:
                                                       SETTINGS
                                                       PLOT
                                                       UTILITY

                                                       3D

                                                       SAVE:


All/Center/Dynamic/Extents/Left/Previous/Window/<Scale(X)>: W
First corner: Other corner: Regenerating drawing.
Command:
```

Figure 89

Warnings The image displayed by a three-dimensional view is always looking back to the 0,0,0 origin point. The VPOINT command does not control the distance from the origin, only the orientation. Use the ZOOM command to change the viewing distance.

The image displayed by a three-dimensional view is not a perspective. It is a parallel projection. AutoCAD does not have a true Perspective View mode.

Tips

To return to a standard two-dimensional view, enter a view point with zero X and Y coordinates (for example, 0,0,1). It is helpful to build this response into your menus.

Some users use only the compass to orient their view point, some only the tripod, and some a combination of both. Use whichever method suits you best.

The current X, Y, and Z view point components are stored in the system variables VPOINTX, VPOINTY, and VPOINTZ, respectively. These values can be read using the SETVAR command or AutoLISP but cannot be altered (they are read-only variables).

In AutoCAD release 9 (with the standard menu ACAD.MNU), you can select the VPOINT command from the Display pull-down menu. A view direction can be selected from an icon menu (and a view height selected from the standard screen menu) or the regular compass and tripod selected by picking the "VPOINT" icon. (See Advanced User Interface and Icon Menus.)

VPOINTX

[ADE-3] [v2.5, v2.6, r9]

Overview

The X-axis component of the current three-dimensional view point is stored in the VPOINTX system variable. It is a read-only variable.

Procedure

You can look at the VPOINTX value using the SETVAR command or (getvar) AutoLISP function, but only the VPOINT command can change its value (see VPOINTY, VPOINTZ, and VPOINT).

Examples

In the following example, the SETVAR command is used to return the current VPOINTX value.

```
Command: SETVAR
Variable name or ?: VPOINTX
VPOINTX = 0.0000 (read only)
```

Tips   The initial VPOINTX value is determined from the prototype drawing (default value is 0.0000). The current value saved with the drawing file.

VPOINTY   [v2.5, v2.6, r9] [ADE-3]

Overview   Like the VPOINTX variable, VPOINTY is a read-only variable that stores the Y-axis component of the current three-dimensional view point.

Procedure   Changing the current three-dimensional view point using the VPOINT command is the only way you can change the VPOINTY value. The current VPOINTY value can be returned using the SETVAR command or the (getvar) AutoLISP function.

Examples   In the following example, the SETVAR command is used to return the current VPOINTY value.

```
Command: SETVAR
Variable name or ?: VPOINTY
VPOINTY = 0.0000 (read only)
```

Tips   The initial VPOINTY value is determined from the prototype drawing (default value is 0.0000). The current value is saved with the drawing.

VPOINTZ   [ADE-3] [v2.5, v2.6, r9]

Overview   The VPOINTZ variable stores the Z-axis component of the current three-dimensional view point, which is determined by the VPOINT command. The three view point variables — VPOINTX, VPOINTY, and VPOINTZ — determine the current three-dimensional viewing point for each AutoCAD drawing. The values stored in these three variables determine how AutoCAD displays the screen whenever you load a drawing or change the three-dimensional view point.

Procedure Changing the current three-dimensional view point using the VPOINT command is the only way you can change the VPOINTZ value. The current VPOINTZ value can be observed using the SETVAR command or the (getvar) AutoLISP function.

Examples In the following example, the SETVAR command is used to return the current VPOINTZ value.

```
Command: SETVAR
Variable name or ?: VPOINTZ
VPOINTZ = 1.0000 (read only)
```

Tips The initial VPOINTZ value is determined from the prototype drawing (default value is 1.0000). The current value is saved with the drawing.

VSLIDE

Options:

file   Views slide
*file   Preloads slide, next VSLIDE will view

Overview The VSLIDE (View Slide) command is used to view a slide file previously created with the MSLIDE command.

Procedure To view a previously created slide file, enter the VSLIDE command. AutoCAD prompts for the name of the slide file you want to view. Enter the name of an existing slide file. You can include a drive designation and a subdirectory name, but do not include a file extension. The file type .SLD is assumed.

```
Command: VSLIDE
Slide file: (enter slide file name)
```

Normally slide files are displayed on the screen at redraw speed, subject to reading the slide file information from the disk. You can preload the slide file into memory by preceding the slide file name with an asterisk (*). This loads the slide file into memory but does not display it on the screen. To display the preloaded slide, enter the VSLIDE command again, but this time simply press Return rather than enter a slide file name. The pre-loaded slide displays on the screen at redraw speed.

Release 9 Only   If you are using AutoCAD release 9, you can combine several slides into a slide library. A special utility, SLIDELIB.EXE, is provided for this purpose (see SLIDES). Slides that have thus been combined into a slide library can still be viewed individually using the VSLIDE command.

To view slides contained in a library, you must respond to the VSLIDE prompt by using the format *library(slide-name)*, where *library* is the name of the slide library file (including any drive designation and subdirectory path) and *slide-name* is the name of the individual slide within the slide library file.

If you have a slide named DOORS in the slide library file named MYSLIDES stored on the drive D in a subdirectory named SAMPLES, you would view this slide with the VSLIDE command, as follows:

```
Command: VSLIDE
Slide file name: D:\SAMPLES\MYSLIDES(DOORS)
```

Examples   The preloading of slide files is particularly useful in creating slide shows for an audience. You can control the slide show by a command script using AutoCAD's script function (see SCRIPT). In this way, each slide can be preloaded and will display much quicker. The command script can incorporate the DELAY command so that slides remain on the screen long enough for your audience to see them (see DELAY). The script can even be made continuous by including the RSCRIPT command as the last command in the script file (see RSCRIPT). Below is a sample command script for a self-running slide show. The file is saved with the file extension .SCR and is activated using AutoCAD's SCRIPT command.

PROGRAM	EXPLANATION
VSLIDE SLIDE1	Begins slide show.
VSLIDE *SLIDE2	Preloads slide 2.
DELAY 2000	Pauses to allow reading of slide 1.
VSLIDE	Displays slide 2.
VSLIDE *SLIDE3	Preloads slide 3.
DELAY 2000	Pauses to allow reading of slide 2.
VSLIDE	Displays slide 3.
DELAY 2000	Pauses to allow reading of slide 3.
RSCRIPT	Repeats the entire cycle.

579

The ability to view a slide file can also be used to remind you of the appearance of part of your drawing or to make it easier to select a pre-drawn block for insertion into your drawing. If you find that you frequently have to refer to a portion of your drawing (or even another drawing), you can save the image as a slide file and quickly pop it onto your screen whenever necessary. If you have a collection of "canned" details, which you sometimes get confused trying to decide upon, you could save the image of the collection of details along with text entries of their block names as a slide file. Then when you can't remember the name of the block or what it looks like or can't decide which one to insert, you can view the block slide file and then make your decision. AutoCAD includes a slide of the different methods for displaying point entities (named POINTS.SLD) for just this purpose.

You can also use the ability to view slide files to create a drawing as a help screen for a particularly difficult operation. Save the drawing as a slide file, complete with whatever descriptive text you find helpful. Next time you have trouble remembering how the command sequence works, view the slide file.

Warnings   Slides cannot be edited. Only the drawings from which they were created can be edited.

The current layer and colors present in a drawing have no effect on a slide. A slide is simply a "snapshot" of a drawing.

Tips   To clear a slide from the display, simply enter the REDRAW command.

VSMAX
[v2.6, r9]

Overview   The VSMAX system variable contains the drawing coordinate point of the upper right corner of the current virtual screen. It is a read-only variable.

Procedure   The VSMAX value can be returned using the SETVAR command or the (getvar) AutoLISP function. It can be changed only by regenerating the drawing and thus setting a new virtual screen.

Examples   In the following example, the SETVAR command is used to return the VMAX value.

```
Command: SETVAR
Variable name or ?: VSMAX
VSMAX = 13.1995,9.6689 (read only)
```

Warnings   VSMAX is not available in AutoCAD version 2.5.

Tips   If the Fast Zoom mode has been disabled, the VSMAX value will be the same as the Display shows value for the upper right corner returned by the STA-TUS command. A portion of the STATUS screen is shown in Figure 90.

   The VSMAX value is not saved. A new value is determined whenever the drawing is regenerated.

```
Display shows     X:    0.0000    14.4381
                  Y:    0.0000     9.0000
```

Figure 90

VSMIN

[v2.6, r9]

Overview   The VSMIN system variable contains the drawing coordinate point of the lower left corner of the current virtual screen. It is a read-only variable.

Procedure   The VSMIN value can be returned using the SETVAR command or the (getvar) AutoLISP function. It can be changed only by regenerating the drawing and thus setting a new virtual screen.

Examples   In the following example, the SETVAR command is used to return the VSMIN value.

```
Command: SETVAR
Variable name or ?: VSMIN
VSMIN = 0.0000,0.0000 (read only)
```

Warnings   VSMIN is not available in AutoCAD version 2.5.

Tips   If the Fast Zoom mode has been disabled, the VSMIN value will be the same as the Display shows value for the lower left corner returned by the STATUS command. A portion of the STATUS screen is shown in Figure 90 (see VSMAX).

The VSMIN value is not saved. A new value is determined whenever the drawing is regenerated.

WBLOCK

[all versions]

Options

Name	Writes specified block definition
=	Block name same as file name
*	Writes entire drawing
Blank	Writes selected objects

Overview   The WBLOCK (Write Block) command writes all or part of a drawing to a disk file. The disk file created by the WBLOCK command is a new drawing file. Once written to disk, this Write Block can be inserted into any other AutoCAD drawing using the INSERT command (as can any AutoCAD drawing) or edited using AutoCAD's main menu task 2 (Edit Existing Drawing).

Procedure   To write all or a portion of the current drawing to a disk file, enter the WBLOCK command. AutoCAD first prompts for the name of the file. You can enter any valid file name, using up to eight characters. You can include a drive designation and a subdirectory name, but do not include a file extension. The file type .DWG is assumed. The WBLOCK command next prompts for the block name.

```
Command: WBLOCK
File name: (enter output file name)
Block name:
```

There are four responses available. If you enter a block name, the WBLOCK command writes the entities that make up an existing block to a file. A block with the specified name must already exist in the current drawing.

Responding with an equal sign (=) has the same effect as entering a block name, except that the name of the block is the same as the name of the file already entered.

If you respond to the block name prompt with an asterisk (*), the entire drawing is written out as a file. This is similar to saving a drawing except the WBLOCK command does not include unreferenced block definitions. This would be identical to first using AutoCAD's PURGE command to eliminate unreferenced block definitions and then saving the drawing (see PURGE).

If you respond to the block name prompt simply by pressing Return, the WBLOCK command responds in a similar fashion to AutoCAD's BLOCK command. You are prompted to select objects and an insertion base point. The selected objects are then removed from the drawing. You can restore objects thus removed by using the OOPS command.

```
Command: WBLOCK
File name: (enter output file name)
Block name: ( RETURN )
Select objects: (do so) (press Return when finished)
Insertion base point: (select point)
```

Examples   One possible use for the WBLOCK command is to create a base floor plan that will be plotted at a small scale. Once the base plan is completed, use WBLOCK to block out a portion of the base plan that will need to be plotted at a larger scale to show additional detail. This way, you don't need to redraw information. You can use what you have already drawn as the basis of your second drawing.

This is done quite often in many disciplines and represents one of the particularly powerful features of computer-aided drafting. There is no point in drawing anything more than once. If one drawing is similar to a previous drawing, you can start with the previous drawing as a base and then edit it accordingly.

In the following example and drawing, a stair section has been drawn previously. The WBLOCK command is used to isolate the first three risers of the stair (using the window method of Entity Selection). This creates a new drawing named DETAIL, which will serve as the basis for a large-scale detail of the base of the stair.

```
Command: WBLOCK
File name: DETAIL
```

```
Block name: ( RETURN )
Insertion base point: CENTER
of (select point A)
Select objects: W
First corner: (select point B)
Other corner: (select point C)
Select objects: ( RETURN )
```

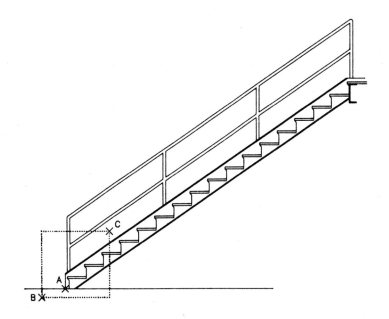

Tips   Remember that the point you specify as the insertion base point will be
used as the base point by the INSERT command.

Remember that you don't have to apply the WBLOCK command to one
drawing in order to insert it into another drawing. The WBLOCK command simply creates a drawing file on disk of part or all of the current
drawing. Any AutoCAD drawing file can be inserted into any other Auto-
CAD drawing.

The WBLOCK command uses the current elevation as the Z-coordinate
for the base point of the Block. You can specify a Z-coordinate explicitly
when entering the insertion point.

584

Windowing

[all versions]

Overview   Windowing is a method of selecting objects that can be used with Auto-CAD's editing or inquiry commands and also a means of zooming or saving named screen views. When AutoCAD prompts you to select objects, you can select the desired objects by placing a "window" around them. When using the ZOOM command to zero in on a smaller area of the screen, you can indicate the area you want to zoom in on by drawing a window around it. And when saving a particular view of the screen for later recall, you can indicate the view by placing it in a window.

In each of these cases, you must first tell AutoCAD that you are going to use the windowing method and then manipulate the screen cursor to draw the desired window.

Procedure   Refer to the descriptions of the ZOOM and VIEW commands for a more complete description of using the windowing method for defining the zoomed area or windowed view. In either case, after activating the command, you inform AutoCAD that you are using the windowing method by entering W or *Window*. AutoCAD then prompts you for the first corner and the second corner. You indicate the two corners by moving the screen cursor and pressing the pick button when the cursor is at the desired position. As soon as you select the first corner, AutoCAD displays a rubberband window, which contracts or expands as you move the cursor. When you pick the second corner, the command immediately executes.

When selecting objects using the windowing method, after you have selected the second corner, AutoCAD scans the area within the window. Any entity that falls entirely within the window is selected while entities outside the window or crossing the boundary of the window are not.

Release 9 Only   If you are using AutoCAD release 9, there are three additional object selection options you can use to make a windowing selection. If you enter *Box* in reply to the prompt to select objects, you are immediately prompted to select the first and second corners as usual. After you select the first corner, however, if you move the cursor to the right, AutoCAD immediately assumes you are making a windowing selection. Moving the cursor toward the left indicates a crossing selection

585

method. If your graphics card supports the Advanced Users Interface (see Advanced User Interface), AutoCAD further indicates the windowing method by displaying the rubberband box with a solid linetype as shown in Figure 91. The crossing method of entity selection is indicated by a dashed-line rubberband box (see Crossing).

Entering *Auto* or *AU* in response to the prompt to select objects causes AutoCAD to prompt again for objects. If you select a blank area of the screen, AutoCAD immediately goes into the Box selection mode, as previously described.

Normally AutoCAD continues prompting you to select objects until you press Return a final time. If you respond to the prompt to select objects by entering *Single* or *SI*, AutoCAD prompts you again to select objects. You can then use any valid selection method, such as windowing, box, or auto. But in this case as soon as an object is selected, object selection is complete.

Examples AutoCAD's window is always rectangular. Picking the first corner locks down that corner of the window. You can locate the second corner anywhere else on the screen. For example, you can choose the lower left corner of the window as the first point and then the upper right as

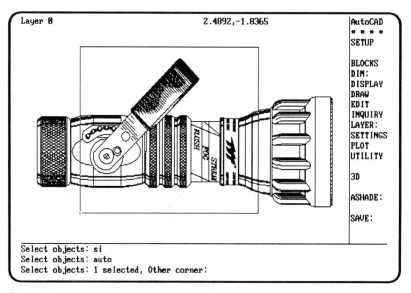

Figure 91

the second corner. Or you can select the upper right as the first corner. No matter what points you choose, the corners are always diagonally opposite.

The following example illustrates the use of the window method of object selection. When selecting objects in this fashion, only those entities that fall entirely within the window are selected. Those that are outside or cross the window are not selected. This can be seen in the following drawing.

```
Select objects: W
First corner: (select point A)
Other corner: (select point B)
Select objects: ( RETURN )
```

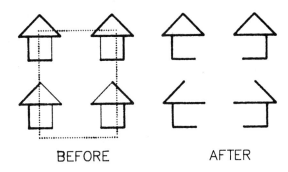

BEFORE   AFTER

Tips   When selecting objects, remember that you can use the window method in combination with other entity selection methods to add or remove any number of objects from the selection set. When an object is selected, it is highlighted. When you have completed your selections, press Return when AutoCAD again prompts you to select objects to cause the command to finish executing.

Windowing should not be confused with the crossing method of object selection. With windowing, only objects totally within a window are selected. With the crossing method, objects within the window or crossing its boundary are selected. (See Crossing.)

Some commands, such as TRIM and EXTEND, do not permit the use of windowing to select objects. On the other hand, the STRETCH command requires the use of either the window or crossing method of object selection.

587

X/Y/Z Filtering

[ADE-3] [v2.6, r9]

Overview X/Y/Z filtering is a method of entering coordinates that allows you to enter partial coordinates; AutoCAD then prompts you for the remaining coordinate information.

Procedure You can use X/Y/Z filtering whenever AutoCAD prompts for a coordinate. Indicate that you want to use X/Y/Z filtering by responding to the prompt with a filter in the form

`.coordinate`

where *coordinate* is one or more of the letters X, Y, and Z. AutoCAD indicates that you are using a filter by prompting of. You must then provide the coordinates that you filtered for. For example, if you respond to a prompt with the filter .XY, AutoCAD expects you to provide the X,Y coordinate pair. If an additional coordinate is required, AutoCAD prompts for it. For example, if the command requires a Z-coordinate in addition to the X and Y already entered, AutoCAD prompts need Z. You must then enter the Z-coordinate.

Examples You can use X/Y/Z filtering with any command that requires a coordinate as input. It is particularly useful in conjunction with the 3DFACE and 3DLINE commands. You can enter absolute or relative coordinates. In the following example below, a 3DLine is drawn from the point 2,3,2 to the point 3,5,5 (the point 1 unit away in the X-direction, 2 units away in the Y-direction, and 3 units away in the Z-direction). These points are built up by first supplying the X- and Y-coordinates and then providing the Z-coordinate.

```
Command: 3DLINE
From point: .XY
of 2,3
(need Z): 2
To point: .XY
of @1,2
(need Z): @0,0,3
To point: ( RETURN )
```

Tips X/Y/Z filtering is very useful in conjunction with a pointing device. You can filter for the X- and Y-coordinates, using the pointing device to select a point in a two-dimensional view of the drawing, and then enter the Z-coordinate from the keyboard.

'ZOOM

Options:

number	Multiplier from original scale
*number*X	Multiplier from current scale
A	All
C	Center
D	Dynamic pan/zoom [ADE-3]
E	Extents (drawing uses)
L	Lower left corner
P	Previous
W	Window

Overview   The ZOOM command changes the degree of magnification at which the current drawing is displayed on the graphics screen. One description of the image you see on AutoCAD's graphics screen is an analogy to being in a blimp. There's a whole world beneath you, which you see through a viewport of limited size. To see more of this world, you can increase your altitude (zoom out). You can now see more area but with less discernable detail. You can lower the blimp (zoom in) to see more detail, but you can't see as large an area. As you work in the drawing editor, creating and editing a drawing, you must constantly move in and out (or up and down if you will) within the drawing. The ZOOM command provides all the tools necessary to do this.

Procedure   The ZOOM command has nine available options to control its functioning. You can select any of these nine options either by entering the magnification scale or by entering the first letter of the desired option.

```
Command: ZOOM
All/Center/Dynamic/Extents/Left/Previous/Window/<Scale(X)>:
```

You can use any of the available ZOOM command options transparently (that is, while another AutoCAD command is active) except All or Extents, subject to certain restrictions. The transparent ZOOM command is activated by preceding the command with an apostrophe, as in 'ZOOM. Only ZOOM commands that will not cause the drawing to be regenerated can be executed transparently.

Examples

Zoom Scale At its simplest, if you think of an AutoCAD drawing as having one absolute scale factor (1 to 1, that is, one drawing unit equals one real-world unit), entering a number in response to the ZOOM command simply changes the magnification factor of the display. Thus, a magnification factor of 2 makes the drawing appear at twice its absolute size. A magnification factor of .5 causes the drawing to appear at half its absolute size.

```
Command: ZOOM
All/Center/Dynamic/Extents/Left/Previous/Window/<Scale(X)>: (number)
```

You can also enter a scale factor that alters the magnification relative to the current screen magnification, as opposed to the absolute screen magnification. To zoom relative to the current magnification, enter a number followed by *X*.

```
Command: ZOOM
All/Center/Dynamic/Extents/Left/Previous/Window/<Scale(X)>: (numberX)
```

For example, entering a scale factor of *2X* makes objects on the screen appear twice their current size. A scale factor of .5X reduces the drawing to half its current size, displaying more of the drawing but with less detail.

In either of these two cases, the center of the graphics display remains constant. The screen image appears at greater or lesser magnification, but the physical center point of the view always remains at the center of the screen.

ZOOM—All The ZOOM—All option causes AutoCAD to display the drawing either to its limits or to its extents, whichever is greater (see LIMITS and EXTENTS).

```
Command: ZOOM
All/Center/Dynamic/Extents/Left/Previous/Window/<Scale(X)>: A
```

The ZOOM—All command causes the drawing to be regenerated. If the drawing extents have changed since the last time the drawing was regenerated, the drawing is regenerated twice, and AutoCAD displays the message

```
** Second regeneration caused by change in drawing extents.
```

ZOOM—Center The ZOOM—Center option allows you to choose the physical center of the display. You can enter this center point as a coordinate or use the screen cursor to select a point on the screen. You can then specify the height of the displayed image in drawing units (similar to the absolute zoom scale factor previously described). The current display height is shown as the default. If you do not enter a new value, the screen image remains at the current magnification (the image is simply panned over to the new center point). Entering a smaller value increases the magnification. Entering a larger number decreases the magnification. If you enter a height followed by X, you alter the magnification relative to the current displayed image rather than relative to absolute drawing units. This is identical to the *numberX* method described previously. You can also specify the height by using the screen cursor to select two points on the screen. The height is determined by the distance between the two points.

```
Command: ZOOM
All/Center/Dynamic/Extents/Left/Previous/Window/<Scale(X)>: C
Center point: (select center point)
Magnification or Height <current>: (enter magnification or height)
```

ZOOM — Dynamic [ADE-3 Only]   The ZOOM—Dynamic option actually accomplishes both pans and zooms. ZOOM—Dynamic causes the currently generated portion of the drawing to be displayed on the screen along with a view box. You can resize and move this view box about the displayed image by using the pointing device. When you press Return, the area within the view box is redrawn on the graphics screen.

```
Command: ZOOM
All/Center/Dynamic/Extents/Left/Previous/Window/<Scale(X)>: D
```

When you execute the Dynamic option, the current screen image clears and a special dynamic image begins to be presented, filling the screen at redraw speed. You do not need to wait until the entire drawing image is presented. As soon as the view box is sized and positioned as you want it, press Return to complete the ZOOM—Dynamic option.

The dynamic zoom screen contains some special symbols. The current drawing extents are contained in a solid white rectangle. The view that was on the graphics screen just prior to beginning the ZOOM—Dynamic command is marked by a dotted box (displayed in green on color monitors). Four corner markers (displayed in red on color monitors) indicate the limits of the virtual screen.

A view box, initially the same size as the dotted current view box, can be moved about on the screen with the pointing device. When this view box is displayed with an X in its center, it means you are panning the view box about the screen. The magnification of the image to be displayed remains the same. If you press the pick button, the X is replaced with an arrow displayed on the right side of the view box. Now as you move the pointing device, the size of the view box changes. This indicates that the magnification of the resulting image will change. Press the pick button again to toggle back and forth between Pan and Zoom mode as many times as you want. When you have selected the view you want to display, press Return. The view box represents the edges of the graphics screen once the ZOOM command is completed.

As long as the view box remains within the red virtual screen limits, the ZOOM–Dynamic command will be executed at redraw speed. To further indicate when a selected view box image will cause regeneration of the drawing, an hourglass symbol is displayed in the lower left corner of the dynamic zoom screen whenever the view box is moved outside this virtual screen area.

The screens in Figure 92 illustrate the ZOOM–Dynamic command. The first screen shows the view box with an X in its center (indicating you are panning the view box). The second screen shows an arrow (indicating a change in magnification). Notice the hourglass symbol in the second screen (indicating that a drawing regeneration would be required).

ZOOM–Extents   The ZOOM–Extents option causes AutoCAD to display the drawing out to its extents, making the image fill the screen to as great a magnification factor as possible (see Extents). The image displayed is only that occupied by drawing objects, even if the drawing limits exceed this drawn area (unlike the ZOOM–All option). The ZOOM–Extents command always causes the drawing to be regenerated. If the drawing extents have changed since the last time the drawing was zoomed to the extents, the ZOOM–Extents command causes a second drawing regeneration.

```
Command: ZOOM
All/Center/Dynamic/Extents/Left/Previous/Window/<Scale(X)>: E
```

ZOOM–Left   The ZOOM–Left option is similar to ZOOM–Center. The difference is that instead of specifying the center point of the display, you must specify the lower left corner. This corner can be selected by entering a coordinate or selecting a point on the screen. The selected point will be moved to the lower left corner of the screen at the completion of

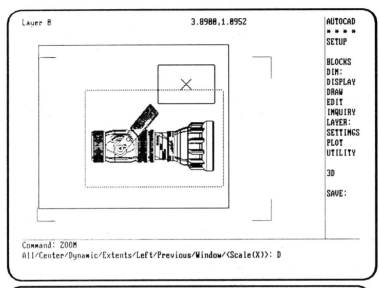

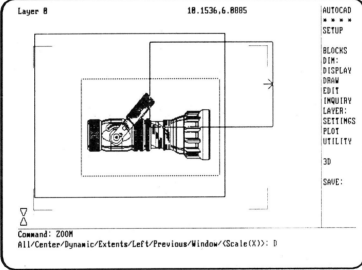

Figure 92

the ZOOM command. You are also prompted for the magnification or the height. As with the ZOOM—Center command, entering a number specifies the absolute height of the screen image in drawing units. Rather than enter a number, you can indicate two points on the screen, the height being the distance between them. If you enter a number followed immediately by an X, as in 2X, the magnification factor will be relative to the current screen image. The current absolute magnification factor is displayed as the default value. Pressing Return causes the drawing to simply pan over so that the selected point is at the lower left corner and the magnification remains the same.

```
Command: ZOOM
All/Center/Dynamic/Extents/Left/Previous/Window/<Scale(X)>: L
Lower left corner point: (select point)
Magnification or Height <current>: (enter magnification or height)
```

Zoom — Previous   The ZOOM—Previous option simply returns the screen image to its previous view. AutoCAD remembers the image presented on the screen by the ZOOM, PAN, and VIEW—Restore commands. You can use the ZOOM—Previous option to move back through the five previous views. Each time, the exact screen area and magnification factor are restored to the screen, but new objects added to the drawing appear and erased objects are removed.

```
Command: ZOOM
All/Center/Dynamic/Extents/Left/Previous/Window/<Scale(X)>: P
```

Zoom — Window   The ZOOM—Window option allows you to specify the area you want to enlarge on the screen by drawing a rectangular window around it. AutoCAD prompts for the two opposite corners of this window (similar to the window method of entity selection). The center of the selected window becomes the new center point of the display, and the image is enlarged to fill the display as completely as possible (subject to the ratio of height and width of the graphics screen to that of the selected window). Once you select one corner of the window, a rubberband window is displayed to aid in selecting the other corner. This is illustrated in the following example and shown in Figure 93.

```
Command: ZOOM
All/Center/Dynamic/Extents/Left/Previous/Window/<Scale(X)>: W
First corner: (select point A)
Other corner: (select point B)
```

Normally you use the screen cursor to select the two corners. This causes the area of the current display within the window to be enlarged. But if you enter coordinates, you can use the ZOOM — Window option to reduce the magnification scale by selecting points that are outside the current display.

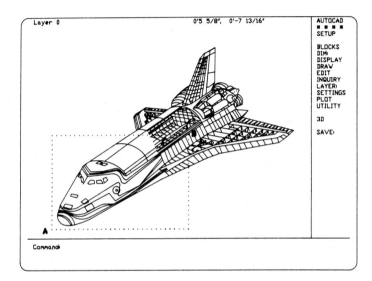

Before
ZOOM

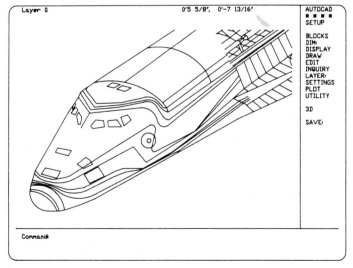

After
ZOOM

Figure 93

Warnings   You cannot use the ZOOM command transparently unless the VIEWRES command is used to turn Fast Zoom mode on. It also cannot be used transparently while the VPOINT, PAN, ZOOM, or VIEW command is active.

ZOOM—Dynamic does not perform a zoom at redraw speed when looking at a three-dimensional image. To remind you of this, a three-dimensional box image is displayed in the lower left corner of the dynamic zoom screen (see Figure 94).

The ZOOM—Dynamic command is designed to work with a pointing device rather than keyboard pointing keys. Therefore, pressing Return redisplays the current view-box image unless you first press one of the keyboard arrow keys.

Whenever a ZOOM command causes the drawing to be regenerated, the drawing extents reported by the STATUS command (drawing uses) are updated. If this value has changed since the drawing was last regenerated, a second regeneration of the drawing is necessary.

The transparent ZOOM command is not available in AutoCAD version 2.5.

Tips   Since the ZOOM command can be used transparently in version 2.6, you should go back and edit your AutoCAD menus to include the apostrophe

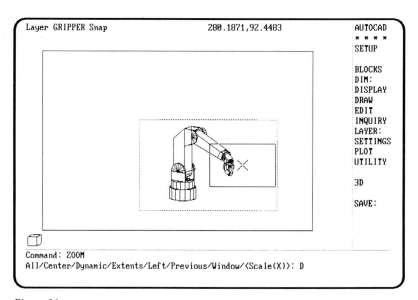

Figure 94

before every ZOOM command. If the command cannot be used transparently in a given instance, AutoCAD will tell you when a transparent ZOOM command cannot be executed.

Often filling the screen with a ZOOM—All or a ZOOM—Extents places part of the drawing too close to the edge of the screen to allow you to select objects. To eliminate this problem, after you perform a ZOOM—All or a ZOOM—Extents, do a ZOOM .85X. This reduces the screen image an additional 15 percent, giving you a working margin around the displayed drawing.

It is helpful to save both the extents image and the 15-percent smaller image as named views (see VIEW). Thus you avoid having to repeat this process multiple times (with the associated delays of the required drawing regeneration each time) and can restore these views transparently later as named views.

Normally you should work with REGENAUTO off. Do this so that if you try to zoom to a view that would normally force a regeneration, AutoCAD will warn you

```
About to regen, proceed? <Y>:
```

At that point you can answer no, and then attempt the ZOOM command using a slightly different window or magnification factor or use ZOOM—Dynamic to accurately control the zoomed view.

In AutoCAD release 9 (with the standard menu ACAD.MNU), you can select the transparent ZOOM—Previous, ZOOM—Window, or ZOOM—Dynamic command from the Display pull-down menu. (See Advanced User Interface.)

Numeric
Entries

3DFACE

[ADE-3] [v2.6, r9]

Overview   The 3DFACE command creates a 3DFace entity, a section of a plane, in three-dimensional space. You must describe the plane by providing the X-, Y-, and Z-coordinates of three or more corners. 3DFaces are always displayed as wire frames and cannot be extruded.

Procedure   To create a 3DFace, activate the 3DFACE command. AutoCAD prompts for the first point, the second point, the third point, and the fourth point. After you provide a fourth point, the command continues to prompt for points by alternating prompts for third point and fourth point to allow you to build up a complex three-dimensional object.

You can enter absolute 3D coordinates or relative coordinates. When entering points, you can build up a three-dimensional point coordinate by defining each X-, Y-, and Z-coordinate individually. This method is referred to as X/Y/Z filtering (see X/Y/Z Filtering).

Examples   You can draw a simple three-dimensional box (with 1-unit square sides) using the 3DFACE command. This box would appear as a wire frame when you view it using the VPOINT command, but the planes would be treated as solid planes by the HIDE command. The resulting box appears with an open top and bottom, since you defined only four planes. You can add a top and a bottom using the 3DFACE command again and object snapping to the four upper corners (and again snapping to the four lower corners to make a solid cube).

The following example draws the three-dimensional box in the following drawing. Each step (and the resulting figure at each stage) is shown.

```
Command: 3DFACE
First point: 0,0,0
Second point: @0,0,1
Third point: @1,0,0
Fourth point: @0,0,-1 (completes first side)
Third point: @0,1,0
Fourth point: @0,0,1 (completes second side)
Third point: @-1,0,0
Fourth point: @0,0,-1 (completes third side)
Third point: @0,-1,0
```

Fourth point: **@0,0,1** (completes fourth side)
Third point: (RETURN) (ends the command)

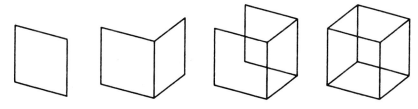

Warnings   Because each corner of a 3DFACE can have different Z-coordinates, it is possible to create nonplanar objects.

Tips   AutoCAD version 2.6 and release 9 include a file of AutoLISP routines that utilize the 3DFACE command to create more complex shapes, such as cones, domes, and spheres. This file, 3D.LSP, is on the Support disk (Support2 with release 9). In addition, if you are using AutoCAD release 9, you can select these 3D AutoLISP routines from the icon menu displayed by picking 3D Objects from the Options pull-down menu.

3DLINE   [ADE-3] [v2.6, r9]

Overview   The 3DLINE command creates 3DLine entities, which basically are lines in three-dimensional space. The lines can have varying X-, Y-, and Z-coordinates.

Procedure   The 3DLINE command operates identically to the regular LINE command except that Z-coordinates are also accepted. The command prompts for a "from point" and then prompts continuously for the "to point." Separate 3DLine segments are drawn between the specified points.

You can create 3DLines by enering distinct X-, Y-, and Z-coordinates (absolute coordinates), by providing relative coordinates (using the at-symbol method; for example, @0,0,1 would draw a line from the current location to a point one unit away in the Z-direction), by screen pointing, or by using the X/Y/Z filtering method (see X/Y/Z Filtering). If you construct a 3DLine using screen pointing, the current elevation is used for the Z-coordinate.

3DLines cannot be extruded and are always drawn with a solid (continuous) linetype.

Examples

The following command sequence creates a wire frame box similar to the box in the 3DFACE example. Note that the box drawn using 3DLines always remains a wire frame, even after using the HIDE command, since the lines always remain lines in space. They are not joined to represent planes. Each step (and the resulting drawing at each stage) is shown in the following example and drawings.

```
Command: 3DLINE
From point: 0,0,0
To point: @0,0,1
To point: @1,0,0
To point: @0,0,-1
To point: C (closes the 3DLine to draw first side)
Command: ( RETURN )
3DLINE From point: ( RETURN )
To point: @0,1,0
To point: @0,0,1
To point: @0,-1,0 (completes second side)
To point: ( RETURN ) (ends command)
Command: ( RETURN ) (restarts command)

3DLINE From point: 0,1,0
To point: @1,0,0
To point: @0,-1,0 (completes bottom)
To point: ( RETURN )  (ends command)
Command: ( RETURN )  (restarts command)
3DLINE From point: 1,1,0
To point: @0,0,1
To point: @0,-1,0 (completes third side)
To point: ( RETURN )  (ends command)
Command: ( RETURN )  (restarts command)
3DLINE From point: 1,1,1
To point: @-1,0,0 (completes fourth side)
To point: ( RETURN )  (ends command)
```

Tips

You can use object snap with 3DLines but only to snap to the end point or the midpoint of 3DLines.

602

Appendices

Appendix A: AutoCAD System Variables

The following table comprises all AutoCAD system variables, arranged alphabetically. The default value is provided along with notations for read-only variables, the variable type (string, integer, real, point, 3D point), where the variable is saved, the required ADE level, and what version of AutoCAD the variable was introduced in (when listed, this indicates that the variable is not present in earlier versions). Note that the dimensioning variables were not available as system variables until AutoCAD version 2.6. Version 2.5 users can access dimensioning variables only from the Dimensioning mode. Also note that this list does not include the undocumented user variables (see System Variables). In the Default Value column, quotes (" ") denote a blank string value.

NAME	DEFAULT VALUE		TYPE	SAVED IN	ADE	VERSION
ACADPREFIX	" "	(read only)	string	—	—	2.6
ACADVER	"9.0"	(read only)	string	—	—	r9
AFLAGS	0		integer	—	2	2.5
ANGBASE	0		real	drawing	1	2.5
ANGDIR	0		integer	drawing	1	2.5
APERTURE	10		integer	config	2	—
AREA	0.0000	(read only)	real	—	—	—
ATTDIA	0		integer	drawing	3	r9
ATTMODE	1		integer	drawing	2	—
ATTREQ	1		integer	drawing	2	r9
AUNITS	0		integer	drawing	1	—
AUPREC	0		integer	drawing	1	—
AXISMODE	0		integer	drawing	1	—
AXISUNIT	0.0000,0.0000		integer	drawing	1	—
BLIPMODE	1		integer	drawing	—	—
CDATE	19871015.234806400	(read only)	real	—	—	2.5
CECOLOR	"BYLAYER"	(read only)	string	drawing	—	2.5

NAME	DEFAULT VALUE		TYPE	SAVED IN	ADE	VERSION
CELTYPE	"BYLAYER"	(read only)	string	drawing	—	2.5
CHAMFERA	0.0000		real	drawing	1	—
CHAMFERB	0.0000		real	drawing	1	—
CLAYER	"0"	(read only)	string	drawing	—	—
CMDECHO	1		integer	—	3	2.5
COORDS	0		integer	drawing	1	2.5
DATE	2447084.99251192	(read only)	real	—	—	2.5
DIMALT	0		integer	drawing	1	2.5
DIMALTD	2		integer	drawing	1	2.5
DIMALTF	25.4000		integer	drawing	1	2.5
DIMAPOST	" "	(read only)	integer	drawing	1	2.6
DIMASO	1		integer	drawing	1	2.6
DIMASZ	0.1800		real	drawing	1	—
DIMBLK	" "	(read only)	string	drawing	1	2.5
DIMCEN	0.0900		real	drawing	1	—
DIMDLE	0.0000		real	drawing	1	2.5
DIMDLI	0.3800		real	drawing	1	—
DIMEXE	0.1800		real	drawing	1	—
DIMEXO	0.0625		real	drawing	1	—
DIMLFAC	1.0000		real	drawing	1	2.5
DIMLIM	0		integer	drawing	1	—
DIMPOST	" "	(read only)	string	drawing	1	2.6
DIMRND	0.0000		real	drawing	1	2.5
DIMSCALE	1.0000		real	drawing	1	—
DIMSE1	0		integer	drawing	1	—
DIMSE2	0		integer	drawing	1	—
DIMSHO	0		integer	drawing	1	2.6
DIMTAD	0		integer	drawing	1	—

APPENDIX A: AUTOCAD SYSTEM VARIABLES

NAME	DEFAULT VALUE		TYPE	SAVED IN	ADE	VERSION
DIMTIH	1		integer	drawing	1	—
DIMTM	0.0000		real	drawing	1	—
DIMTOH	1		integer	drawing	1	—
DIMTOL	0		integer	drawing	1	—
DIMTP	0.0000		real	drawing	1	—
DIMTSZ	0.0000		real	drawing	1	—
DIMTXT	0.1800		real	drawing	1	—
DIMZIN	0		integer	drawing	1	2.5
DISTANCE	0.0000	(read only)	real	—	—	2.5
DRAGMODE	2		integer	drawing	2	—
DRAGP1	10		integer	drawing	2	2.5
DRAGP2	25		integer	drawing	2	2.5
DWGNAME	" "	(read only)	string	—	—	2.5
DWGPREFIX	" "	(read only)	string	—	—	2.5
ELEVATION	0.0000		real	drawing	3	—
EXPERT	0		integer	—	—	2.5
EXTMAX	$-1.0000E+20$ $-1.0000E+20$	(read only)	point	drawing	—	—
EXTMIN	$1.0000E+20$ $1.0000E+20$	(read only)	point	drawing	—	—
FILLETRAD	0.0000		real	drawing	1	—
FILLMODE	1		integer	drawing	—	—
GRIDMODE	0		integer	drawing	—	—
GRIDUNIT	0.0000,0.0000		point	drawing	—	—
HIGHLIGHT	1		integer	—	3	2.5
INSBASE	0.0000,0.0000		point	drawing	2	—
LASTANGLE	0	(read only)	real	—	—	2.5
LASTPOINT	0.0000,0.0000		point	—	—	2.5
LASTPT3D	0.0000,0.0000,0.0000		3D point	—	3	2.6

NAME	DEFAULT VALUE		TYPE	SAVED IN	ADE	VERSION
LIMCHECK	0		integer	drawing	—	—
LIMMAX	12.0000,9.0000		point	drawing	—	—
LIMMIN	0.0000,0.0000		point	drawing	—	—
LTSCALE	1.0000		real	drawing	—	—
LUNITS	2		integer	drawing	1	—
LUPREC	4		integer	drawing	1	—
MENUECHO	0		integer	—	—	2.5
MENUNAME	"acad"	(read only)	string	drawing	—	r9
MIRRTEXT	1		integer	drawing	2	2.5
ORTHOMODE	0		integer	drawing	—	—
OSMODE	0		integer	drawing	2	—
PDMODE	0		integer	drawing	—	2.5
PDSIZE	0.0000		real	drawing	—	2.5
PERIMETER	0.0000	(read only)	real	—	—	2.5
PICKBOX	3		integer	config	—	2.5
POPUPS	1	(read only)	integer	drawing	3	r9
QTEXTMODE	0		integer	drawing	—	—
REGENMODE	1		integer	drawing	—	—
SCREENSIZE	572.0000,292.0000	(read only)	point	—	—	2.5
SKETCHINC	0.1000		real	drawing	1	—
SKPOLY	0		integer	drawing	3	2.5
SNAPANG	0		real	drawing	2	—
SNAPBASE	0.0000,0.0000		point	drawing	2	—
SNAPISOPAIR	0		integer	drawing	2	—
SNAPMODE	0		integer	drawing	—	—
SNAPSTYL	0		integer	drawing	2	—
SNAPUNIT	1.0000,1.0000		point	drawing	—	—
SPLFRAME	0		integer	drawing	3	r9

APPENDIX A: AUTOCAD SYSTEM VARIABLES

NAME	DEFAULT VALUE		TYPE	SAVED IN	ADE	VERSION
SPLINESEGS	8		integer	drawing	3	r9
TDCREATE	2447084.99164606	(read only)	real	drawing	—	2.5
TDINDWG	0.00195475	(read only)	real	drawing	—	2.5
TDUPDATE	2447084.99164606	(read only)	real	drawing	—	2.5
TDUSRTIMER	0.00195671	(read only)	real	drawing	—	2.5
TEMPPREFIX	" "	(read only)	string	—	—	2.6
TEXTEVAL	0		integer	—	3	2.6
TEXTSIZE	0.2000		real	drawing	—	—
TEXTSTYLE	"STANDARD"	(read only)	string	drawing	—	2.6
THICKNESS	0.0000		real	drawing	3	—
TRACEWID	0.0500		real	drawing	—	—
VIEWCTR	7.2190,4.5000	(read only)	point	drawing	—	—
VIEWSIZE	9.0000	(read only)	real	drawing	—	—
VPOINTX	0.0000	(read only)	real	drawing	3	2.5
VPOINTY	0.0000	(read only)	real	drawing	3	2.5
VPOINTZ	1.0000	(read only)	real	drawing	3	2.5
VSMAX	14.4381,9.0000	(read only)	point	—	—	2.6
VSMIN	0.0000,0.0000	(read only)	point	—	—	2.6

Appendix B: Scale Factors Chart

FINISHED-SHEET SCALE	GOING DOWN TO A 1X FILE	GOING UP FROM A 1X FILE
$\frac{1}{32}'' = 1'$	0.0026041	384
$\frac{1}{16}'' = 1'$	0.0052083	192
$\frac{1}{8}'' = 1'$	0.0104166	96
$\frac{3}{16}'' = 1'$	0.015625	64
$\frac{1}{4}'' = 1'$	0.0208333	48
$\frac{3}{8}'' = 1'$	0.03125	32
$\frac{1}{2}'' = 1'$	0.041666	24
$\frac{3}{4}'' = 1'$	0.0625	16
$1'' = 1'$	0.08333	12
$1\frac{1}{2}'' = 1'$	0.125	8
$3'' = 1'$	0.25	4
$1'' = 10'$	0.0083333	120
$1'' = 20'$	0.0041666	240
$1'' = 30'$	0.0027777	360
$1'' = 40'$	0.0020833	480
$1'' = 50'$	0.0016666	600
$1'' = 60'$	0.0013888	720
$1'' = 70'$	0.0011904	840
$1'' = 80'$	0.0010416	960
$1'' = 90'$	0.0009259	1080
$1'' = 100'$	0.0008333	1200
$\frac{1}{100}$	0.01	100
$\frac{1}{80}$	0.0125	80
$\frac{1}{64}$	0.015625	64

For example, to insert a detail onto a sheet that will plot at $1 = 1$ so that the detail appears at $\frac{1}{4}'' = 1'$, insert the detail drawing at an insertion scale of .0208333.

APPENDIX B: SCALE FACTORS CHART

FINISHED-SHEET SCALE	GOING DOWN TO A 1X FILE	GOING UP FROM A 1X FILE
$\frac{1}{40}$	0.025	40
$\frac{1}{32}$	0.03125	32
$\frac{1}{20}$	0.05	20
$\frac{1}{16}$	0.0625	16
$\frac{1}{10}$	0.1	10
$\frac{1}{8}$	0.125	8
$\frac{1}{4}$	0.25	4
$\frac{3}{8}$	0.375	2.6666666
$\frac{1}{2}$	0.5	2
$\frac{5}{8}$	0.625	1.6
$\frac{3}{4}$	0.75	1.3333333
$\frac{7}{8}$	0.875	1.1428671
1	1	1
2	2	0.5
4	4	0.25
10	10	0.1
20	20	0.05

Appendix C: Quick Reference to Entries

ENTRY	CATEGORY	VERSIONS
ACAD	DOS command	[all versions]
ACADPREFIX	system variable	[v2.6, r9]
ACADVER	system variable	[r9]
ADE	program feature	[all versions]
ADI	program feature	[v2.5, v2.6, r9]
Advanced User Interface	program feature	[ADE-3] [r9]
AFLAGS	system variable	[ADE-2] [v2.5, v2.6, r9]
ANGBASE	system variable	[ADE-1] [v2.5, v2.6, r9]
ANGDIR	system variable	[ADE-1] [v2.5, v2.6, r9]
Angles	program feature	[all versions]
APERTURE	command & system variable	[ADE-2] [all versions]
ARC	command	[all versions]
AREA	command & system variable	[all versions]
ARRAY – Polar	command	[all versions]
ARRAY – Rectangular	command	[all versions]
Associative Dimensions	dimensioning feature	[ADE-1] [v2.6, r9]
@ (At symbol)	program feature	[all versions]
ATTDEF (Attribute Define)	command	[ADE-2] [all versions]
ATTDIA	system variable	[ADE-2] [r9]
ATTDISP (Attribute Display)	command	[ADE-2] [all versions]
ATTEDIT (Attribute Edit)	command	[ADE-2] [all versions]
ATTEXT (Attribute Extract)	command	[ADE-2] [all versions]
ATTMODE	system variable	[ADE-2] [all versions]
ATTREQ	system variable	[ADE-2] [r9]
AUNITS	system variable	[ADE-1] [all versions]
AUPREC	system variable	[ADE-1] [all versions]

APPENDIX C: QUICK REFERENCE TO ENTRIES

ENTRY	CATEGORY	VERSIONS
AutoLISP	program feature	[ADE-3] [all versions]
AXIS	command	[ADE-1] [all versions]
AXISMODE	system variable	[ADE-1] [all versions]
AXISUNIT	system variable	[ADE-1] [all versions]
Backup	program feature	[all versions]
BASE	command	[all versions]
BLIPMODE	command & system variable	[all versions]
BLOCK	command	[all versions]
BREAK	command	[ADE-1] [all versions]
Button Menu	program feature	[all versions]
CDATE	system variable	[v2.5, v2.6, r9]
CECOLOR	system variable	[v2.5, v2.6, r9]
CELTYPE	system variable	[v2.5, v2.6, r9]
CHAMFER	command	[ADE-1] [all versions]
CHAMFERA	system variable	[ADE-1] [all versions]
CHAMFERB	system variable	[ADE-1] [all versions]
CHANGE	command	[all versions]
CIRCLE	command	[all versions]
CLAYER	system variable	[all versions]
CMDECHO	system variable	[ADE-3] [v2.5, v2.6, r9]
COLOR	command	[v2.5, v2.6, r9]
Configuring AutoCAD	program feature	[all versions]
Coordinates	program feature	[all versions]
COORDS	system variable	[ADE-1] [v2.5, v2.6, r9]
COPY	command	[all versions]
Crossing	program feature	[v2.5, v2.6, r9]

ENTRY	CATEGORY	VERSIONS
DATE	system variable	[v2.5, v2.6, r9]
DBLIST	command	[all versions]
DDATTE	command	[ADE-3] [r9]
'DDEMODES	command	[ADE-3] [r9]
'DDLMODES	command	[ADE-3] [r9]
'DDRMODES	command	[ADE-3] [r9]
DELAY	command	[all versions]
Dialog Box	program feature	[ADE-3] [r9]
DIM	command	[ADE-1] [all versions]
DIM1	command	[ADE-1] [v2.5, v2.6, r9]
DIM—ALIGNED	dimension subcommand	[ADE-1] [all versions]
DIM—ANGULAR	dimension subcommand	[ADE-1] [all versions]
DIM—BASELINE	dimension subcommand	[ADE-1] [all versions]
DIM—CENTER	dimension subcommand	[ADE-1] [all versions]
DIM—CONTINUE	dimension subcommand	[ADE-1] [all versions]
DIM—DIAMETER	dimension subcommand	[ADE-1] [all versions]
DIM—EXIT	dimension subcommand	[ADE-1] [all versions]
DIM—HOMETEXT	dimension subcommand	[ADE-1] [v2.6, r9]
DIM—HORIZONTAL	dimension subcommand	[ADE-1] [all versions]
DIM—LEADER	dimension subcommand	[ADE-1] [all versions]
DIM—NEWTEXT	dimension subcommand	[ADE-1] [v2.6, r9]
DIM—RADIUS	dimension subcommand	[ADE-1] [all versions]
DIM—REDRAW	dimension subcommand	[ADE-1] [all versions]
DIM—ROTATED	dimension subcommand	[ADE-1] [all versions]
DIM—STATUS	dimension subcommand	[ADE-1] [all versions]
DIM—STYLE	dimension subcommand	[ADE-1] [v2.5, v2.6, r9]
DIM—UNDO	dimension subcommand	[ADE-1] [all versions]
DIM—UPDATE	dimension subcommand	[ADE-1] [v2.6, r9]

APPENDIX C: QUICK REFERENCE TO ENTRIES

ENTRY	CATEGORY	VERSIONS
DIM — VERTICAL	dimension subcommand	[ADE-1] [all versions]
DIMALT	dimensioning variable	[ADE-1] [v2.5, v2.6, r9]
DIMALTD	dimensioning variable	[ADE-1] [v2.5, v2.6, r9]
DIMALTF	dimensioning variable	[ADE-1] [v2.5, v2.6, r9]
DIMAPOST	dimensioning variable	[ADE-1] [v2.6, r9]
DIMASO	dimensioning variable	[ADE-1] [v2.6, r9]
DIMASZ	dimensioning variable	[ADE-1] [all versions]
DIMBLK	dimensioning variable	[ADE-1] [v2.5, v2.6, r9]
DIMCEN	dimensioning variable	[ADE-1] [all versions]
DIMDLE	dimensioning variable	[ADE-1] [v2.5, v2.6, r9]
DIMDLI	dimensioning variable	[ADE-1] [all versions]
Dimension Text	program feature	[ADE-1] [v2.5, v2.6, r9]
DIMEXE	dimensioning variable	[ADE-1] [all versions]
DIMEXO	dimensioning variable	[ADE-1] [all versions]
DIMLFAC	dimensioning variable	[ADE-1] [v2.5, v2.6, r9]
DIMLIM	dimensioning variable	[ADE-1] [all versions]
DIMPOST	dimensioning variable	[ADE-1] [v2.6, r9]
DIMRND	dimensioning variable	[ADE-1] [v2.5, v2.6, r9]
DIMSCALE	dimensioning variable	[ADE-1] [all versions]
DIMSE1	dimensioning variable	[ADE-1] [all versions]
DIMSE2	dimensioning variable	[ADE-1] [all versions]
DIMSHO	dimensioning variable	[ADE-1] [v2.6, r9]
DIMTAD	dimensioning variable	[ADE-1] [all versions]
DIMTIH	dimensioning variable	[ADE-1] [all versions]
DIMTM	dimensioning variable	[ADE-1] [all versions]
DIMTOH	dimensioning variable	[ADE-1] [all versions]
DIMTOL	dimensioning variable	[ADE-1] [all versions]
DIMTP	dimensioning variable	[ADE-1] [all versions]

ENTRY	CATEGORY	VERSIONS
DIMTSZ	dimensioning variable	[ADE-1] [all versions]
DIMTXT	dimensioning variable	[ADE-1] [all versions]
DIMZIN	dimensioning variable	[ADE-1] [v2.5, v2.6, r9]
DIST (Distance)	command	[all versions]
DISTANCE	system variable	[v2.5, v2.6, r9]
DIVIDE	command	[ADE-3] [v2.5, v2.6, r9]
DONUT or DOUGHNUT	command	[ADE-3] [v2.5, v2.6, r9]
DRAGMODE	command & system variable	[ADE-2] [all versions]
DRAGP1	system variable	[ADE-2] [v2.5, v2.6, r9]
DRAGP2	system variable	[ADE-2] [v2.5, v2.6, r9]
Drawing	program feature	[all versions]
DTEXT (Dynamic Text)	command	[ADE-3] [v2.5, v2.6, r9]
DWGNAME	system variable	[v2.5, v2.6, r9]
DWGPREFIX	system variable	[v2.5, v2.6, r9]
DXBIN	command	[ADE-3] [all versions]
DXF	program feature	[all versions]
DXFIN	command	[all versions]
DXFOUT	command	[all versions]
ELEV (Elevation)	command	[ADE-3]
ELEVATION	system variable	[ADE-3] [all versions]
ELLIPSE	command	[ADE-3] [v2.5, v2.6, r9]
END	command	[all versions]
Entities	program feature	[all versions]
Entity Selection	program feature	[all versions]
Environment Variables	program feature	[all versions]
ERASE	command	[all versions]
EXPERT	system variable	[v2.5, v2.6, r9]

ENTRY	CATEGORY	VERSIONS
EXPLODE	command	[ADE-3] [v2.5, v2.6, r9]
EXTEND	command	[ADE-3] [v2.5, v2.6, r9]
Extents	program feature	[all versions]
EXTMAX	system variable	[all versions]
EXTMIN	system variable	[all versions]
Fast Zoom Mode	program feature	[v2.5, v2.6, r9]
FILES	command	[all versions]
FILL	command	[all versions]
FILLET	command	[ADE-1] [all versions]
FILLETRAD	system variable	[ADE-1] [all versions]
FILLMODE	system variable	[all versions]
FILMROLL	command	[ADE-3] [v2.6, r9]
Fonts	program feature	[all versions]
Function Keys	program feature	[all versions]
'GRAPHSCR	command	[all versions]
GRID	command	[all versions]
GRIDMODE	system variable	[all versions]
GRIDUNIT	system variable	[all versions]
HATCH	command	[ADE-1] [all versions]
'HELP (or '?)	command	[all versions]
HIDE	command	[ADE-3] [all versions]
HIGHLIGHT	system variable	[ADE-3] [v2.5, v2.6, r9]
Icon Menu	program feature	[ADE-3] [r9]
ID	command	[all versions]

ENTRY	CATEGORY	VERSIONS
IGESIN	command	[ADE-3] [v2.5, v2.6, r9]
IGESOUT	command	[ADE-3] [v2.5, v2.6, r9]
INSBASE	system variable	[all versions]
INSERT	command	[all versions]
ISOPLANE	command	[ADE-2] [all versions]
Last Entity Selection	program feature	[all versions]
Last Point	program feature	[all versions]
LASTANGLE	system variable	[v2.5, v2.6, r9]
LASTPOINT	system variable	[v2.5, v2.6, r9]
LASTPT3D	system variable	[ADE-3] [v2.6, r9]
LAYER	command	[all versions]
LIMCHECK	system variable	[all versions]
LIMITS	command	[all versions]
LIMMAX	system variable	[all versions]
LIMMIN	system variable	[all versions]
LINE	command	[all versions]
LINETYPE	command	[all versions]
LIST	command	[all versions]
LOAD	command	[all versions]
LTSCALE	command & system variable	[all versions]
LUNITS	system variable	[ADE-1] [all versions]
LUPREC	system variable	[ADE-1] [all versions]
MEASURE	command	[ADE-3] [v2.5, v2.6, r9]
MENU	command	[all versions]
Menu Bar	program feature	[ADE-3] [r9]
MENUECHO	system variable	[v2.5, v2.6, r9]

ENTRY	CATEGORY	VERSIONS
MENUNAME	system variable	[r9]
MINSERT	command	[v2.5, v2.6, r9]
MIRROR	command	[ADE-2] [all versions]
MIRRTEXT	system variable	[ADE-2] [v2.5, v2.6, r9]
MOVE	command	[all versions]
MSLIDE (Make Slide)	command	[ADE-2] [all versions]
MULTIPLE	command	[r9]
Object Snap	program feature	[ADE-2] [all versions]
OFFSET	command	[ADE-3] [v2.5, v2.6, r9]
OOPS	command	[all versions]
ORTHO	command	[all versions]
ORTHOMODE	system variable	[all versions]
OSMODE	system variable	[ADE-2] [all versions]
OSNAP	command	[ADE-2] [all versions]
'PAN	command	[all versions]
PDMODE	system variable	[v2.5, v2.6, r9]
PDSIZE	system variable	[v2.5, v2.6, r9]
PEDIT (Polyline Edit)	command	[ADE-3] [all versions]
PERIMETER	system variable	[v2.5, v2.6, r9]
PICKBOX	system variable	[v2.5, v2.6, r9]
PLINE (Polyline)	command	[ADE-3] [all versions]
PLOT	command	[all versions]
Plotting	program feature	[all versions]
POINT	command	[all versions]
POLYGON	command	[ADE-3] [v2.5, v2.6, r9]
POPUPS	system variable	[ADE-3] [r9]

ENTRY	CATEGORY	VERSIONS
Previous Selection Set	program feature	[v2.5, v2.6, r9]
Printer Echo	program feature	[all versions]
Prototype Drawing	program feature	[all versions]
PRPLOT (Printer Plot)	command	[all versions]
Pull-Down Menus	program feature	[ADE-3] [r9]
PURGE	command	[all versions]
QTEXT (Quick Text)	command	[all versions]
QTEXTMODE	system variable	[all versions]
QUIT	command	[all versions]
REDEFINE	command	[ADE-3] [r9]
REDO	command	[v2.5, v2.6, r9]
'REDRAW	command	[all versions]
REGEN	command	[all versions]
REGENAUTO	command	[all versions]
REGENMODE	system variable	[all versions]
RENAME	command	[all versions]
Repeating Commands	program feature	[all versions]
'RESUME	command	[all versions]
ROTATE	command	[ADE-3] [v2.5, v2.6, r9]
RSCRIPT	command	[all versions]
SAVE	command	[all versions]
SCALE	command	[ADE-3] [v2.5, v2.6, r9]
Screen Menu	program feature	[all versions]
SCREENSIZE	system variable	[v2.5, v2.6, r9]
SCRIPT	command	[all versions]

ENTRY	CATEGORY	VERSIONS
SELECT	command	[v2.5, v2.6, r9]
Selection Set	program feature	[v2.5, v2.6, r9]
'SETVAR	command	[v2.5, v2.6, r9]
SH	command	[ADE-3] [v2.5, v2.6, r9]
SHAPE	command	[all versions]
SHELL	command	[ADE-3] [all versions]
SKETCH	command	[ADE-1] [all versions]
SKETCHINC	system variable	[ADE-1] [all versions]
SKPOLY	system variable	[ADE-3] [v2.5, v2.6, r9]
Slides	program feature	[all versions]
SNAP	command	[all versions]
SNAPANG	system variable	[ADE-2] [all versions]
SNAPBASE	system variable	[ADE-2] [all versions]
SNAPISOPAIR	system variable	[ADE-2] [all versions]
SNAPMODE	system variable	[all versions]
SNAPSTYL	system variable	[ADE-2] [all versions]
SNAPUNIT	system variable	[all versions]
SOLID	command	[all versions]
SPLFRAME	system variable	[ADE-3] [r9]
SPLINESEGS	system variable	[ADE-3] [r9]
STATUS	command	[all versions]
Status Line	program feature	[ADE-1] [all versions]
STRETCH	command	[ADE-3] [v2.5, v2.6, r9]
STYLE	command	[all versions]
System Variables	program feature	[all versions]
TABLET	command	[all versions]
Tablet Menu	program feature	[all versions]

ENTRY	CATEGORY	VERSIONS
TDCREATE	system variable	[v2.5, v2.6, r9]
TDINDWG	system variable	[v2.5, v2.6, r9]
TDUPDATE	system variable	[v2.5, v2.6, r9]
TDUSRTIMER	system variable	[v2.5, v2.6, r9]
TEMPPREFIX	system variable	[v2.6, r9]
TEXT	command	[all versions]
TEXTEVAL	system variable	[ADE-3] [v2.6, r9]
'TEXTSCR	command	[all versions]
TEXTSIZE	system variable	[all versions]
TEXTSTYLE	system variable	[all versions]
THICKNESS	system variable	[ADE-3] [all versions]
TIME	command	[v2.5, v2.6, r9]
TRACE	command	[all versions]
TRACEWID	system variable	[all versions]
Transparent Commands	program feature	[v2.5, v2.6, r9]
TRIM	command	[ADE-3] [v2.5, v2.6, r9]
U	command	[v2.5, v2.6, r9]
UNDEFINE	command	[ADE-3] [r9]
UNDO	command	[v2.5, v2.6, r9]
UNITS	command	[ADE-1] [all versions]
'VIEW	command	[ADE-2] [all versions]
VIEWCTR	system variable	[all versions]
VIEWRES	command	[v2.5, v2.6, r9]
VIEWSIZE	system variable	[all versions]
Virtual Screen	program feature	[v2.5, v2.6, r9]
VPOINT	command	[ADE-3] [all versions]

APPENDIX C: QUICK REFERENCE TO ENTRIES

ENTRY	CATEGORY	VERSIONS
VPOINTX	system variable	[ADE-3] [v2.5, v2.6, r9]
VPOINTY	system variable	[ADE-3] [v2.5, v2.6, r9]
VPOINTZ	system variable	[ADE-3] [v2.5, v2.6, r9]
VSLIDE (View Slide)	command	[all versions]
VSMAX	system variable	[v2.6, r9]
VSMIN	system variable	[v2.6, r9]
WBLOCK	command	[all versions]
Windowing	program feature	[all versions]
X/Y/Z Filtering	program feature	[ADE-3] [v2.6, r9]
'ZOOM	command	[all versions]
3DFACE	command	[ADE-3] [v2.6, r9]
3DLINE	command	[ADE-3] [v2.6, r9]

Appendix D: Selected Examples of AutoCAD Drawings

AutoCAD is used by more than 100,000 users for a variety of applications. The drawings on the following pages illustrate some of the many uses of AutoCAD. They also show examples of different computer-aided drafting techniques. Many of the drawings would have been difficult or impractical to create had it not been for the power of AutoCAD. Others were simply drawn much more quickly thanks to the ease with which AutoCAD can replicate parts and simplify editing. Take time to study these drawings. Try to apply the techniques used to the work you do with AutoCAD.

Figure D-1.   The appearance of perspective is cleverly created in this technical illustration of an airplane. This drawing is actually a simple two-dimensional drawing. It was drawn without using AutoCAD's isometric or three-dimensional drawing capabilities. Instead, its artist saved several typical parts as blocks and then inserted them into the drawing at a successively smaller insertion scale to give the appearance of perspective.

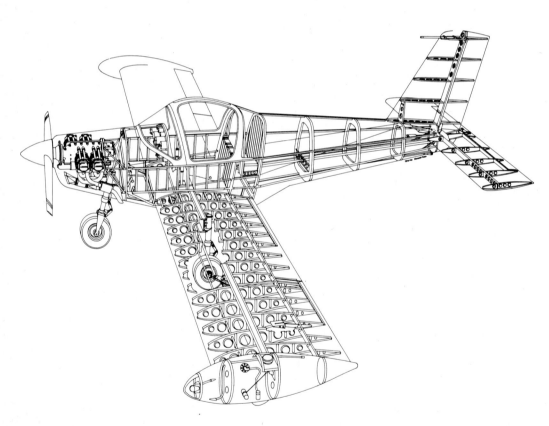

AutoCAD drawing by Chris Knowlton; courtesy of Autodesk, Inc.

Figure D-2. *These electrical and lighting floor plans show how layers are effectively used. The final drawing is a composite, created by inserting the actual floor plan on the predrawn, bordered sheet. The walls and the column grid were drawn on base layers, which were turned on in both drawings. The same drawing file can actually be used to create multiple drawings, by turning the power plan layers on for one drawing and turning the ceiling grid and light fixture layers on for the other. Drawing each contractor's work on a different layer of the same drawing assures that everyone's work will be in alignment, and interference between features can be quickly checked.*

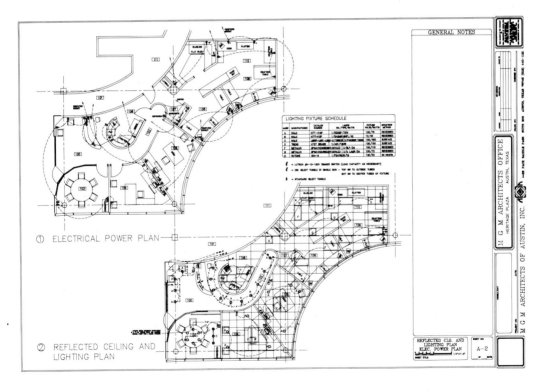

AutoCAD drawing by MGM Architects of Austin, Inc.; courtesy of Autodesk, Inc.

APPENDIX D: SELECTED EXAMPLES OF AUTOCAD DRAWINGS

Figure D-3.   AutoCAD is often used to create maps, as typified by this geology map. An AutoCAD map is much easier to update than a conventional printed map. Polylines were used to draw the contours and the course of the creek. Attributes can be associated with the drill hole markings to allow extraction of additional map data to a database program for further analysis. Features can also be accurately located using absolute coordinates.

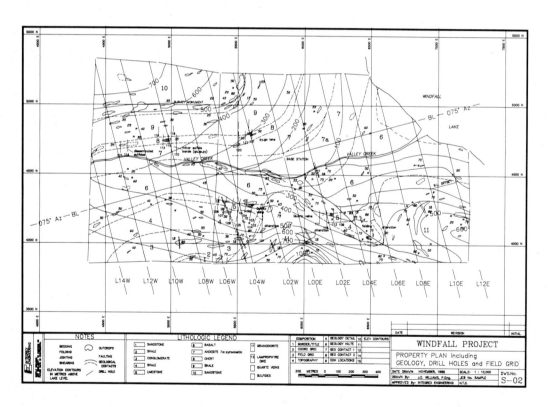

AutoCAD drawing by Integrex Instruments; courtesy of Autodesk, Inc.

626

Figure D-4.   This beautifully rendered elevation of the Carson Mansion was one of the winners of Autodesk's 1987 AEC drawing contest. This seemingly complex drawing was created mainly by copying and mirroring repetitious objects, such as the railings and cornice work. Standard hatch patterns were used to create the window shading, the siding, and the latticework.

AutoCAD drawing by Dave Harland; courtesy of Autodesk, Inc.

Figure D-5. *This complex printed circuit board drawing makes use of the ARRAY command to replicate the same circuit assembly over the entire board. All that needed to be done after the 34 copies were in place was to attach the end assemblies to the connectors at either end of the board.*

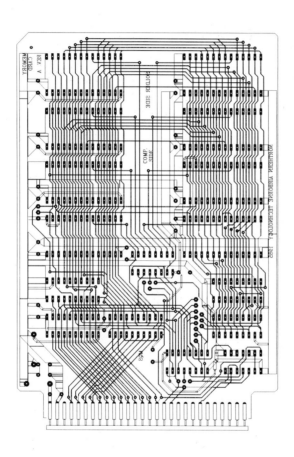

AutoCAD drawing by Northern Airborne Technology Ltd.; courtesy of Autodesk, Inc.

Figure D-6. *This rendered elevation of a private residence is another winner of Autodesk's 1987 AEC drawing contest. The extensive use of shading, which makes this drawing so striking, would have been impractical had this drawing been done by hand. With AutoCAD, simple hatch patterns quickly create the texture of the roof and the appearance of shadows. The architect enclosed the area to be hatched with a polyline, used the HATCH command to add the desired pattern, and then erased the polyline border.*

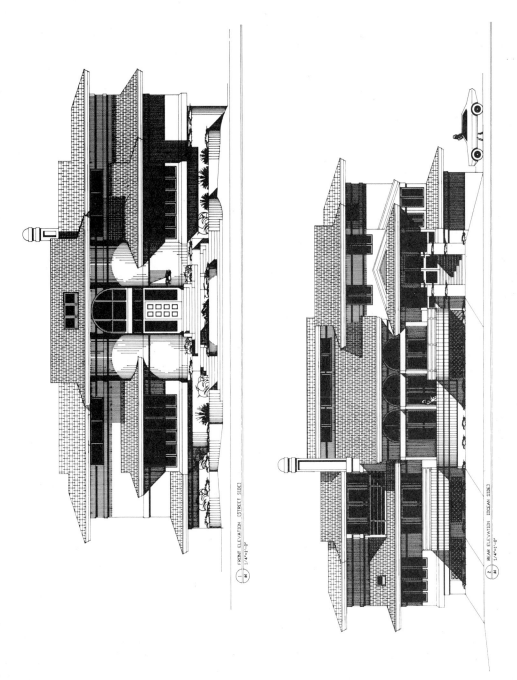

AutoCAD drawing by Jon L. Bourne, Architect AIA; courtesy of Autodesk, Inc.

APPENDIX D: SELECTED EXAMPLES OF AUTOCAD DRAWINGS

Figure D-7. This isometric drawing of a machine tool lathe was created by projecting circles onto the three isometric planes.

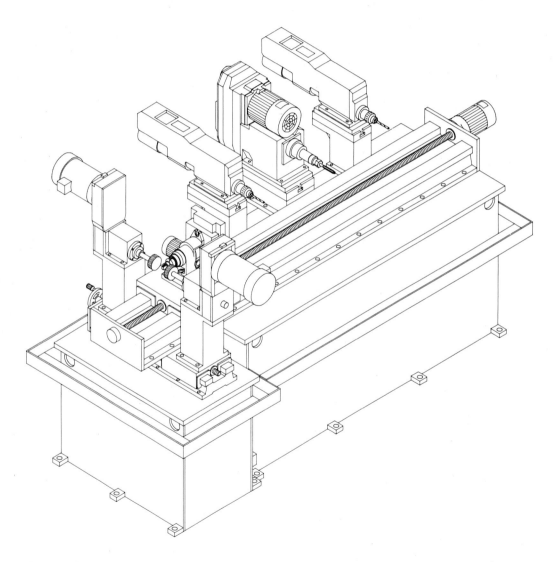

AutoCAD drawing; courtesy of Autodesk, Inc.

Indices

Applications Index

General Index

639